Maths, Science & Technology

A–Z of Essential Terms

David Meagher & Anne Pitkethley

OXFORD
UNIVERSITY PRESS
AUSTRALIA & NEW ZEALAND

Contents

OXFORD
UNIVERSITY PRESS
AUSTRALIA & NEW ZEALAND

253 Normanby Road, South Melbourne, Victoria 3205, Australia

Oxford University Press is a department of the University of Oxford. It furthers the University's objective of excellence in research, scholarship, and education by publishing worldwide in

Oxford New York

Auckland Bangkok Buenos Aires Cape Town Chennai Dar es Salaam Delhi Hong Kong Istanbul Karachi Kolkata Kuala Lumpur Madrid Melbourne Mexico City Mumbai Nairobi São Paulo Shanghai Taipei Tokyo Toronto

OXFORD is a trade mark of Oxford University Press in the UK and in certain other countries

First published 2002
Reprinted 2009, 2010, 2018(D)

National Library of Australia

Cataloguing-in-Publication data:

Meagher, David, 1958–.

A student's A–Z of maths, science and technology.
Upper primary/lower secondary.

ISBN 978 0 19 551505 3.

1. Mathematics—Dictionaries—Juvenile literature.
2. Science—Dictionaries—Juvenile literature.
3. Technology—Dictionaries—Juvenile literature.
I. Pitkethly, Anne, 1946–. II. Title.

500

Edited by Jocelyn Hargrave
Text designed by Olga Lavecchia
Illustrated by David Meagher, Pat Kermode & Currency Communications
Typeset by Currency Communications & Charlotte Reynolds
Printed and bound in Australia by Ligare Book Printers Pty Ltd

Introduction

Why do you need a dictionary?

Like all things in our universe, maths, science and technology are always changing. Keeping up with the latest ideas and methods is an exciting challenge.

Many everyday words have a special meaning when we use them in maths, science and technology. Mathematicians and scientists also use words that are not used by anyone else. Knowing what these words mean will make your maths, science and technology learning much easier.

Using this dictionary

This dictionary is divided into two parts. The first part covers maths. The second part covers science and technology. There is also a special section about units of measurement and maths formulae and symbols on pages 122–127.

We have given you the exact meanings of words as they are used today, as well as examples and illustrations to help you understand how the words are used. Because many of the words are related to other words and ideas, we have also given you links to these other words and ideas in the definition. Here are example definitions from each of the sections with the different parts shown.

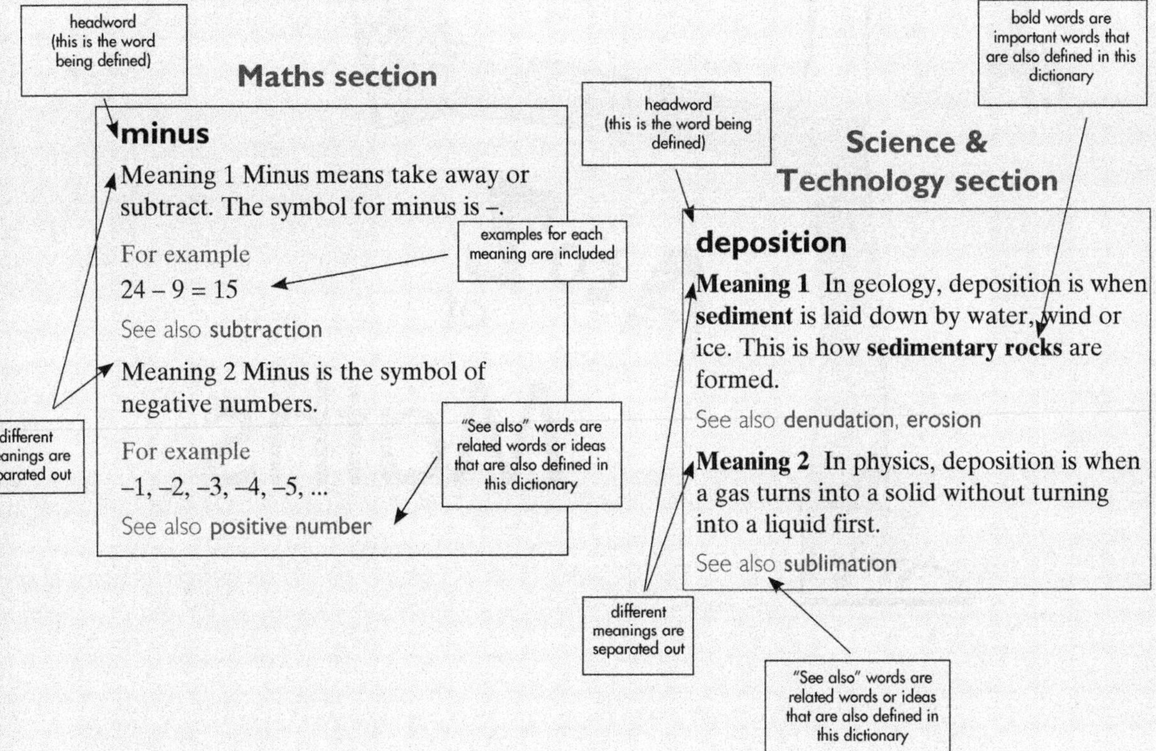

When an entry continues on the next page, a pointer ☞ tells you to turn over.

You can use this dictionary for more than just finding out the meaning of a word. You can also read it like a book to discover things about maths, science and technology that you did not know before. And after all, knowledge and discovery is really what makes maths, science and technology exciting for everyone.

A to Z of
Maths

Aa

abacus

An abacus is a device used for counting and calculating. A frame supports rods on which beads slide. Each rod represents a place value in a counting system. The plural form of *abacus* is abaci. There are different types of abaci.

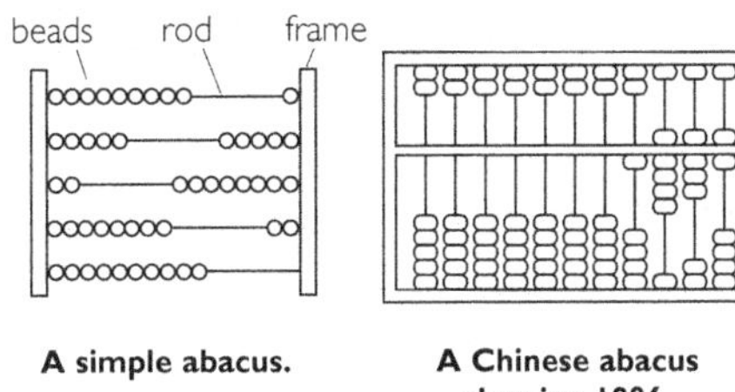

A simple abacus. **A Chinese abacus showing 1986.**

See also **calculator**

acceleration

Acceleration is the rate of change of the quickness of motion.

See also **rate**, **speed**

accurate

Accurate means correct or exact.

See also **approximation**

acute angle

An acute angle is a sharp angle that measures less than 90 degrees.

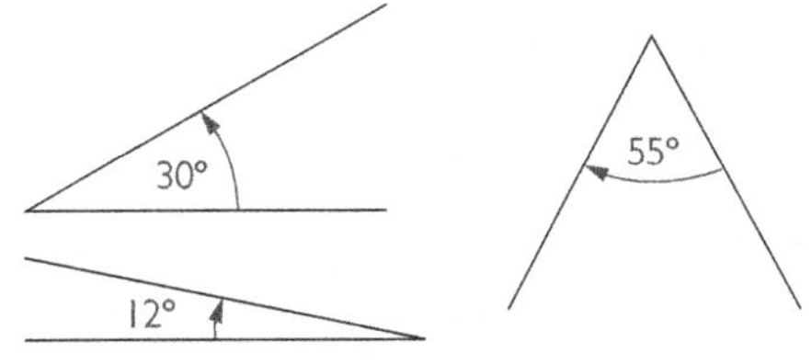

See also **angle**

acute-angled triangle

An acute-angled triangle is a triangle with all angles measuring less than 90 degrees.

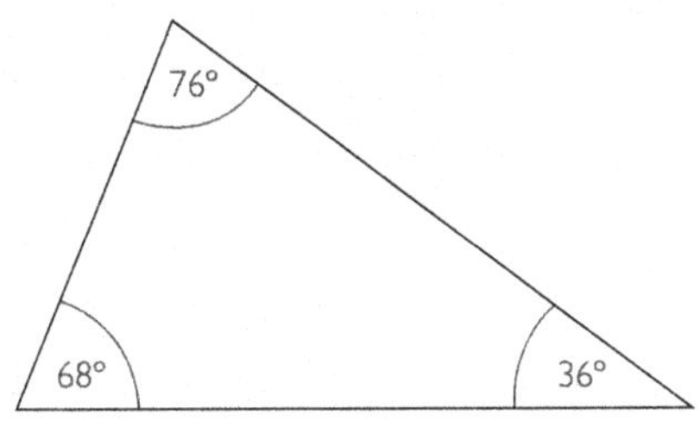

See also **triangle**

add

To add is to combine two or more numbers or quantities to give their sum or total. The symbol for the combination is + (plus).

See also **addend**, **addition**

addend

An addend is any number that is to be added.

For example

$9 + 8 + 3 = 20$

9, 8 and 3 are the addends.

20 is the sum.

See also **addition**, **missing addend**

addition

Addition is the operation of combining two or more numbers or quantities to give their sum or total. The symbol for the operation is + (plus).

For example

1 $5 + 7 = 12$

2 $\frac{1}{4} + \frac{1}{2} = \frac{1}{4} + \frac{2}{4} = \frac{3}{4}$

See also **operation**

adjacent

Adjacent means next to.

For example

1 Adjacent angles are angles that share a common arm and common vertex.

☞

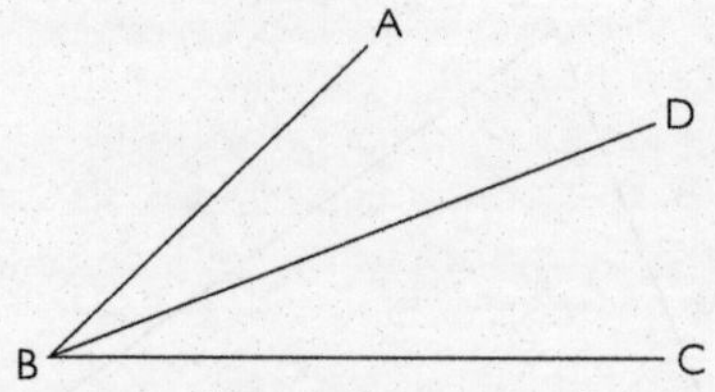

Angle ABD is adjacent to angle DBC, having B as the common vertex.

2 Adjacent sides are sides that share a common vertex.

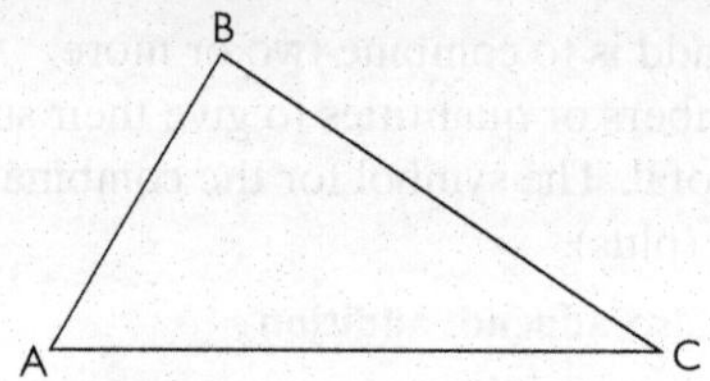

Side AB is adjacent to side AC, having A as the common vertex.

algebra

Algebra is that part of mathematics that deals with properties and relations of numbers and sets, using symbols to represent unknown quantities.

For example

$x + y = 15$

In this equation, letters symbolise the unknown quantities.

See also **al-Khwarizmi**, **pronumeral**, **relation**, **symbol**, **variable**

algebraic

Algebraic means occurring in algebra.

For example

$3a$ and x^2y are algebraic terms.

algorithm

An algorithm is a given series of steps for reaching a solution to a mathematical problem.

For example

The methods of setting out and solving addition, subtraction, multiplication and division problems are called algorithms.

```
  ¹7 ¹8 4          5 1 r 8
+  6  8 9      9 ) 4 6 7
 1 4  7 3          4 5
                   1 7
                     9
                     8

   ⁴ ²3 6
 ×     7 4          ⁶7̸ ¹¹2̸ ¹4
     1 4 4        - 3  4  8
   2 5 2 0          3  7  6
   2 6 6 4
```

See also **al-Khwarizmi**

al-Khwarizmi

al-Khwarizmi was an Arabic mathematician in the 9th century. He wrote a book on equations, *Al-jabr w'al muqabala*, which popularised algebra. The word *algebra* comes from the title. The word *algorithm* comes from his own name.

See also **algebra**, **algorithm**

allied angles

Allied angles are a pair of angles formed by a straight line crossing a pair of parallel lines. The allied angles are between the parallel lines on the same side of the transversal. Pairs of allied angles add up to 180 degrees. Another name for allied angles is cointerior angles.

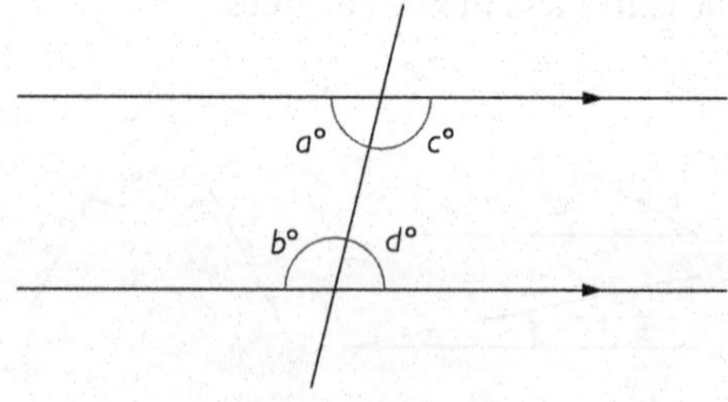

$a°$ and $b°$ are allied angles.
$c°$ and $d°$ are allied angles.

See also **alternate angles**, **corresponding angles**

alternate angles

Alternate angles are a pair of equal angles formed by a straight line crossing a pair of parallel lines. The alternate angles are between the parallel lines, one on either side of the transversal.

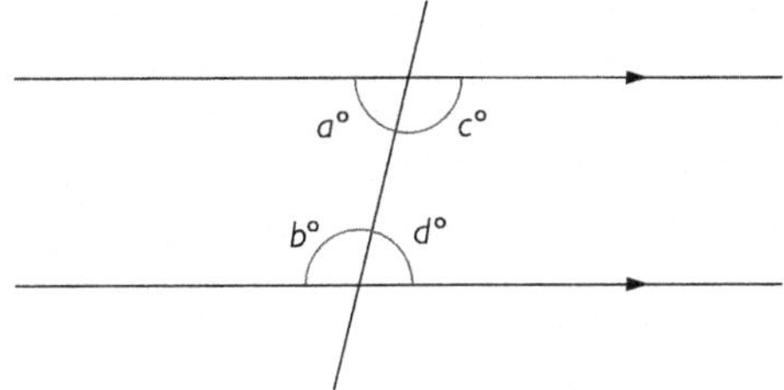

$a°$ and $d°$ are alternate angles.
$b°$ and $c°$ are alternate angles.

See also **allied angles**, **corresponding angles**

altitude

Altitude is the perpendicular distance from the base of a figure to the top.

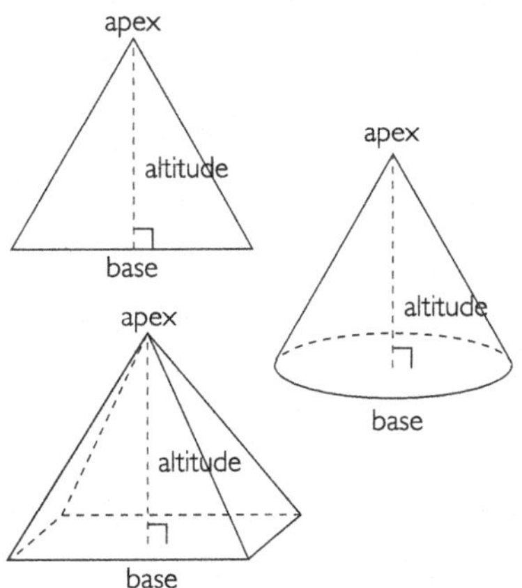

See also **apex**, **perpendicular**, **vertex**

a.m.

a.m. is the period between 12 o'clock at night and 12 o'clock in the middle of the day that is used with 12-hour time. The letters *a.m.* are an abbreviation of the Latin words *ante meridiem*.

For example

The time is 3.45 a.m.

See also **p.m.**, **twelve-hour time**

amount

Amount means how much.

For example

If I have $20, what amount of money would I have left after spending $12?

See also **quantity**

analysis

Analysis means breaking down something complex into its parts.

For example

Data analysis is the process of finding patterns in a set of data.

angle

An angle is the amount of turn between two straight lines that share a common starting point. The amount of turn is measured in degrees. There are 360 degrees in a complete turn, or circle.

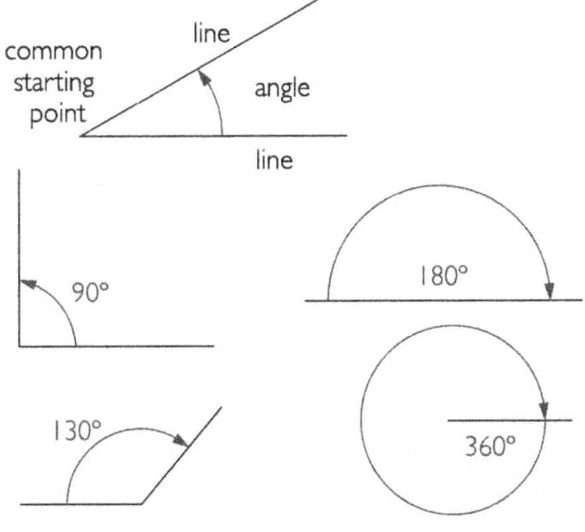

See also **acute angle**, **obtuse angle**, **ray**, **reflex angle**, **revolution**, **right angle**, **straight angle**, **vertex**

A

anticlockwise

Anticlockwise is the direction opposite to the direction of movement of the hands of an analogue clock.

See also **clockwise**

apex

The apex is the vertex that is furthest from the base of a plane or solid figure.

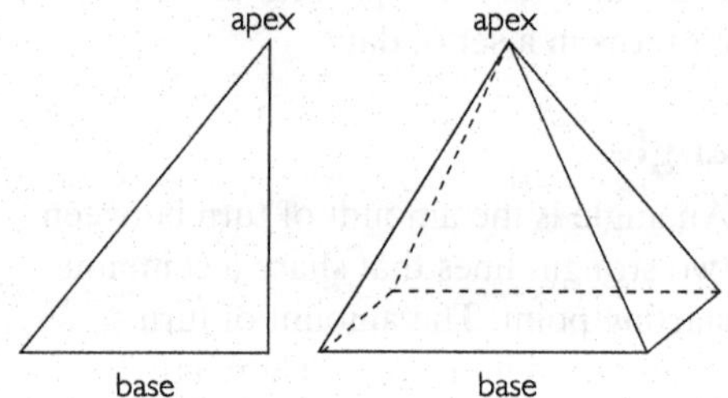

See also **altitude**, **base**, **vertex**

approximation

An approximation is an answer that is not exact. An approximation is almost accurate. The symbol ≈ stands for "is approximately equal to".

For example

$$99 \times 5 \approx 100 \times 5$$
$$= 500$$

See also **accurate**, **front-end estimation**

Arabic numerals

Arabic numerals are the symbols 0, 1, 2, 3, 4, 5, 6, 7, 8 and 9. Arabic numerals were introduced into Western use from Egypt around the 12th century. They may have originated with the Hindus in India around 200 BC. The ten digits form the basis of the base-ten counting system.

See also **base**, **digit**, **numeral**, **place value**

arc

An arc is part of a circle or other curved line. Two points on a circle divide the circle into two arcs. The shorter arc is called the minor arc. The longer arc is called the major arc. If the two arcs are the same length, they are called semicircles.

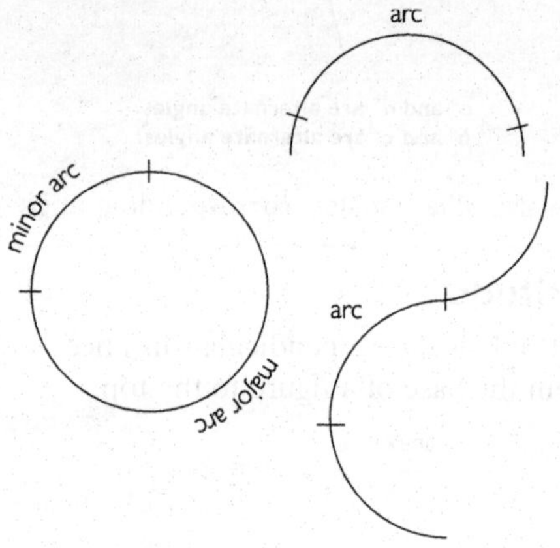

See also **circle**, **curve**

Archimedes

Archimedes was a Greek mathematician and inventor who lived 287–212 BC. His discoveries in mathematics included the law of the lever and the formula for the volume of a sphere. He also found volumes by displacement and developed a spiral device for raising water.

area

Area is the size of a surface. It is the amount of space within the boundary of a plane figure. Metric units of area are the square centimetre (cm^2), square metre (m^2), hectare (ha) and square kilometre (km^2).

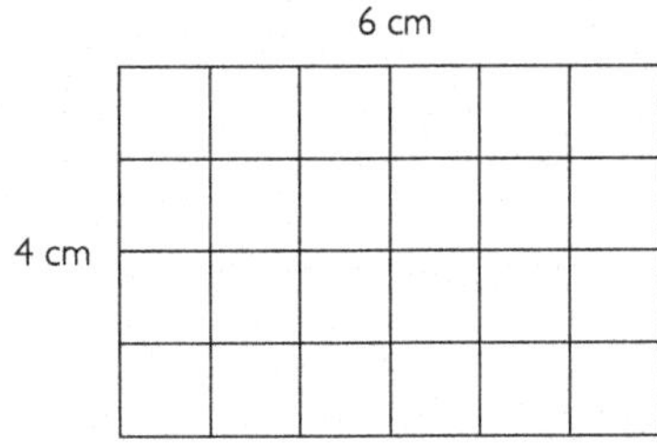

4 x 6 = 24
The area of this shape is 24 cm².

See also **metric system**, **size**, **surface**

arithmetic

Arithmetic is the study of numbers. It involves computations with whole numbers, fractions and decimals using the operations of addition, subtraction, multiplication and division. The word *arithmetic* comes from the Greek word for number, *arithmos*.

See also **computation**, **operation**, **real number**

arithmetic mean

The arithmetic mean of a group of numbers is a measure of the "middle" of the group. Arithmetic mean is another name for mean, which is commonly called average.

See also **average**, **mean**, **median**, **mode**

arithmetic progression

An arithmetic progression is a sequence of numbers in which the difference between each number and the previous number is always the same.

For example

1. The sequence 2, 4, 6, 8, 10 has a common difference of 2.
2. The sequence 50, 40, 30, 20, 10 has a common difference of –10.

See also **difference**, **geometric progression**, **sequence**

array

An array is a regular arrangement of objects or numbers in rows and columns. An array visually represents the grouping and sharing of numbers.

• • •
• • •

3 in each row
2 rows

$$\begin{array}{r} 3 \\ \times\ 2 \\ \hline 6 \end{array}$$

• •
• •
• •

2 in each row
3 rows

$$\begin{array}{r} 2 \\ \times\ 3 \\ \hline 6 \end{array}$$

• • • •
• • • •
• • • •
• • • •

16 shared among 4 is 4

$$\begin{array}{r} 4 \\ 4\overline{)16} \end{array}$$

See also **division**, **grouping**, **multiplication**, **share**

arrow diagram

In an arrow diagram, arrows are used to show a relationship between objects or numbers.

For example

This arrow diagram shows the favourite fruits of five children.

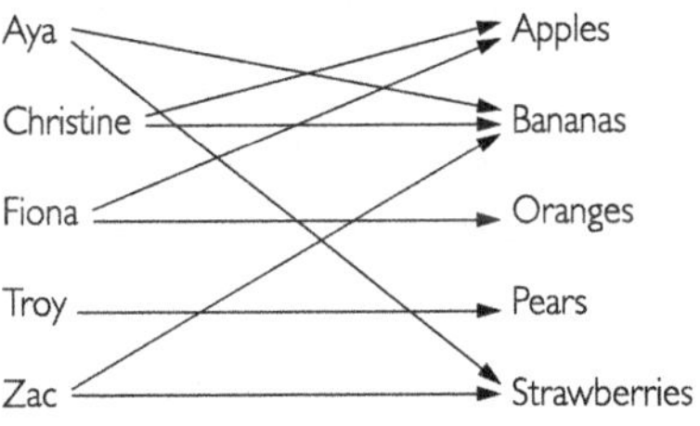

See also **mapping**, **one-to-one correspondence**, **relation**

ascending order

Ascending order is from smallest to largest.

For example

1. The numbers 5, 10, 15, 20 are arranged in ascending order.
2. The lengths 27.5 cm, 57.8 cm, 80.2 cm, 1 m are in ascending order.

See also **descending order**

A

associative

An operation is called associative if it does not matter how the quantities are grouped when they are combined.

Both addition and multiplication are associative.

For example

1 $3 + (8 + 5) = (3 + 8) + 5$
2 $3 \times (8 \times 5) = (3 \times 8) \times 5$

Neither subtraction nor division is associative.

For example

1 $3 - (2 - 1) \neq (3 - 2) - 1$
2 $8 \div (4 \div 2) \neq (8 \div 4) \div 2$

See also **commutative**, **distributive**, **operation**

assumption

An assumption is something taken to be true. An assumption is either assumed without proof, or assumed as a basis for reasoning.

asymmetry

Asymmetry is no symmetry at all.

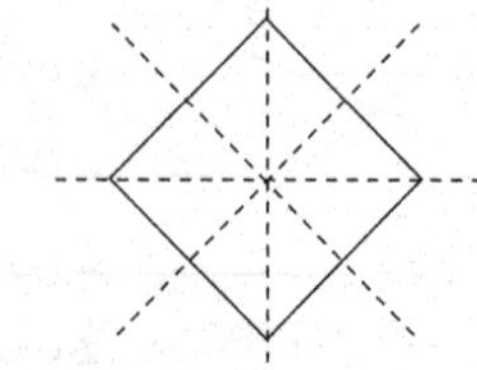

This figure has four lines of symmetry.

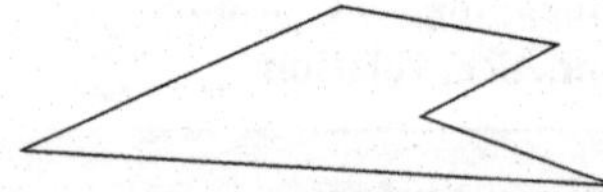

This figure has asymmetry.

See also **symmetry**

attribute

An attribute is a characteristic of an object. Size, colour, shape and thickness are attributes that can be used in the classification of objects.

See also **classify**

average

Average is a single measure that tries to show the "middle" of a group of numbers or quantities. Three such measures are used: mean, median and mode.

The word *average* is commonly used to refer to the arithmetic mean, which is found by adding all the values and dividing by how many values there are.

See also **mean**, **median**, **mode**

axis

Meaning 1 An axis is a line through the centre of a figure or solid. The line usually divides the figure or solid symmetrically. The plural form of *axis* is axes.

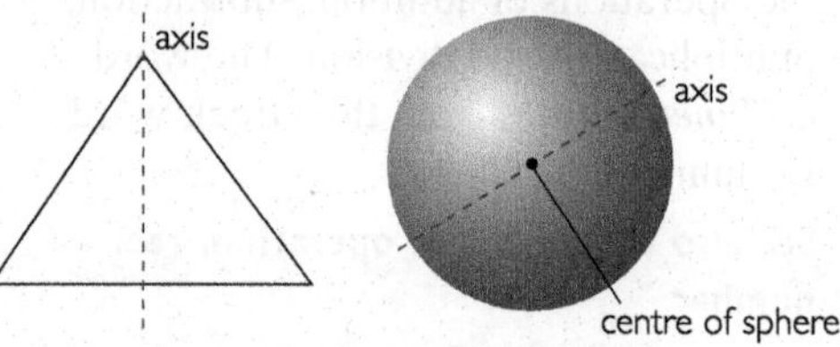

See also **line of symmetry**, **mirror line**, **symmetry**

Meaning 2 An axis is a fixed, scaled line that acts as a reference for a graph. Of the two axes of a graph, one is the vertical axis and the other is the horizontal axis. Points on the graph may be named using these axes.

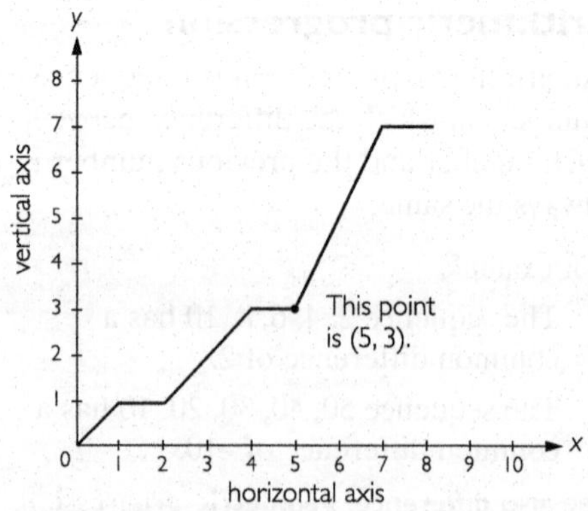

See also **coordinates**, **graph**, **scale**

Bb

balance

Meaning 1 The balance is the difference between the totals of the credits and the debits.

For example

The bank account has a balance of $876.54.

Meaning 2 Some types of scales are called balance scales. They measure the mass or the weight of an object.

For example

1 A beam balance is used to compare the mass of an object against standard masses. When the mass of the object and the standard masses are equal, the beam is balanced.

2 A spring balance measures the weight of an object. The stretch of the coiled spring and the pull of the object are balanced.

See also **mass**, **scales**, **weight**

Meaning 3 To balance is to have an equal distribution of amounts or quantities.

For example

1 The two bags of stones balanced the beam balance.

2 The two sides of an equation are balanced.

3 Income and expenditure were balanced in the budget.

See also **equal**

bar graph

A bar graph is a graph that uses bars to represent information. The heights or lengths of the bars are scaled to show size or quantity. Another name for a bar graph is a bar chart.

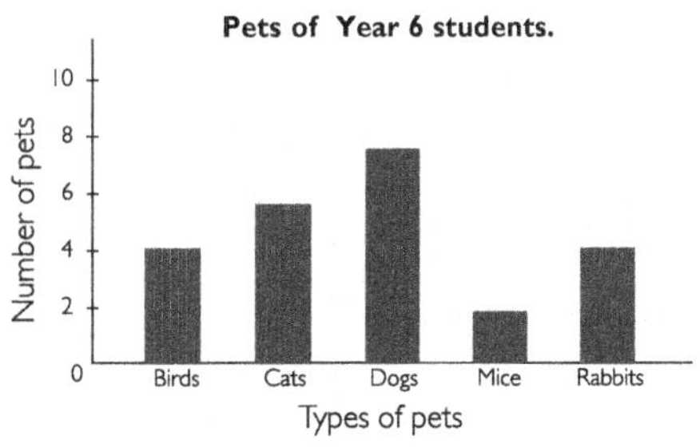

See also **data**, **graph**, **scale**

base

Meaning 1 A base is the line or surface on which a shape stands.

For example

1 A plane shape stands on a line.

2 A solid shape stands on a surface.

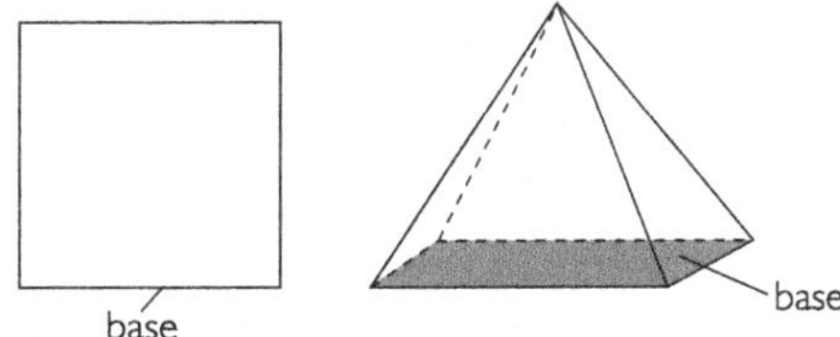

See also **line**, **surface**

☞

Meaning 2 The base of a place-value number system is the size of the group on which the system is built. It corresponds to the number of symbols, including zero, that are used to represent the numbers.

For example

1. The ten Arabic numerals are the foundation of the base-ten or decimal system in which place values are powers of 10.

Base-ten system.

Hundreds	Tens	Ones	Tenths
10 x 10 x bigger	10 x bigger		10 x smaller

2. Two symbols, 0 and 1, are the foundation of the base-two or binary system.

See also **place value**

Meaning 3 The base is the number to which a power is applied.

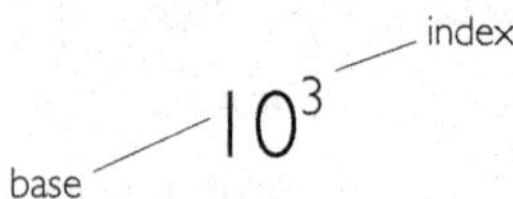

See also **power**

base line

Meaning 1 A base line is the horizontal axis of a graph.

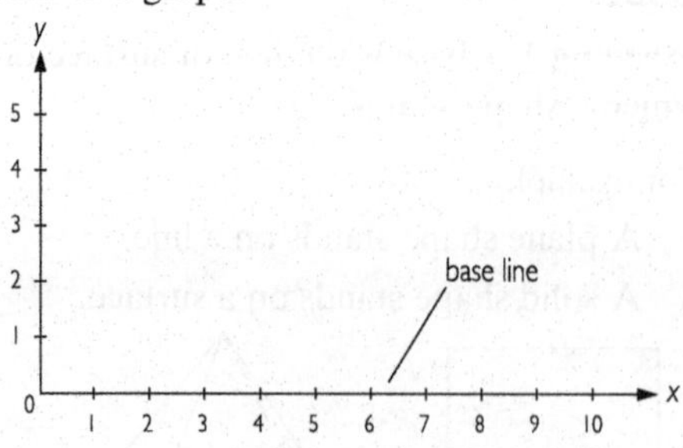

See also **axis, graph, horizontal**

Meaning 2 A base line is a line from which heights of objects are compared.

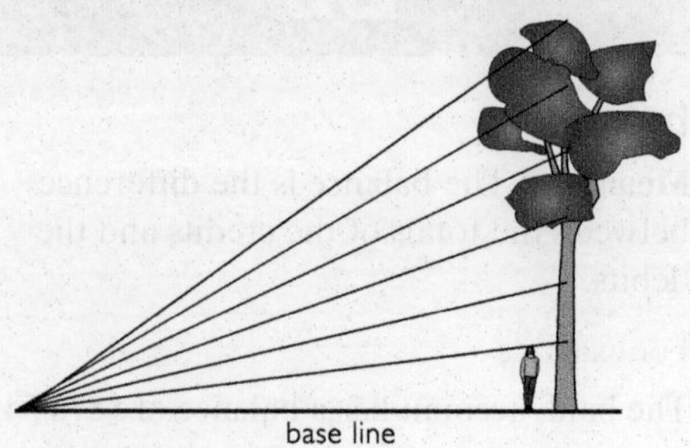

See also **height**

basic facts

Basic facts are number facts that arise from the operations of addition, subtraction, multiplication and division performed with the numbers 0 to 9. These basic facts are usefully committed to memory.

For example

The multiplication tables are basic facts.

See also **fraction facts, operation**

bearing

A bearing is the angle on the ground measured from the north or south direction that fixes the direction of an object.

For example

The bearing of the tallest tree is north 20 degrees east (N20°E).

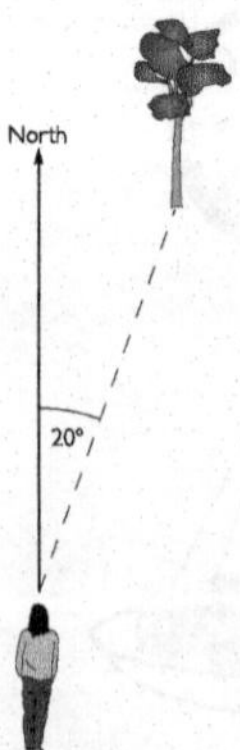

See also **compass points, direction**

bilateral symmetry

Bilateral symmetry is symmetry about a line. The line divides the shape or object into two parts that reflect each other. Bilateral symmetry is also known as line symmetry.

For example

The letters A and M have bilateral symmetry about the vertical line down the middle of the letter.

See also **line of symmetry**, **mirror line**, **reflection**, **symmetry**

billion

A billion is the number that is one thousand times one million:

1 000 000 000 or 10^9

See also **place value**

binary system

The binary system is the number system that uses groups of two as the foundation of place value. Only two digits, 0 and 1, are used to represent the numbers.

For example

The numbers that are usually written as 0, 1, 2, 3 and 4 are written as 0, 1, 10, 11 and 100 in binary notation.

Decimal notation	*Binary notation*			
	Eights	*Fours*	*Twos*	*Ones*
0				0
1				1
2			1	0
3			1	1
4		1	0	0

place values

See also **decimal system**, **number system**, **place value**

bird's-eye view

A bird's-eye view is the view seen from above, such as in plans and maps. It is sometimes called a plan view.

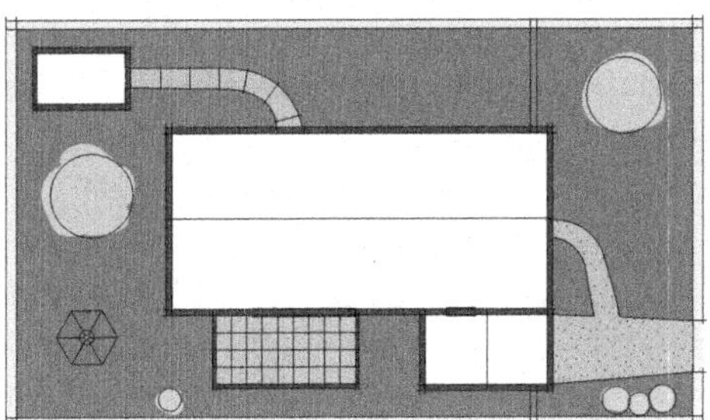

See also **map**, **plan**

bisect

To bisect is to cut or divide into two equal parts.

For example

1 A line can bisect an angle.

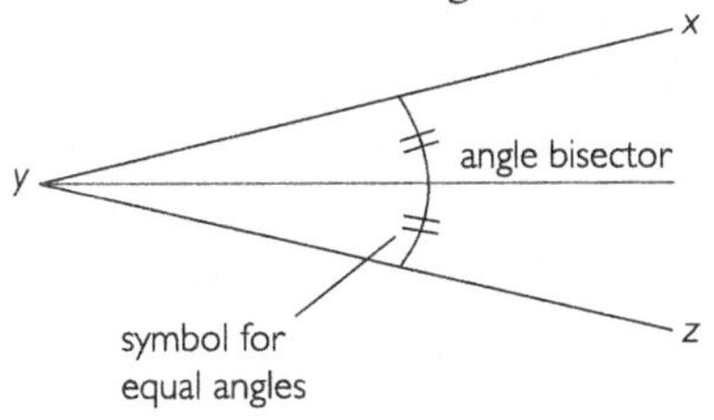

2 A line can bisect a line segment.

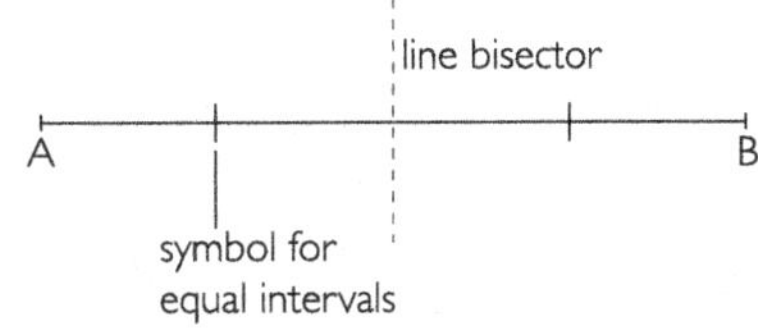

See also **equal**

box plot

A box plot is a summary graph. The box represents 50% of the data. The lines at each end of the box represent the top 25% and the bottom 25% of the data. The graph shows the centre, the spread and the overall range of distribution of the data.

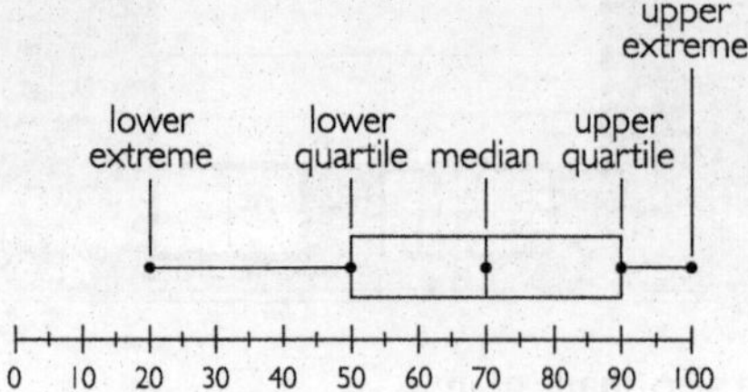

See also **distribution, graph, median, range, spread**

brackets

Brackets are the symbols used to group elements together. Brackets indicate which elements are to be combined first. The elements may be enclosed within different types of brackets:

parentheses ( )

square brackets [ ]

braces { }

For example

(5 + 3) × (8 ÷ 4) = 8 × 2 = 16

See also **element, order of operations**

breadth

Breadth is another name for width.

See **width**

Cc

calculate

To calculate is to work out a number or quantity using mathematical methods.

For example

1. Calculate the total number of people attending the concerts during the season.
2. Calculate how much money is spent on groceries each week.

calculation

Meaning 1 Calculation is the act of calculating.

For example

Problems may be worked out by mental or written calculation.

Meaning 2 A calculation is the result of calculating.

See also **calculate, computation, solution**

calculator

A calculator is a machine that performs mathematical operations. Electronic calculators commonly include a keyboard and a display.

See also **abacus**

calendar

Meaning 1 A calendar is a system for dividing up a year into intervals.

For example

In the Gregorian calendar, now in use, a year normally has 365 days, but has 366 days each leap year. There are other systems, including the Chinese lunar calendar and the Jewish lunisolar calendar.

See also **time, year**

Meaning 2 A calendar is made up of tables that show the days, weeks and months of a year.

2003 CALENDAR.

January

S	S	M	T	W	T	F
				1	2	3
4	5	6	7	8	9	10
11	12	13	14	15	16	17
18	19	20	21	22	23	24
25	26	27	28	29	30	31

February

S	S	M	T	W	T	F
1	2	3	4	5	6	7
8	9	10	11	12	13	14
15	16	17	18	19	20	21
22	23	24	25	26	27	28

March

S	S	M	T	W	T	F
1	2	3	4	5	6	7
8	9	10	11	12	13	14
15	16	17	18	19	20	21
22	23	24	25	26	27	28
29	30	31				

April

S	S	M	T	W	T	F
			1	2	3	4
5	6	7	8	9	10	11
12	13	14	15	16	17	18
19	20	21	22	23	24	25
26	27	28	29	30		

See also **table**

calibration

Calibration is the marking of a scale on a measuring instrument.

For example

1 A measuring jug is calibrated in millilitres.

2 A protractor is calibrated in degrees.

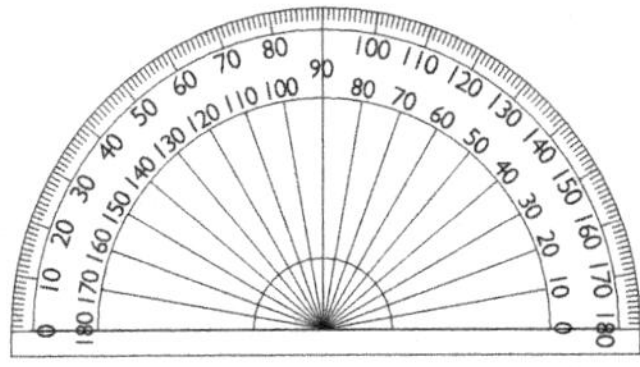

See also **graduation, scale**

capacity

Capacity is the amount a container can hold, and refers to a measure of things that can be poured. The standard metric measures of capacity are the millilitre (mL), litre (L), kilolitre (kL), megalitre (ML), cubic centimetre (cm^3) and cubic metre (m^3).

For example

1 The capacity of the milk carton is 1 litre.

2 The capacity of the eyedropper is 5 millilitres.

See also **standard units of measurement, volume**

cardinal compass points

The cardinal compass points are the four key directions on a compass: north, south, east and west.

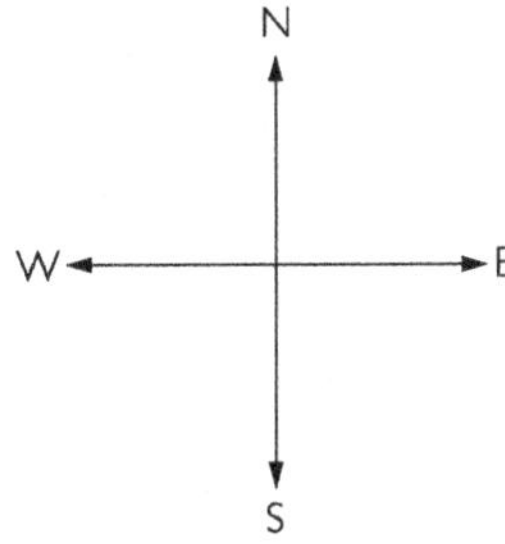

See also **intermediate compass points**

C

cardinal number

A cardinal number is a number that can be used to say how many elements there are in a set. All the counting numbers are cardinal numbers.

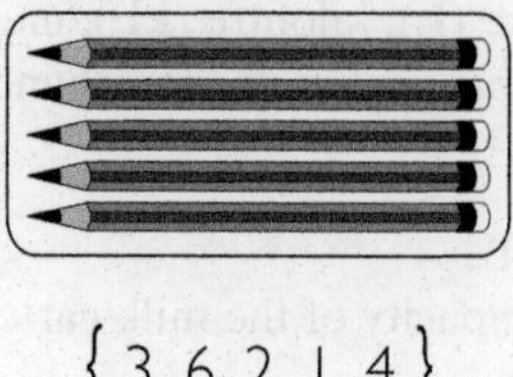

{3, 6, 2, 1, 4}

Each set has five elements.
Five is a cardinal number.

See also **element, one-to-one correspondence, ordinal number**

Carroll diagram

A Carroll diagram is a diagram that is used to classify objects. The diagram was named after Lewis Carroll, an English mathematician and author whose real name was Charles Dodgson.

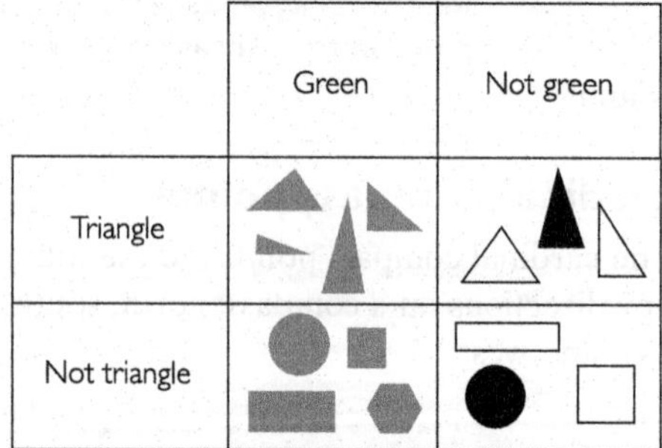

See also **classify, Venn diagram**

Cartesian coordinates

Cartesian coordinates are a method for describing position. They were used by René Descartes, a 17th-century French mathematician and philosopher.

Meaning 1 In two-dimensional space, Cartesian coordinates are pairs of numbers used to describe a point, with reference to two axes, horizontal and vertical, that intersect at the origin.

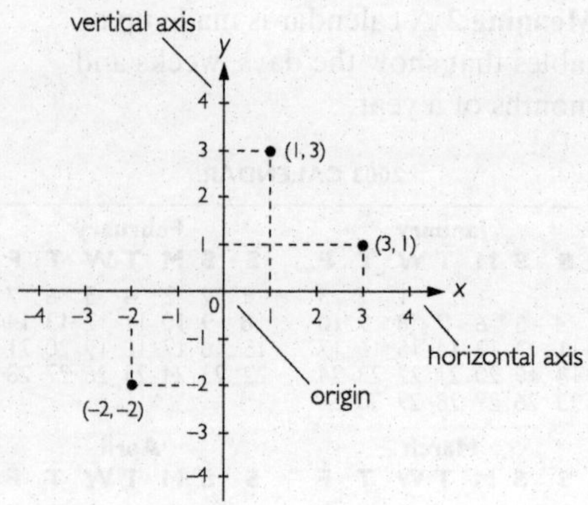

The first coordinate of a point gives its horizontal position. The second coordinate of a point gives its vertical position.

See also **axis, coordinates, origin, two-dimensional**

Meaning 2 In three-dimensional space, Cartesian coordinates are three numbers used to describe a point, with reference to three mutually perpendicular axes that intersect at the origin.

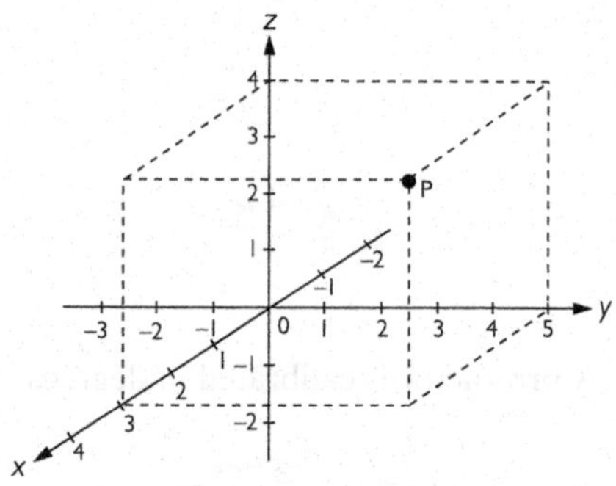

The coordinates of point P are (3, 5, 4).

See also **axis, origin, perpendicular, three-dimensional**

Cartesian graph

A Cartesian graph is a diagram of a set of ordered pairs of numbers, or a set of ordered triples of numbers.

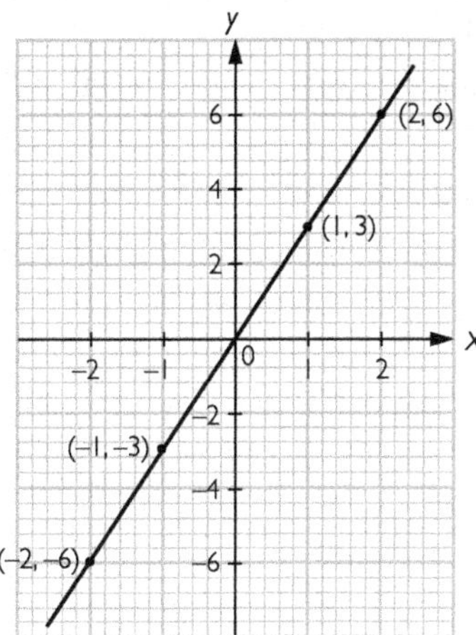

Graph of ordered pairs of the form (*x*, 3*x*).

See also **Cartesian coordinates**, **graph**, **ordered pair**

Celsius scale

The Celsius scale is a scale for measuring temperature. On the Celsius scale, 0 degrees is approximately the freezing point of water and 100 degrees is approximately the boiling point of water. The scale is named after Anders Celsius, a Swedish astronomer and mathematician who lived in the 18th century.

See also **degree**, **scale**, **temperature**

---

cent

A cent is the name of a unit in the decimal money systems of Australia, the USA and Hong Kong, for instance. One cent is one-hundredth of a dollar. The symbol for cent is c. In Australia, the one-cent coin is no longer used.

100 cents = $1.00

For example

The ticket cost 50 cents.

See also **decimal currency**, **dollar**

centimetre

A centimetre is a unit of length. It is equal to one-hundredth of a metre. The symbol for centimetre is cm.

100 centimetres = 1 metre

For example

The length of the ruler is 30 centimetres.

See also **length**, **metre**, **standard units of measurement**

---

centre

The centre is the middle point.

For example

1 The centre of a circle is the point that is the same distance from all the points of the circle.

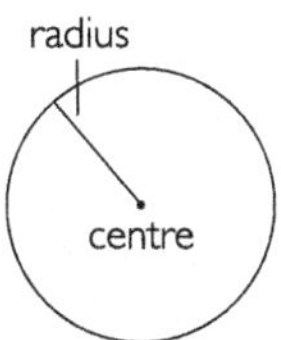

2 The centre of a regular polygon is the point that is the same distance from all the vertices.

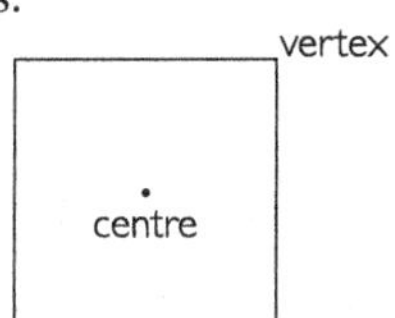

3 When an object is rotated about a fixed point, the fixed point is called the centre of rotation.

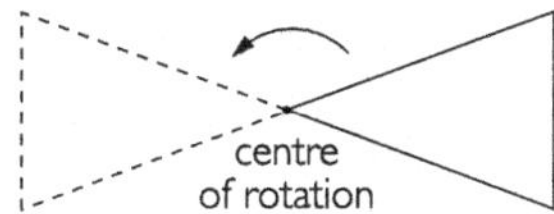

chance

Chance is the likelihood, or probability, of an event happening.

For example

When a die is thrown, the chance of turning up a three is 1 in 6 because there are six possible sides to turn up.

See also **likelihood, probability**

chart

Meaning 1 A chart is information shown in a table.

Quality / Size	Super down	Down	85% down 15% feather	75% down 25% feather	Polyester
Single bed	$90	$75	$60	$45	$30
Double bed	$128	$100	$75	$56	$44
Queen-size bed	$139	$115	$88	$62	$52
King-size bed	$150	$124	$99	$68	$58

See also **table**

Meaning 2 A chart is a graph.

For example

A bar chart can be used to show the different hair colours of a group of students.

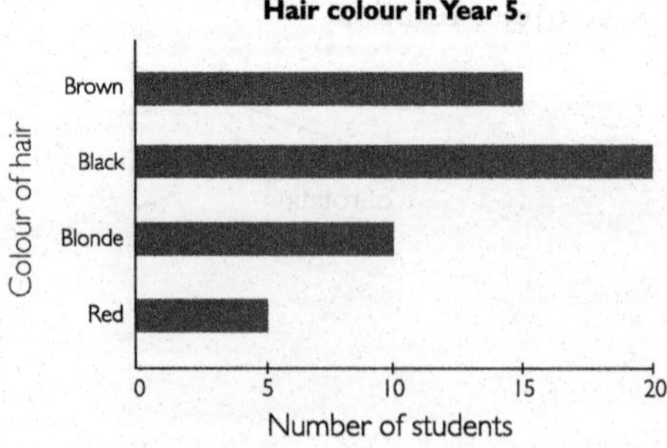

See also **bar graph**

check

To check is to make sure a solution is correct.

For example

To check a division calculation, use the inverse operation: multiplication.

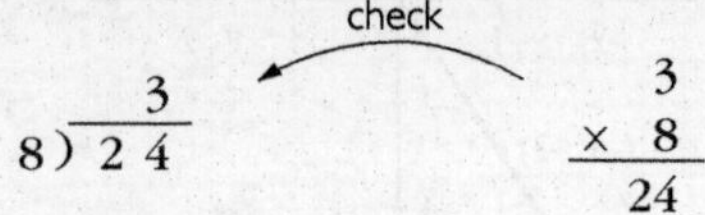

See also **accurate**

chord

A chord is a straight line that joins two points on a circle. The chord divides the region inside a circle into two segments. A chord that passes through the centre of a circle is called a diameter. Diameters are the longest chords.

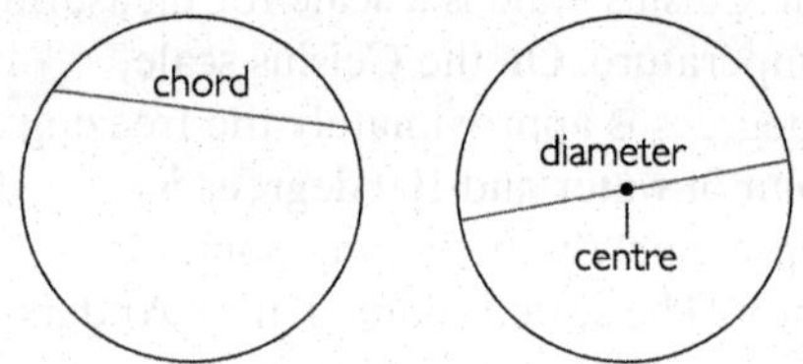

See also **circle, segment**

chronological order

Events arranged in order of when they happened are in chronological order.

For example

A time line is in chronological order.

See also **order, time**

circle

A circle is the set of all points in the plane that are a fixed distance from a given point. The distance from the given point is called the radius. The given point is called the centre of the circle.

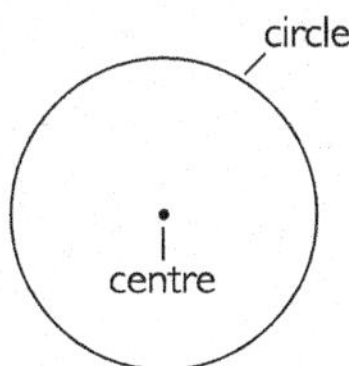

See also **ellipse**, **plane**, **point**, **set**

circumference

The circumference of a circle is the distance around the circle. It is the perimeter of the circle. The circumference of a circle with radius r is $2\varpi r$.

See also **perimeter**, **pi**

classification

Classification is the process of sorting objects into groups according to attributes. Classification involves seeing similarities and differences in objects.

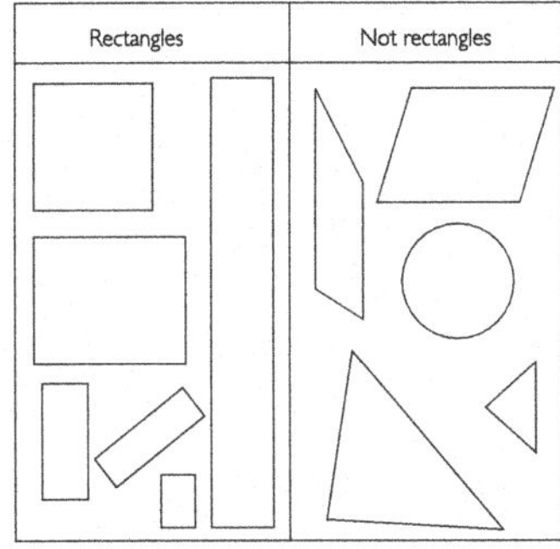

See also **attribute**

classify

To classify is to arrange into groups according to attributes.

For example

Classify polygons according to the number of sides of the shape.

See also **attribute**

class interval

A class interval is one of the divisions into which a set of values has been grouped. Each class generally represents the same number of values.

For example

Earnings per week has been grouped into seven class intervals, all but one with 100 possible values:

> \$0–\$99, \$100–\$199, \$200–\$299, \$300–\$399, \$400–\$499, \$500–\$599, over \$600

clock

A clock is an instrument for measuring and showing time.

For example

Types of clocks include the candle clock, sundial, analogue clock and digital clock.

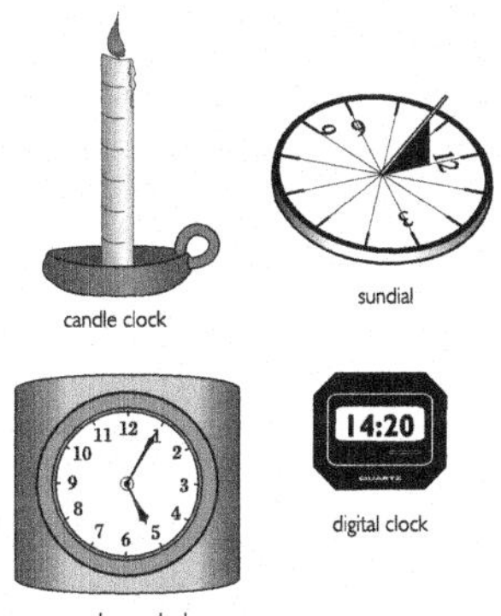

See also **time**

clockwise

Clockwise is the direction in which the hands of an analogue clock move.

See also **anticlockwise**

closed curve

A closed curve is a curved line that begins and ends at the same point.

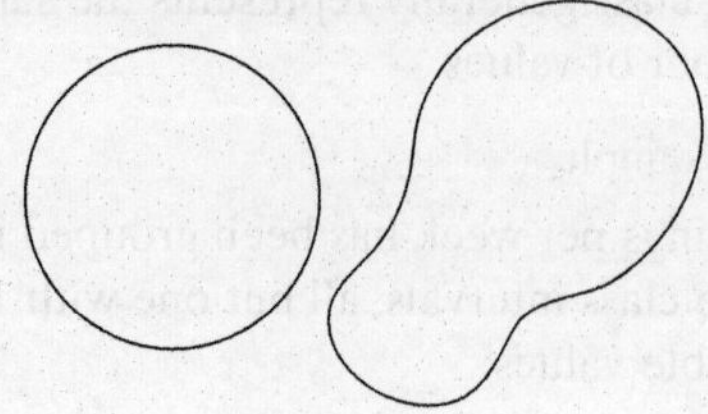

See also **curve**, **region**

coefficient

A coefficient is a number that is placed before a pronumeral to show that the pronumeral is to be multiplied by that number.

For example

In the expression $2x + 3$, the coefficient of x is 2.

See also **pronumeral**

cointerior angles

See **allied angles**

collinear

Collinear means lying on the same straight line.

For example

Points that lie in a straight line are collinear.

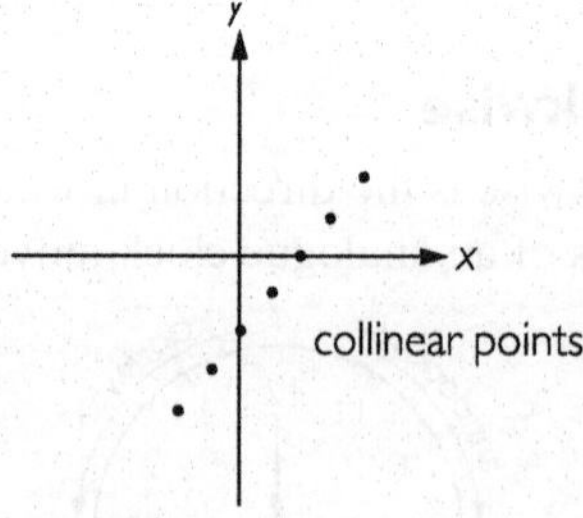

See also **line**

column

Meaning 1 A column is a vertical arrangement of figures.

357

181

343

889

543

Meaning 2 A column is a vertical line or bar, as used in a column graph.

See also **vertical**

column graph

In a column graph, columns are used to represent information. The heights of the columns vary according to the number of items or values represented.

Tennis players at school.

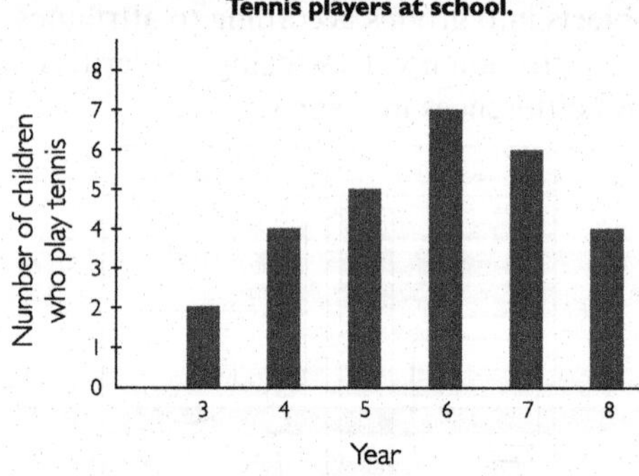

See also **bar graph**, **column**, **graph**

common denominator

A common denominator is a number that is a common multiple of the denominators of two or more common fractions. Put a different way, it is a number that is divisible by each of the denominators exactly.

For example

To simplify $\frac{1}{2} + \frac{1}{3}$, we must find a common denominator of $\frac{1}{2}$ and $\frac{1}{3}$. This number must be a multiple of both 2 and 3.

A common denominator of $\frac{1}{2}$ and $\frac{1}{3}$ is 6.

The number 6 is divisible by both 2 and 3 exactly.

$\frac{1}{2} + \frac{1}{3} = \frac{3}{6} + \frac{2}{6} = \frac{5}{6}$

See also **denominator**, **lowest common denominator**, **multiple**

common factor

A common factor is a factor that a set of numbers have in common.

For example

1 The number 5 is a common factor of 15 and 25.

2 The numbers 1, 2, 5 and 10 are the common factors of 20 and 30, with 10 the highest common factor.

See also **factors**, **highest common factor**

common fraction

A common fraction is a fraction of which the numerator and the denominator are whole numbers. A common fraction is also known as a vulgar fraction or a simple fraction.

For example

$\frac{1}{2}$, $\frac{2}{3}$ and $\frac{3}{4}$ are common fractions.

See also **denominator**, **fraction**, **numerator**

commutative

An operation is called commutative if it does not matter what order the quantities are combined in.

Both addition and multiplication are commutative.

For example

1 $5 + 4 = 4 + 5$

2 $5 \times 4 = 4 \times 5$

Neither subtraction nor division is commutative.

For example

1 $8 - 3 \neq 3 - 8$

2 $6 \div 2 \neq 2 \div 6$

See also **associative**, **operation**

compare

To compare is to identify similarities and differences.

compass

A compass is an instrument used to determine direction. The needle on the compass points to magnetic north. Compare *compass* with *pair of compasses*.

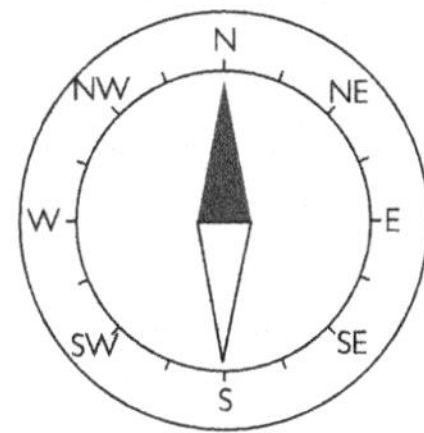

See also **compass points**, **direction**

compass points

Compass points identify direction. The names of the major compass points are north, south, east and west. These directions are at 90 degrees to one another.

See also **compass**, **direction**

complement

The complement is the quantity or amount that completes the whole.

For example

1 What must be added to \$16 to make \$20?

2 The number 2 complements 8 to make 10.

See also **whole**

C

complementary angles

Complementary angles are two acute angles that together add up to 90 degrees.

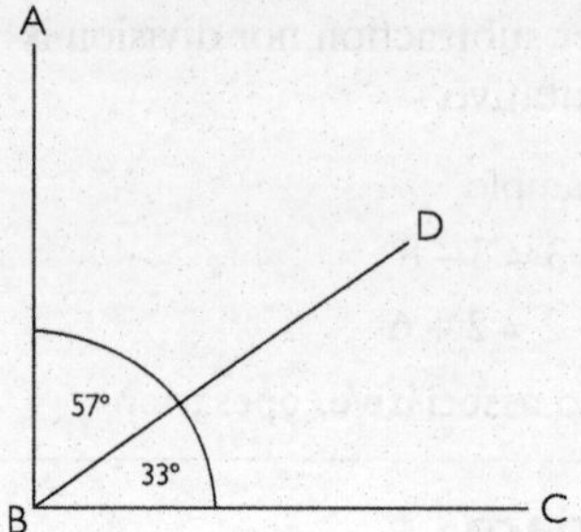

ABD and CBD are complementary angles.

See also **acute angle, supplementary angles**

complete

Meaning 1 Complete is having all its parts.

For example

The diagram is complete.

See also **part**

Meaning 2 To complete is to make whole.

For example

Complete the equation.

See also **whole**

complex fraction

A complex fraction is a fraction of which the numerator and/or the denominator are common fractions or mixed numbers.

For example

1 $\frac{\frac{1}{4}}{4}$

2 $\frac{2}{\frac{3}{4}}$

3 $\frac{1\frac{1}{2}}{2\frac{3}{4}}$

See also **common fraction, denominator, mixed number, numerator**

complex shape

A complex shape is a shape composed of parts in either two or three dimensions. The outline or surface crosses over itself. A complex shape is complicated.

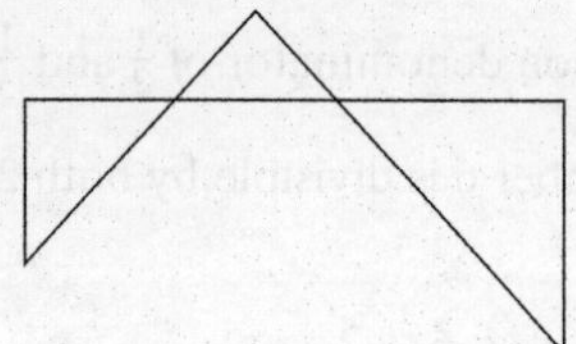

A complex shape in two dimensions.

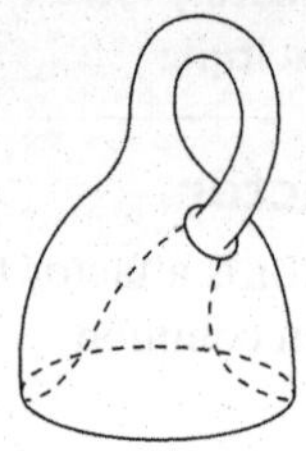

A Klein bottle is a complex shape in three dimensions.

See also **simple shape**

composite number

A composite number is a counting number that has factors other than itself and 1.

For example

16 is a composite number because it has 1, 2, 4, 8 and 16 as its factors.

$1 \times 16 = 16$
$2 \times 8 = 16$
$4 \times 4 = 16$

See also **factors, prime number**

computation

Meaning 1 Computation is the act of working out, or calculating, by mental or written methods or with the use of calculating tools.

See also **calculate**

Meaning 2 A computation is the result of working out a problem.

See also **calculation**

concave

A concave shape is a shape that is curved inwards. A concave polygon has an interior angle of more than 180 degrees.

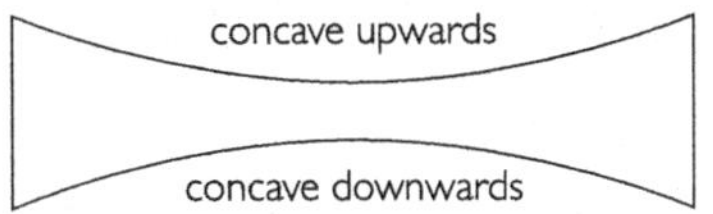

See also **convex**

concentric

Two circles are concentric if they have the same centre. Similarly, two spheres are concentric if they have the same centre.

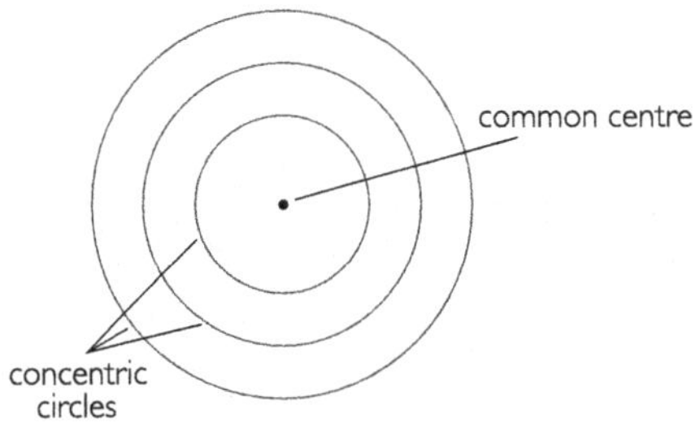

See also **centre**

concurrent lines

Concurrent lines are lines that pass through a common point.

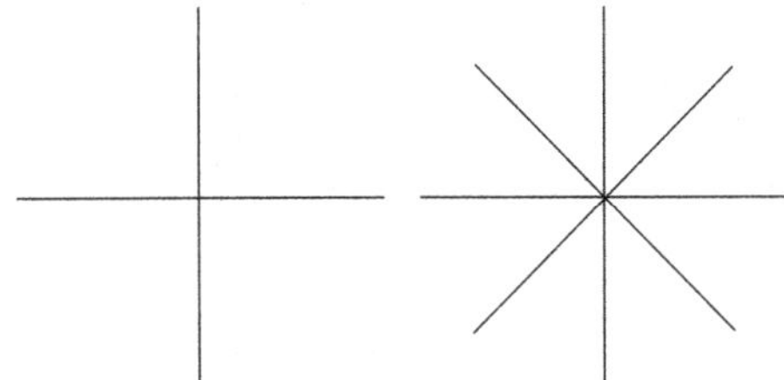

See also **intersection**

cone

A cone is a solid shape that has a circular base and one vertex. The lines that join the vertex to the points around the base form a curved surface.

One particular type of cone is a right cone. In a right cone, the lines that join the vertex to the points around the base are all the same length, and the line from the centre of the base to the vertex is at right angles to the base.

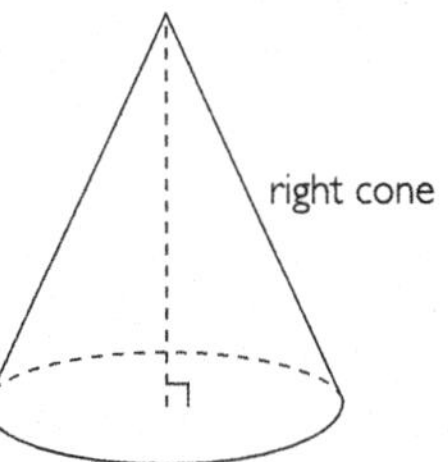

See also **solid**, **three-dimensional**

congruent

Geometric figures that have the same shape and size are congruent. The figures can be in two or three dimensions.

For example

1 Two circles are congruent if they have the same radius.

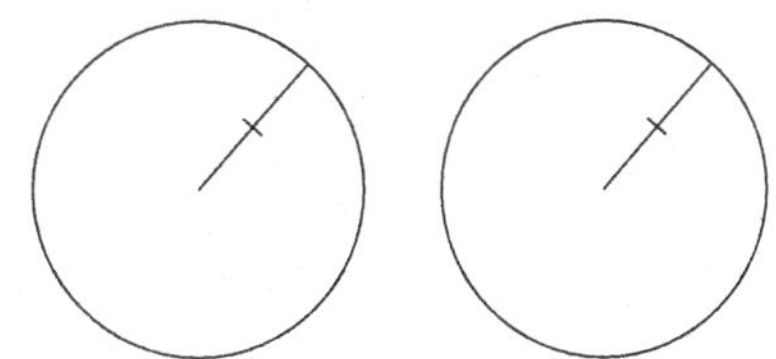

2 Triangles with all sides and angles matching are congruent triangles.

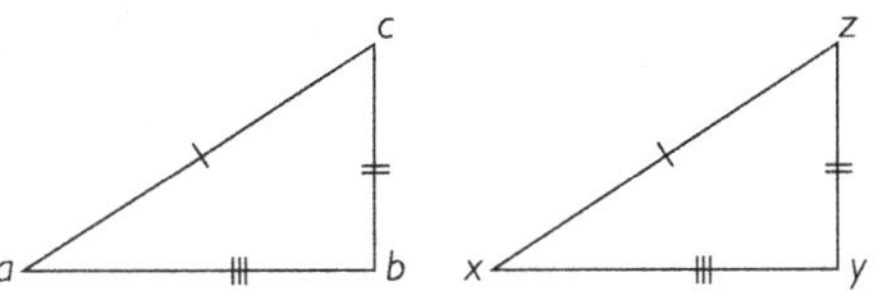

3 Two solid shapes that can coincide in space are congruent.

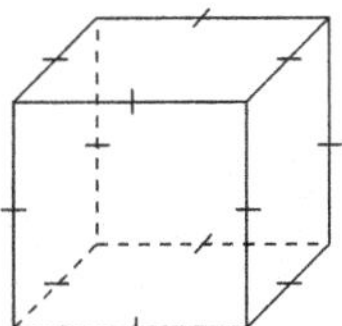
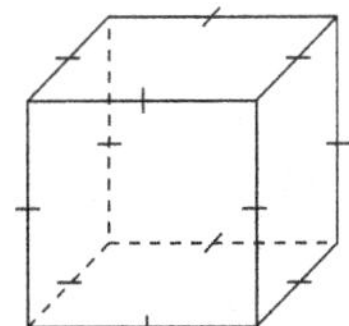

See also **equal**

consecutive numbers

Consecutive numbers are numbers that follow on from each other in sequence.

For example

The numbers 345, 346, 347, 348 are consecutive numbers.

See also **sequence**

conservation

Conservation is when the quantity or amount remains the same however it is arranged. Conservation applies to number, length, area, mass and volume.

For example

1 A group of objects remains the same in number even when their positions are altered.

2 A piece of string with a length of 10 centimetres remains 10 centimetres in length even when the piece of string is knotted.

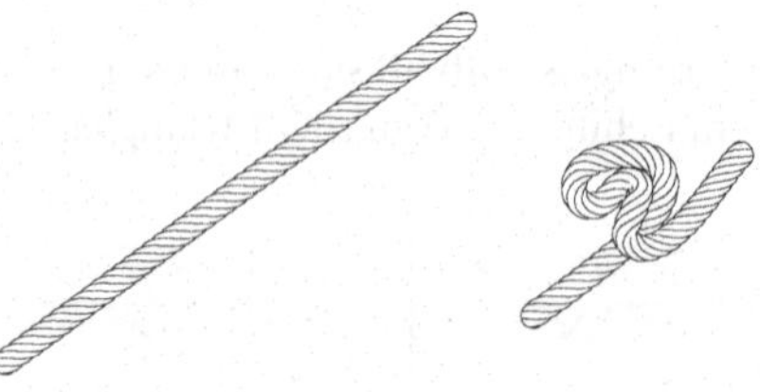

3 A lump of clay with a mass of 1 kilogram remains 1 kilogram in mass even when the clay is moulded.

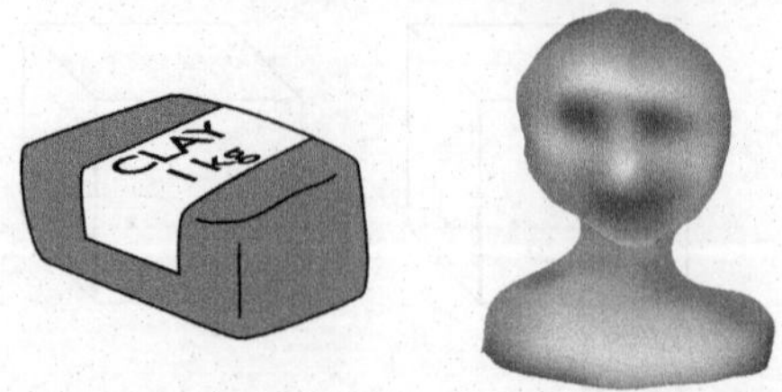

4 1 litre of liquid remains 1 litre whether it is in a bottle or a jug.

constant

A constant is a number or symbol, the value of which does not change.

For example

5 is a constant in the expression $5x$.

See also **value**, **variable**

constant rate of change

A constant rate of change is a uniform trend. Linear functions can be used to model situations where a constant rate of change occurs.

For example

When travelling at a steady speed, the distance travelled increases uniformly with time.

See also **trend**

continuous

Continuous means not separate.

For example

1 A piece of string, a sheet of paper and a block of modelling clay are continuous materials.

2 A curve is a continuous line.

3 Speed is a continuous variable.

4 Height is a continuous quantity.

See also **discrete**

converge

To converge is to come together to a point or value.

For example

1 To show perspective in drawings, parallel lines are made to converge at the horizon.

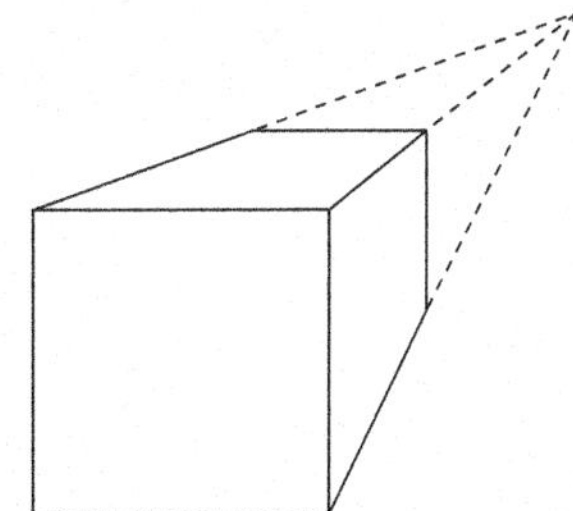

2 An infinite sequence of numbers can converge towards a limit. The following sequence converges to 0:

$1, \frac{1}{2}, \frac{1}{4}, \frac{1}{8}, \ldots$

See also **point**, **value**

convert

To convert is to change into something equivalent.

For example

1 To change the units in which a quantity is measured. The distance 2.8 kilometres converts to 2800 metres.

2 To convert between a common fraction, a decimal fraction and a percentage: $\frac{1}{2} = 0.5 = 50\%$.

See also **transformation**

convex

A shape is convex if it does not bend in on itself.

For example

1 Some optical lenses are convex.

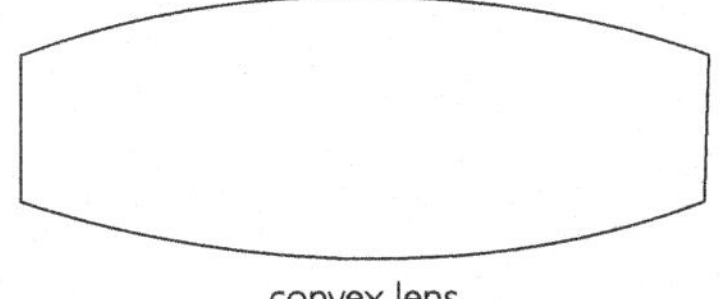

convex lens

2 A polygon is convex if all the vertices point outwards.

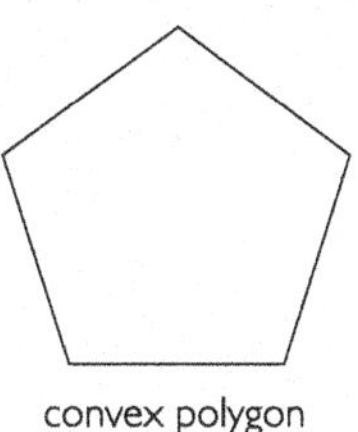

convex polygon

See also **concave**

coordinates

Coordinates are an ordered pair of numbers and/or letters that locate a point on a plane in relation to a frame of reference. By convention, the first coordinate refers to the horizontal position; the second coordinate refers to the vertical position.

For example

Coordinates locate points on graphs, maps and in street directories.

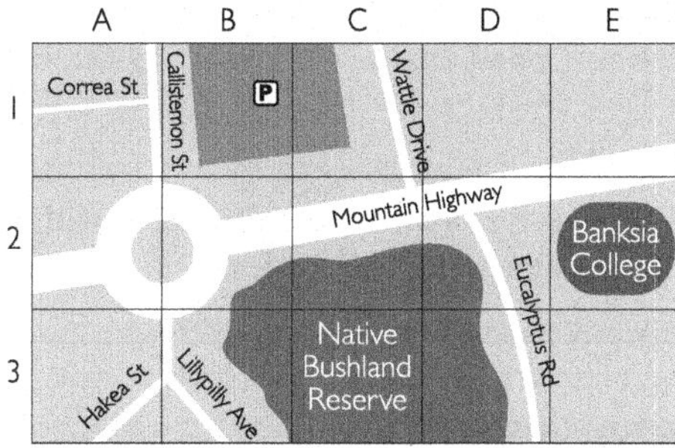

Banksia College is located at E2.

See also **Cartesian coordinates**, **Cartesian graph**, **grid reference**, **ordered pair**

corresponding angles

Corresponding angles are angles in the same relative positions.

For example

1 Corresponding angles are formed when a pair of lines is crossed by a straight line or transversal. If the lines

☞

are parallel, then the corresponding angles will be equal.

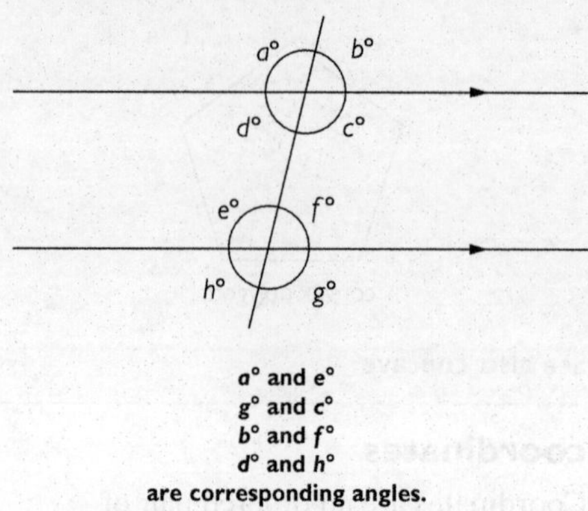

a° and e°
g° and c°
b° and f°
d° and h°
are corresponding angles.

2 Two congruent shapes have corresponding angles.

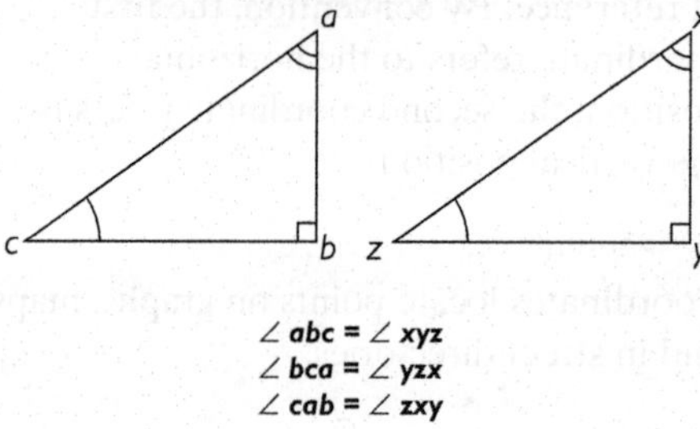

∠ abc = ∠ xyz
∠ bca = ∠ yzx
∠ cab = ∠ zxy

See also **allied angles**, **alternate angles**

count

To count is to match number words with objects in a systematic order.

See also **one-to-one correspondence**

counting-back

Counting-back is to count backwards from a given number.

For example

Counting-back from ten is 10, 9, 8, 7, 6, 5, ...

See also **count, counting-on**

counting number

The counting numbers are 1, 2, 3, 4, 5, ... A counting number is a positive whole number. Counting numbers are also called natural numbers.

See also **positive number**, **whole number**

counting-on

Counting-on is to count forwards from a given number.

For example

Counting-on from ten is 10, 11, 12, 13, 14, 15, 16, ...

See also **count**, **counting-back**

cross-section

Meaning 1 A cross-section is the face that is formed by a plane cutting through a solid.

For example

The diagrams show cross-sections of a cone and a pyramid.

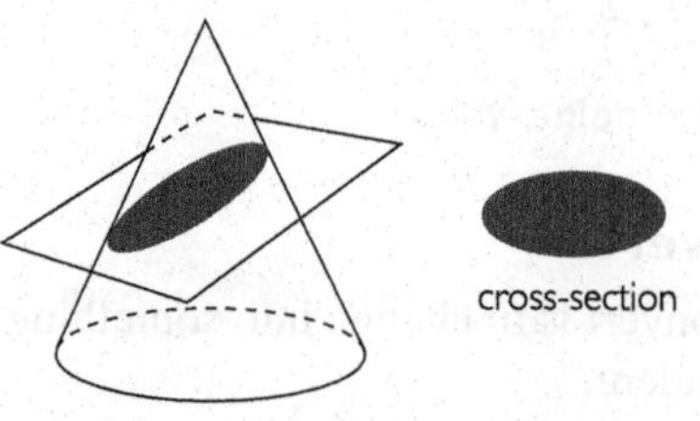

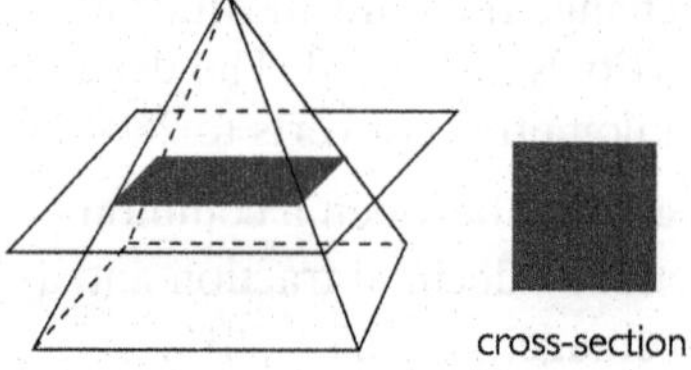

Meaning 2 A cross-section is a sample that is representative of a group.

For example

A cross-section of the population is representative of the whole population.

See also **sample**

cube

Meaning 1 A cube is a polyhedron with six square faces. The cube is one of the five Platonic solids.

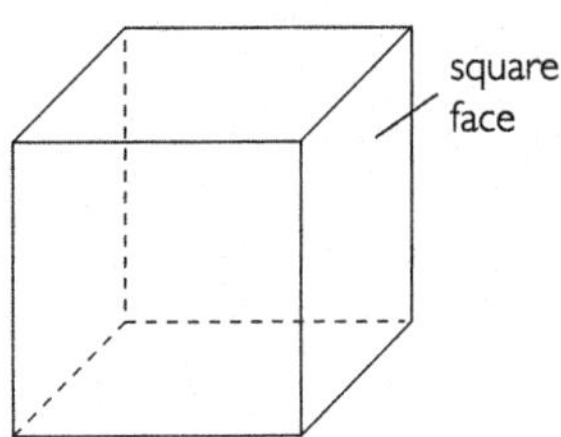

See also **Platonic solids**, **polyhedron**, **square**

Meaning 2 The cube of a number is the third power of the number.

For example

The cube of 2 is $2^3 = 2 \times 2 \times 2 = 8$.

See also **index**, **power**

cubic unit

A cubic unit is a unit for measuring volume. A cubic unit is a unit of length cubed. Cubic units include the cubic centimetre (cm^3) and cubic metre (m^3).

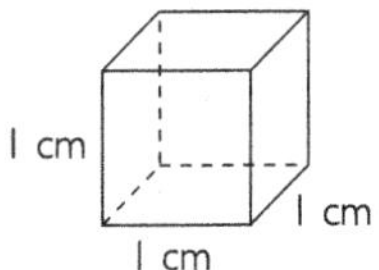

Cubic centimetre.

1 cubic centimetre = 1 millilitre

See also **SI system**, **standard units of measurement**, **unit**, **volume**

cuboid

A cuboid is a polyhedron with six rectangular faces. A cuboid is a type of rectangular prism.

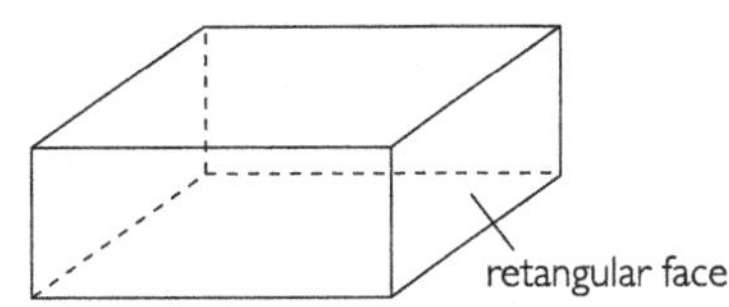

See also **polyhedron**, **prism**, **rectangle**

curve

Meaning 1 A curve is a line. A curve can be bending or straight, and it can be open or closed.

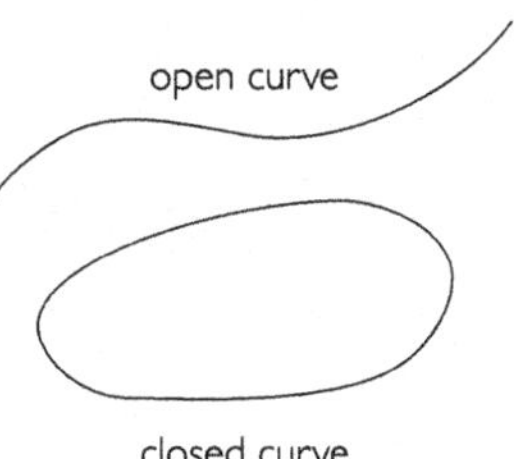

See also **closed curve**, **open curve**

Meaning 2 A curve is a line drawn on a graph that represents a continuous variation in quantity.

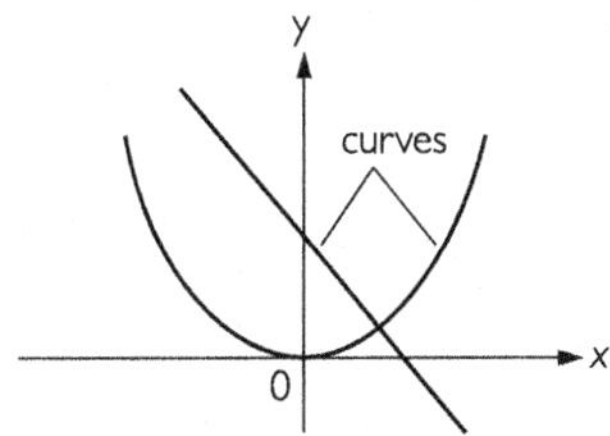

See also **graph**, **line graph**

cylinder

A cylinder is a solid shape made up from two discs of the same size joined together by a curved surface.

For example

Soft drink cans are usually in the shape of cylinders.

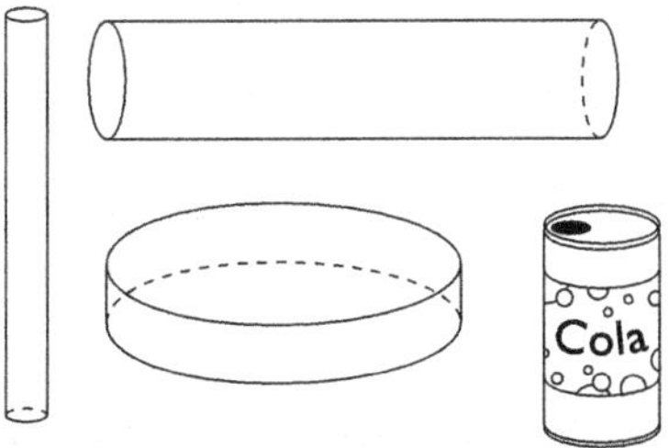

See also **solid**

D

Dd

data

Data is a collection of facts derived from direct observation in response to a question. A set of data is recorded as numbers, measurements or symbols. Worked data can be presented using statistics, such as average, and also using tables or graphs.

For example

What is the most popular make of car in the car park today? The set of data is the number of cars of each make observed throughout the day. The worked data gives the answer to the question.

CARS OBSERVED DURING THE DAY	
DAEWOO	|||
SUBARU	卌	
MAZDA	卌 |||
HOLDEN	卌 卌 ||||
FORD	卌 卌 ||||
TOYOTA	卌 卌 |||
HYUNDAI	卌 卌 ||
MITSUBISHI	卌 卌

MAKES OF CAR	NUMBER
DAEWOO	3
SUBARU	6
MAZDA	8
HOLDEN	14
FORD	14
TOYOTA	13
HYUNDAI	12
MITSUBISHI	10

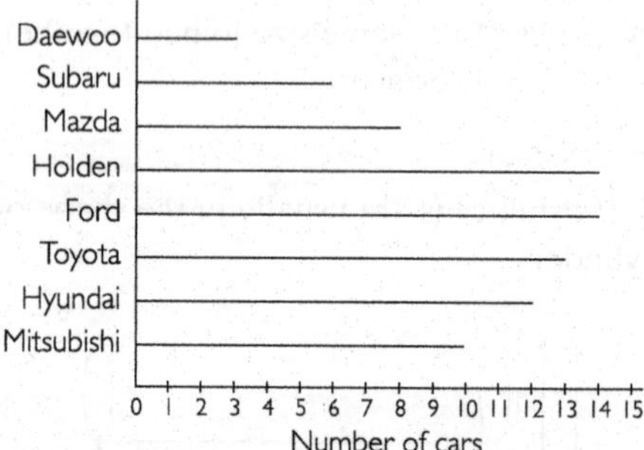

day

A day is a unit of time. It is the 24-hour period during which the Earth makes one revolution on its axis. There are seven days in a week: Monday, Tuesday, Wednesday, Thursday, Friday, Saturday and Sunday.

See also **calendar, hour, week**

decagon

A decagon is a polygon with ten straight sides. In a regular decagon, all the sides have the same length and all the angles are equal.

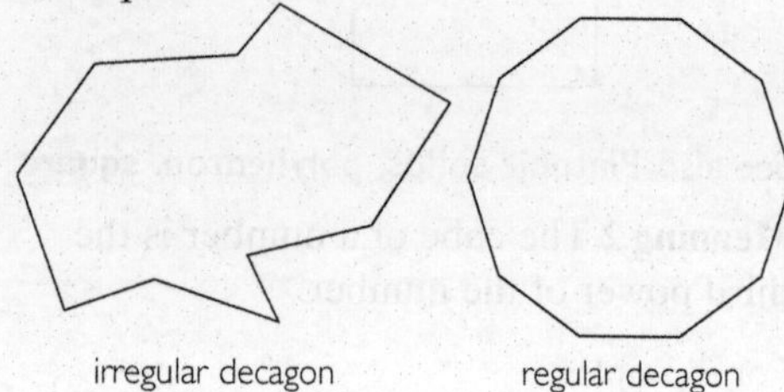

See also **polygon, regular polygon**

decahedron

A decahedron is a polyhedron with ten faces.

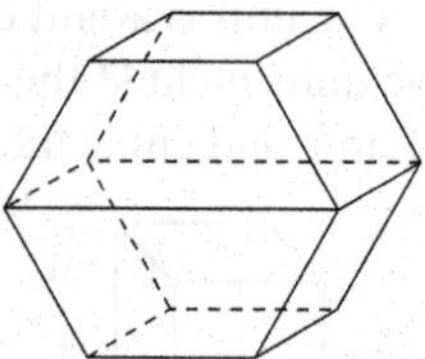

See also **polyhedron**

decimal

Decimal means based on the number 10.

See also **decimal currency, decimal fraction, decimal number, decimal place**

decimal currency

Decimal currency is a system of money based on multiples of ten.

For example

In Australia, 100 cents equals 1 dollar.

See also **decimal**

decimal fraction

A decimal fraction is a fraction written using a decimal point. It is written as tenths, hundredths, thousandths and so on, following the decimal point.

For example

0.5, 0.25 and 0.136 are decimal fractions.

See also **decimal point, decimal system, fraction, place value**

decimal number

A decimal number is any number containing a decimal point.

For example

3.5 and 786.25 are decimal numbers.

See also **decimal point, decimal system, real number**

decimal place

Decimal place means the position of a digit in a decimal fraction, that is, tenths, hundredths, thousandths and so on. Commonly, decimal fractions are rounded to one, two or three decimal places.

For example

0.729 rounded to two decimal places is 0.73 because 9 thousandths is close to one hundredth.

See also **decimal fraction, place value**

decimal point

A decimal point is the dot that separates the whole-number part and the fraction part in a decimal number.

See also **decimal number, fraction, whole number**

decimal system

The decimal system is the usual system for writing numbers. Numbers are written using the ten digits 0, 1, 2, 3, 4, 5, 6, 7, 8, 9 and place value. Each place value is a power of ten. The decimal system is also called the base-ten system.

See also **Arabic numerals, base, binary system, place value, power**

decomposition

Decomposition means breaking into parts. Decomposition is the foundation of regrouping.

For example

1 To simplify a subtraction problem:

$$\begin{array}{r} 53 \\ -\ 18 \\ \hline \end{array} \quad \begin{array}{r} 50 + 3 \\ -\ 10 + 8 \\ \hline \end{array} \quad \begin{array}{r} 40 + 13 \\ -\ 10 + \ \ 8 \\ \hline 30 + \ \ 5 \end{array} \quad \begin{array}{r} 53 \\ -\ 18 \\ \hline 35 \end{array}$$

2 To simplify addition of fractions:

$$1\tfrac{1}{2} + \tfrac{3}{4} = 1 + \tfrac{1}{2} + \tfrac{1}{2} + \tfrac{1}{4}$$

$$= 1 + 1 + \tfrac{1}{4}$$

$$= 2\tfrac{1}{4}$$

See also **part, regroup, rename**

decrease

To decrease is to make less in amount, size or quantity.

For example

1 Decrease the amount of water by half.

2 Decrease the length of the line by 10%.

3 Decrease the number of counters by five.

See also **increase**

D

degree

Meaning 1 A degree is a unit for measuring angles. There are 360 degrees in a complete turn, or circle. The symbol for degree is °. Each degree is divided into 60 minutes, and each minute is divided into 60 seconds.

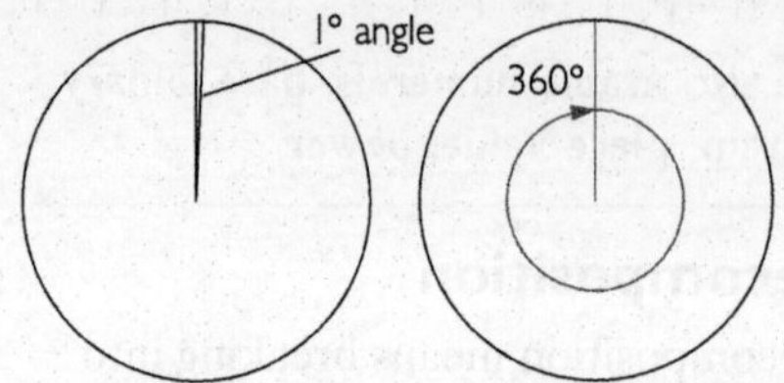

See also **angle, protractor**

Meaning 2 A degree is a unit for measuring temperature. On the Celsius scale, water freezes at approximately 0 degrees and boils at approximately 100 degrees. The symbol for degree is °.

See also **temperature**

denominator

The denominator is the part of a common fraction that is below the fraction line. The denominator indicates how many equal parts make up the whole.

For example

$\frac{1}{2}$ names one part out of two equal parts of a whole.

The denominator of $\frac{1}{2}$ is 2.

See also **common fraction, fraction, numerator, part, whole**

depth

Depth is the distance downwards, inwards or backwards.

For example

1 Depth is the distance below the surface of water.

2 Depth is one dimension of a three-dimensional figure.

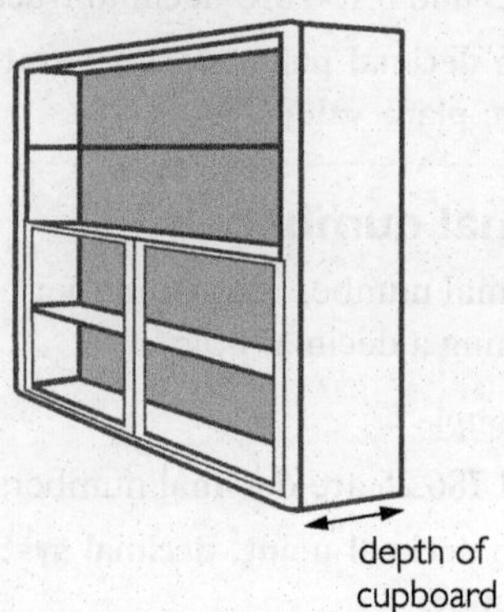

See also **height, length, width**

derived unit

A derived unit is a special kind of unit of measurement. A derived unit is made up of two or more different standard units of measurement.

For example

1 Centimetres per week measures plant growth.

2 Kilometres per hour measures the speed of a car.

3 Litres per second measures water flow.

4 Kilograms per cubic metre measures density.

5 Metres per second per second measures the acceleration of a falling object.

See also **rate, SI system, standard units of measurement**

descending order

Descending order is from largest to smallest.

For example

1 The numbers 20, 15, 10, 5 are arranged in descending order.

2 The volumes 2 L, 1.5 L, 0.75 L are arranged in descending order.

See also **ascending order**

diagonal

Meaning 1 A diagonal is a straight line drawn between two vertices of a polygon or polyhedron that are not adjacent.

For example

A diagonal divides a rectangle into two right-angled triangles.

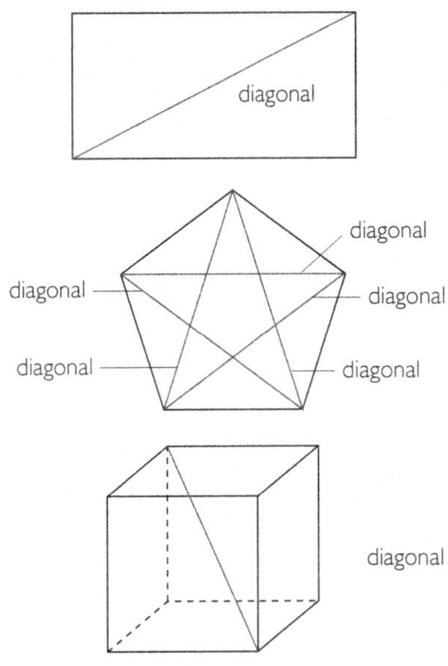

See also **adjacent**, **polygon**, **polyhedron**, **vertex**

Meaning 2 A diagonal plane cuts through a solid figure from one edge to an opposite edge.

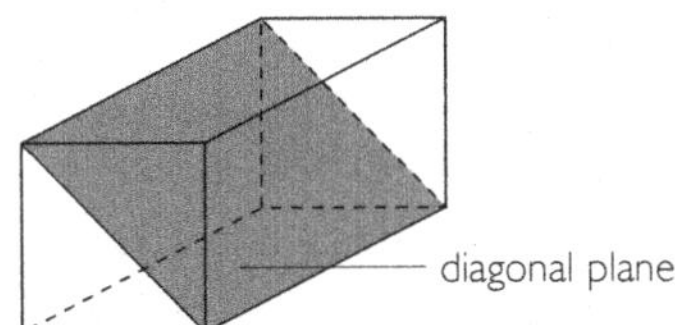

See also **edge**, **plane**

diameter

Meaning 1 A diameter is any straight line that joins two points on a circle and passes through the centre. Similarly, a line joining two points on a sphere and passing through the centre is the diameter of a sphere.

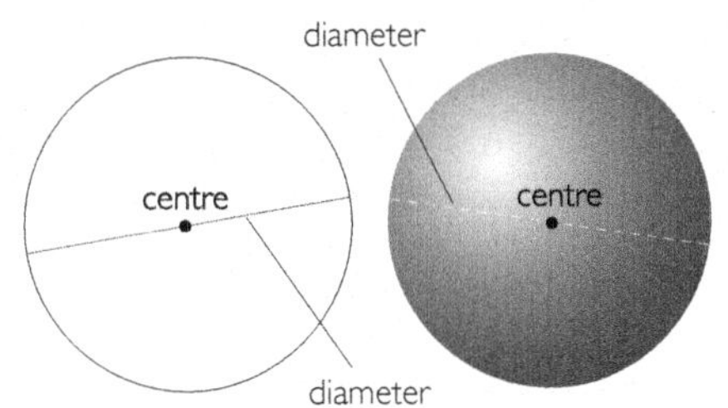

Meaning 2 A diameter is the length of a line that joins two points on a circle and passes through the centre. Similarly, the length of a line joining two points on a sphere and passing through the centre is the diameter of a sphere.

For example

The diameter of the circle is 4 centimetres.

See also **circle**, **point**, **radius**, **sphere**

die

A die is a regular solid, usually a cube, marked with dots or numerals. A die is thrown to generate a random number. The repeated throwing of a die generates a random sequence of numbers, as may be needed in games of chance. The plural form of the word *die* is dice.

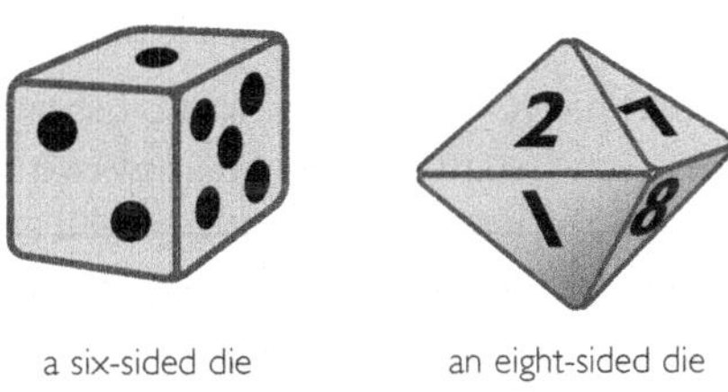

a six-sided die   an eight-sided die

See also **random device**

difference

Difference is the amount by which one quantity is greater or less than another.

For example

The difference between 8 and 3 is 5, which can be shown as $8 - 3 = 5$.

See also **greater than**, **less than**, **subtraction**

digit

A digit is one of the Arabic numerals: 0, 1, 2, 3, 4, 5, 6, 7, 8, 9.

For example

7942 is a four-digit number.

See also **numeral**

dilation

Dilation means enlargement. It is a transformation that maps a shape from a fixed point, using a scale factor. Dilation is also known as dilatation.

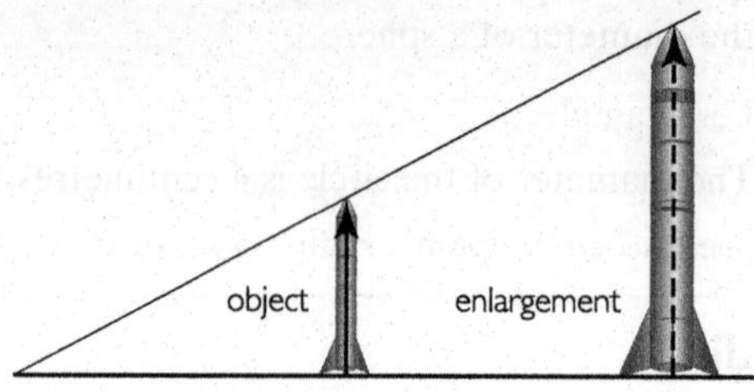

See also **scale**, **transformation**

dimension

Dimension means a measure of size or direction in space.

For example

A point has no dimension; a line has one dimension (length); a polygon has two dimensions (length and width); and a solid has three dimensions (length, height and width).

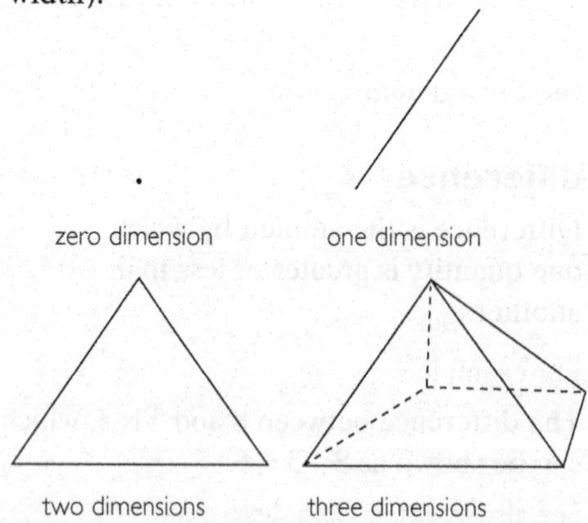

See also **space**

directed number

A directed number is a whole number for which a positive or negative direction is indicated, using a + or – symbol. A number line can show the distance of a number from zero in both positive and negative directions. A directed number is also called an integer.

For example

–3, –2, –1, +1, +2 and +3 are all directed numbers.

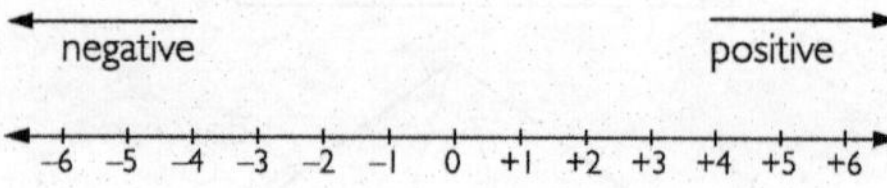

See also **integer**, **negative number**, **positive number**

direction

Direction is where a moving object is headed, or it is where something is pointing or looking.

For example

1. Left, right, up, down, forwards, backwards, clockwise and anticlockwise all indicate direction.
2. The compass points indicate direction.
3. The symbol + or – in front of a whole number indicates the direction of that number from zero.

direct proportion

Direct proportion is a constant relationship between two variables.

For example

Number pairs in a constant ratio are in direct proportion.

If lemons cost 20 cents each, the number bought and the cost can be represented by the number pairs (1, 20), (2, 40), (3, 60), …

Number pairs in direct proportion make a straight-line graph.

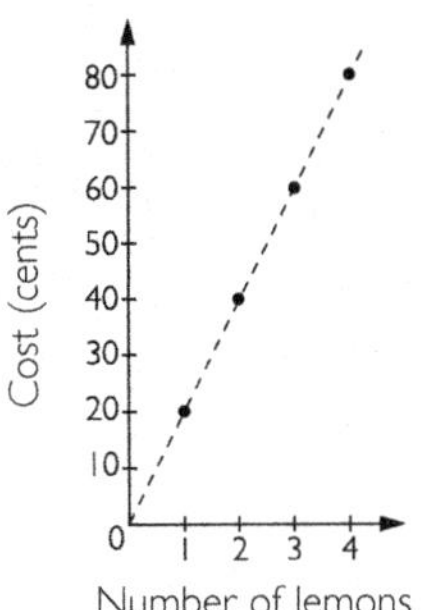

See also **constant**, **variable**

discrete

Discrete means separate or distinct. Discrete is the opposite of continuous.

For example

Counters are discrete objects. A tally of the number of counters of each colour is discrete data.

See also **continuous**

distance

Distance is the length between two points.

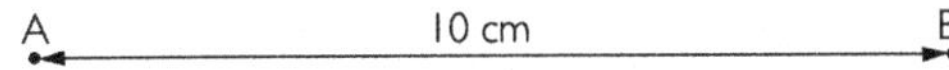

The distance between A and B is 10 cm.

See also **length**, **point**

distribution

A distribution is a set of numbers representing a group of observations or measurements.

For example

The distribution of child heights in a class is:

115 cm, 123 cm, 124 cm, 126 cm, 127 cm, 129 cm, 131 cm, 135 cm, 142 cm

When the number of children of each height is indicated, it becomes a frequency distribution.

Distribution.

Student height in cm								
115	123	124	126	127	129	131	135	142

Frequency distribution.

Student height in cm	*Number of students*
115–119	1
120–124	2
125–129	3
130–134	1
135–139	1
140–144	1

distributive

An operation is distributive if it does not matter whether it is carried out before or after another operation. This allows brackets to be expanded.

For example

1 The distributive property is used to simplify algebraic expressions.

Multiplication is distributive over addition.

$$\begin{aligned} 3(a + 2) &= 3 \times (a + 2) \\ &= 3 \times a + 3 \times 2 \\ &= 3a + 6 \end{aligned}$$

Multiplication is distributive over subtraction.

$$\begin{aligned} 2(a - b) &= 2 \times a - 2 \times b \\ &= 2a - 2b \end{aligned}$$

Division is distributive over addition.

$$\frac{(4a + 2)}{2} = \frac{4a}{2} + \frac{2}{2} = 2a + 1$$

Division is distributive over subtraction.

$$\frac{(9b - 6)}{3} = \frac{9b}{3} - \frac{6}{3} = 3b - 2$$

2 The distributive property is used to simplify calculations. It is the basis of renaming, regrouping, decomposition.

$$\begin{aligned} 6 \times 14 &= 6 \times (10 + 4) \\ &= 6 \times 10 + 6 \times 4 \\ &= 60 + 24 \\ &= 84 \end{aligned}$$

See also **associative**, **commutative**

diverge

To diverge is to get further apart.

For example

1 The lines *oa* and *ob* diverge.

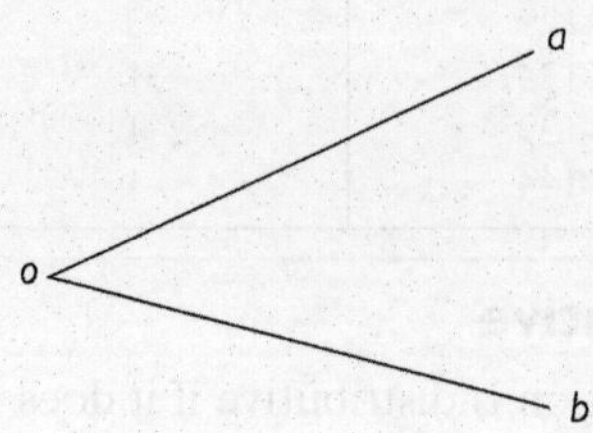

2 The sequence 1, 2, 4, 8, ... diverges since the difference between the numbers in the sequence is increasing.

See also **converge**

dividend

The dividend is the number that is to be divided.

For example

In $48 \div 12 = 4$, the dividend is 48.

See also **division**, **divisor**, **quotient**

division

Division is the operation of finding out how many times one number goes into another number by either grouping or sharing. Division is the inverse of multiplication. Two symbols are used for division. They are the ÷ and / symbols.

For example

The problem $12 \div 3$ can be thought of as "How many groups of 3 make up 12?" or as "If 12 is shared into 3 groups, how many in each group?"

Grouping

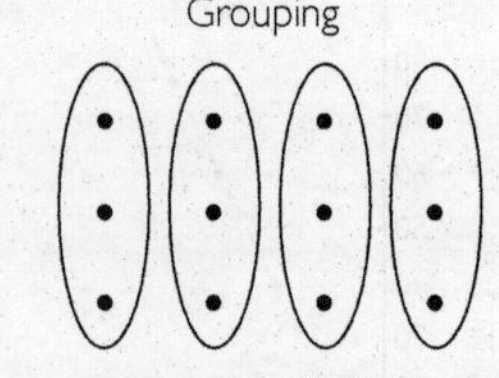

How many boxes of 3 can be made from 12 balls?

$12 \div 3 = 4 \quad 3\overline{)12}$ (quotient 4)

Sharing

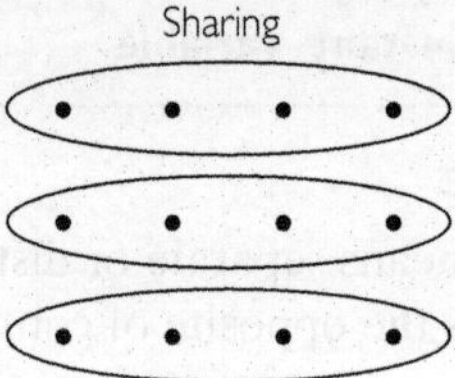

Share 12 balls among 3 boxes. How many in each box?

$12 \div 3 = 4 \quad 3\overline{)12}$ (quotient 4)

See also **grouping**, **operation**, **partition**, **quotition**, **share**

divisor

The divisor is the number that does the dividing.

For example

In $48 \div 12 = 4$, the divisor is 12.

See also **dividend**, **division**, **quotient**

dodecagon

A dodecagon is a polygon with twelve straight sides. In a regular dodecagon, all the sides have the same length and all the angles are equal.

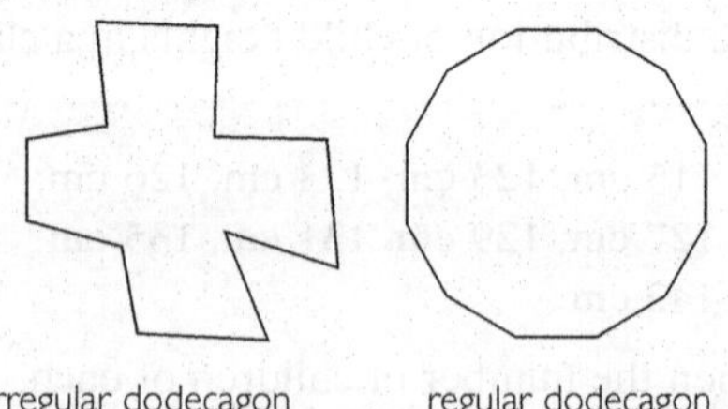

See also **polygon**, **regular polygon**

dodecahedron

A dodecahedron is a polyhedron with twelve faces. The faces of a regular dodecahedron are congruent, regular pentagons. The regular dodecahedron is one of the five Platonic solids.

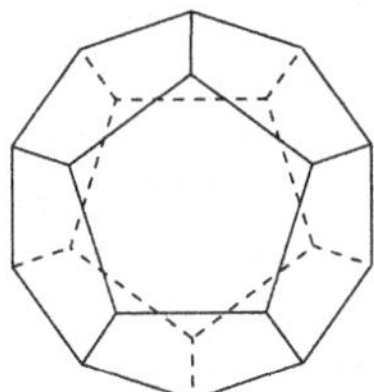

regular dodecahedron

See also **Platonic solids**, **polyhedron**

dollar

A dollar is the name of a unit in the decimal money systems of Australia, the USA and Hong Kong, for instance. The symbol for dollar is $.

1 dollar = 100 cents

For example

The lunch cost $5.

See also **cent**, **decimal currency**

dot paper

Dot paper is paper printed with a regular dot pattern that may be square or isometric. Dot paper is used as a guiding framework for drawings and graphs.

For example

Dot paper is useful in making three-dimensional and scaled drawings of shapes.

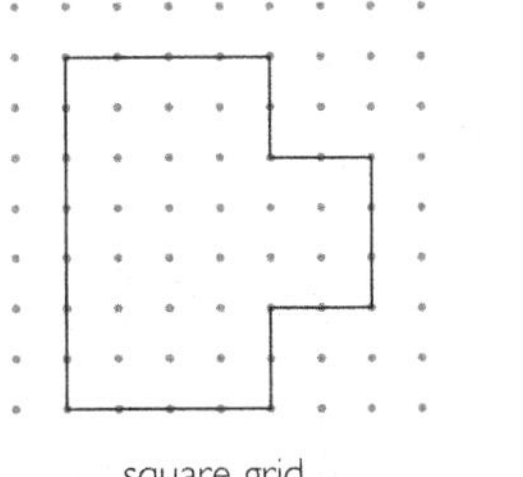

square grid

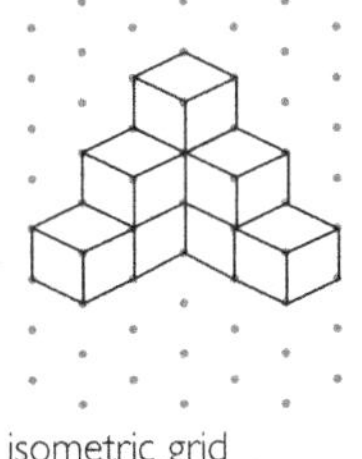

isometric grid

See also **isometric drawing**, **pattern**

dot plot

A dot plot is a graph. It uses a scale to represent data. Each item is shown by a dot on the scale. A dot plot is sometimes called a line plot.

For example

Maximum daily temperatures can be shown on a dot plot.

Maximum daily temperatures in March.

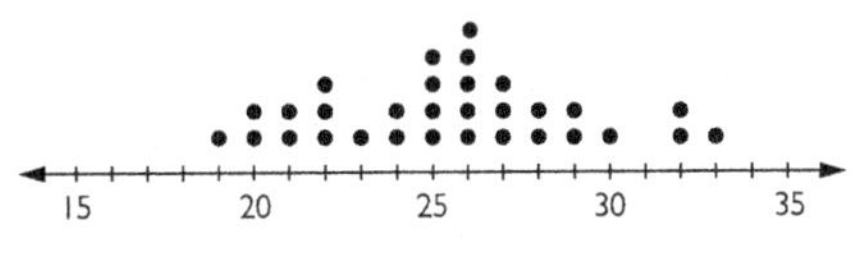

See also **data**, **graph**

double

Double means twice as much or twice as many. The quantity is multiplied by two. The opposite of doubling is halving.

For example

1 Double 5 is 10.

2 8 kilograms is double 4 kilograms.

See also **twice**

dozen

A dozen is a group of twelve.

For example

A dozen eggs is twelve eggs.

duration

Duration is the length of time that anything continues. Duration is the time elapsed.

See also **elapsed time**, **time**

Ee

edge

An edge is where two faces meet.

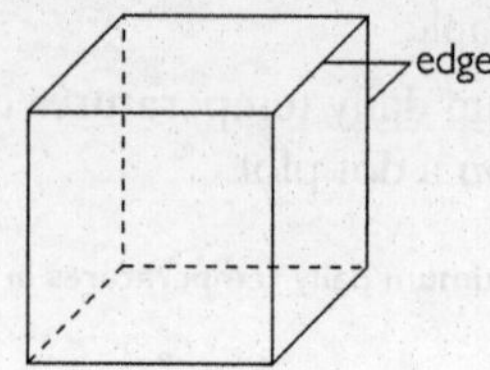

A cube has 12 edges.

See also **Euler, intersection, polyhedron**

elapsed time

Elapsed time is time passed. It is how long it takes to do something. Another name for elapsed time is duration.

For example

The time it takes to run around the oval is elapsed time.

See also **duration, time**

element

Meaning 1 An element is a member of a set.

For example

1 The set of the days in the week has 7 elements: Monday, Tuesday, Wednesday, Thursday, Friday, Saturday, Sunday.

2 The set of natural numbers {1, 2, 3, 4, 5, 6, ...} has an infinite number of elements.

See also **set**

Meaning 2 An element is any one of the units that make up an array.

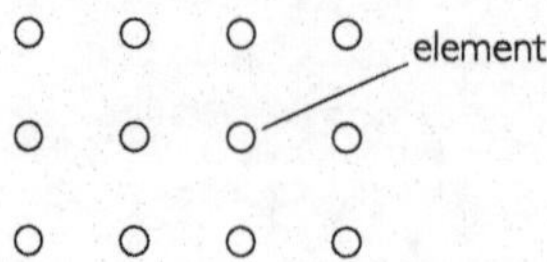

See also **array**

elevation

Elevation is the angle between a horizontal plane and an object above.

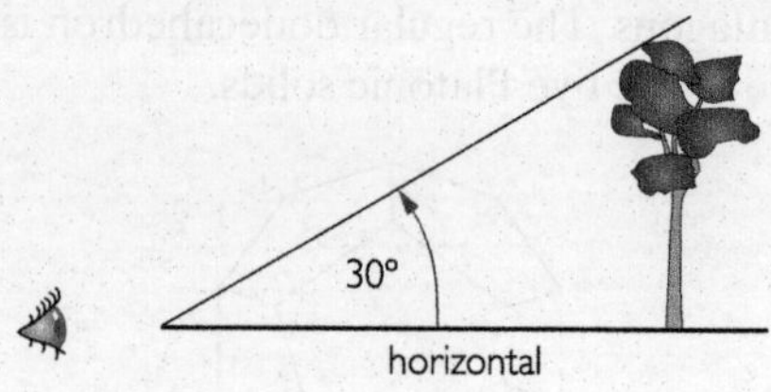

See also **horizontal, plane**

ellipse

An ellipse is a particular closed, plane curve.

For example

1 An ellipse is formed when a cone is cut through by a plane.

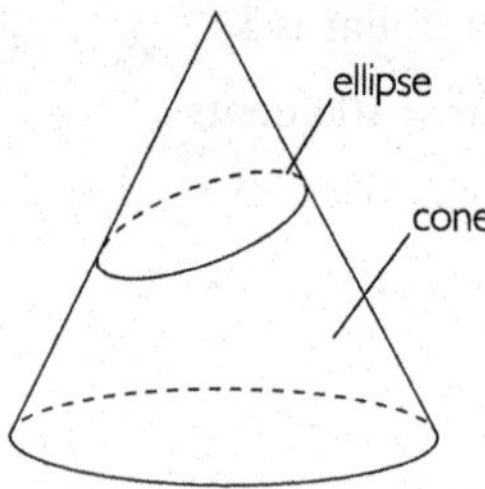

See also **circle, cross-section, oblique**

2 An ellipse can be drawn by fixing the ends of a piece of string and then pulling the string around by the pencil.

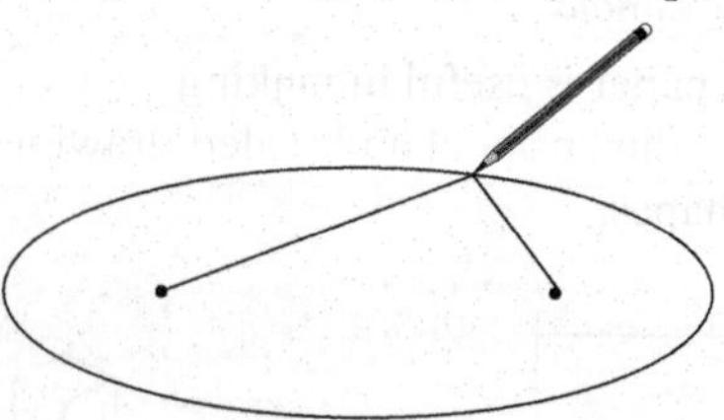

See also **path, point**

end point

An end point is one of the two points that define an interval.

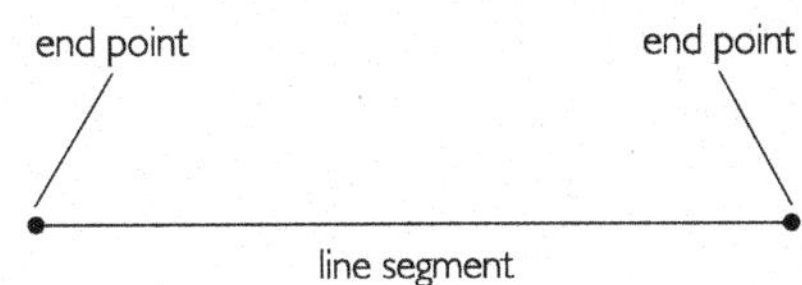

See also **interval**, **point**

enlargement

Meaning 1 Enlargement is the use of a scale factor to increase or reduce the size of a shape. The scale factor can be more than or less than 1. Enlargement produces an image of the shape. Enlargement is one type of transformation.

For example

1 The method of squares uses grids of different sizes as a guide for the enlargement of a drawing.

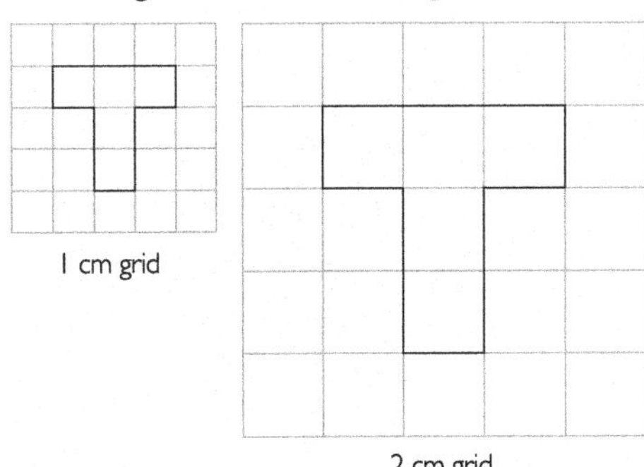

The letter T has been enlarged using the scale factor 2.

2 A photographic enlarger uses light rays through a lens to project an enlarged image.

Meaning 2 An enlargement is the image of a shape that has been increased or reduced in size.

See also **dilation**, **direct proportion**, **ratio**, **scale**, **transformation**

equal

Equal means the same in value, quantity or amount.

For example

1 $1.00 is equal in value to 100 cents.

2 $5 + 4 = 9$

3 Two events may have an equal chance of happening.

4 Four can be shared into two equal parts, or into four equal parts.

5 The flour and rice are equal in mass.

See also **amount**, **quantity**, **value**

equality

Equality is the state of being equal.

See also **equal**, **inequality**

equal sign

The equal sign (=) is the symbol that expresses equality.

For example

$2 + 6 = 8$

See also **equality**, **symbol**, **unequal**

equation

An equation is a mathematical statement that one expression is equal to another.

For example

1 $8 + 2 = 6 + 4$

2 $y = 3x + 1$

3 $2(3x + 4) = 6x + 8$

4 $x^2 + 2x + 1 = 0$

See also **equal**, **equal sign**, **linear equation**, **pronumeral**, **simultaneous equations**

equilateral

Equilateral means having all sides of the same length.

☞

For example

1 Regular polygons are equilateral.

2 A triangle with sides of equal length is called an equilateral triangle.

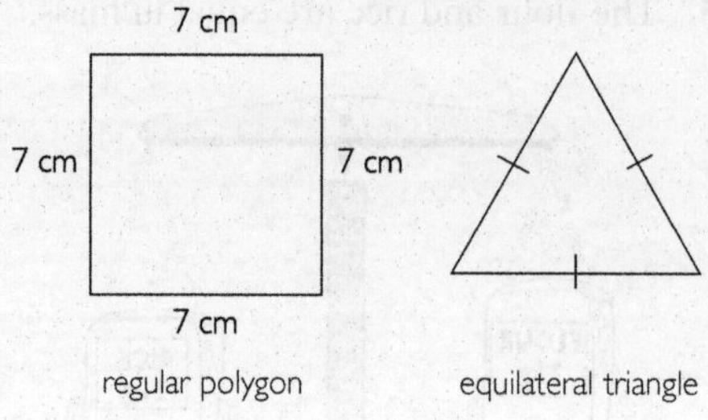

See also **equal, side**

equivalent equations

Equivalent equations are equations that hold the same information.

For example

The three equations below are equivalent:

$$2x + 6 = 10$$
$$2x = 4$$
$$x = 2$$

Each of these equations holds the same information about x.

equivalent fractions

Equivalent fractions are fractions that name the same amount.

For example

1 $\frac{1}{2}$, $\frac{2}{4}$, $\frac{4}{8}$ and $\frac{8}{16}$ are equivalent fractions.

2 $\frac{1}{4}$, 0.25, 25% and 1:4 are equivalent fractions.

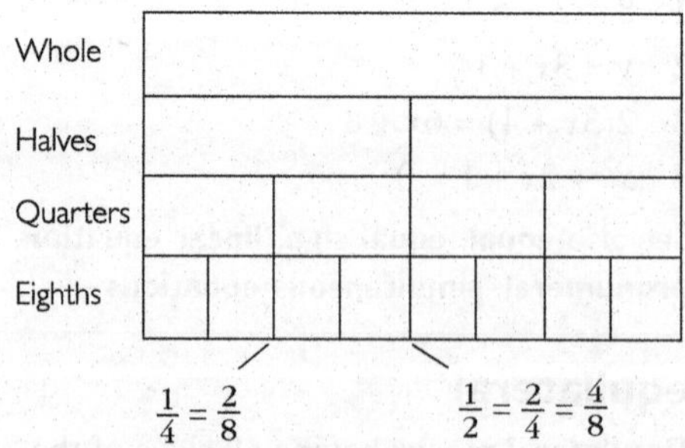

See also **amount, decimal fraction, equal, fraction, percentage, ratio**

Eratosthenes

Eratosthenes was a Greek mathematician who lived at about the same time as Archimedes. He lived 276–194 BC. He had a method for finding prime numbers in a sequence of consecutive numbers. It is called the sieve of Eratosthenes. He also devised a method for working out the distance around the Earth, by using his knowledge of geometry.

See also **Archimedes, prime number, sieve of Eratosthenes**

Escher

Maurits Escher was a Dutch artist whose works included many mathematical ideas. He lived 1898–1972. He used space, scale, tessellations, transformations and the Möbius strip in his drawings.

M.C. Escher's 'Circle Limit I' © 2002 Cordon Art B.V., Baarn, Holland. All rights reserved.

estimate

Meaning 1 To estimate is to make a reasonable guess, or to calculate approximately.

Meaning 2 An estimate is a reasonable guess or an approximate calculation.

See also **accurate, approximation**

estimation strategies

Estimation strategies are techniques for quickly finding an approximate answer to a problem. They can also be used to check whether a solution to a problem is reasonable. The strategies are based on the use of place value, renaming and rounding.

For example

1 When finding 42×39, think 42×40 and subtract 42 from the answer.

2 To get an approximate answer to \$15.99 + \$12.05, work out \$16.00 + \$12.00.

3 When finding $173.25 \div 5.2$, work out 17 tens shared among 5.

See also **approximation**, **front-end estimation**

Euclidian geometry

Euclidean geometry is the geometry of the Greek mathematician Euclid, who lived around 300 BC. Euclid's geometry deals with points, lines and planes, and has served as a model of formal geometry up to the present day.

See also **fractal geometry**, **geometry**, **non-Euclidean geometry**, **topology**

Euler

Leonard Euler was a Swiss mathematician who lived 1707–1783. He was the first mathematician to study networks. He also discovered the formula $F + V - E = 2$ for a polyhedron, where F is the number of faces, V is the number of vertices and E is the number of edges.

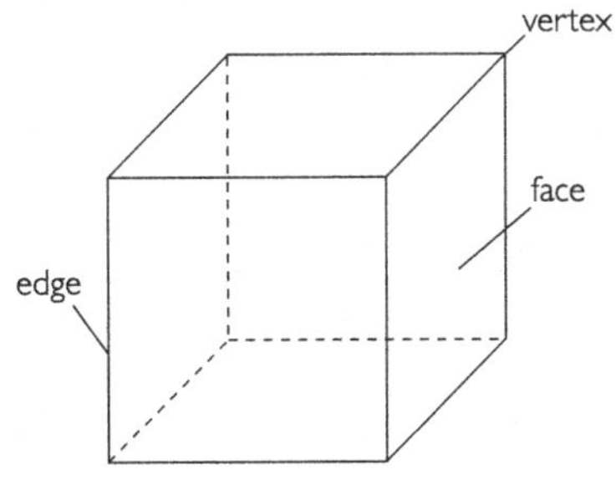

The cube has 6 faces, 8 vertices, 12 edges: 6 + 8 – 12 = 2.

See also **network**, **polyhedron**

evaluate

To evaluate is to find the value of.

For example

Evaluate the algebraic expression $3x + 2y$, knowing that $x = 2$ and $y = 3$.

The expression becomes $3 \times 2 + 2 \times 3 = 6 + 6 = 12$.

See also **value**

even number

An even number is a whole number that can be divided equally by two.

For example

2, 4, 6, 8, 10, 12 are all even numbers because they are all divisible by 2 exactly.

See also **odd number**, **whole number**

event

An event is something that happens, as in statistics and probability.

For example

A toss of a coin is an event.

See also **probability**, **statistics**

exact

Exact means accurate or correct. Exact is not approximate.

See also **accurate**, **approximation**

expanded notation

To write in expanded notation is to rewrite in fuller form. Expanded notation is also known as extended notation.

For example

1 Numbers can be expanded using place value. For instance:

$$1259 = 1000 + 200 + 50 + 9 = (1 \times 1000) + (2 \times 100) + (5 \times 10) + 9$$

$$2431 = (2 \times 10^3) + (4 \times 10^2) + (3 \times 10) + 1$$

2 Algebraic expressions can be expanded using the distributive laws. For instance:

$$3(a + b) = 3a + 3b$$
$$= a + a + a + b + b + b$$

See also **distributive**, **place value**, **power**

exponent

The exponent is the number that indicates the power to which a number is raised. The exponent is written as a superscript after the number.

For example

3 is the exponent in 4^3, which means $4 \times 4 \times 4$.

See also **index**, **power**

expression

An expression consists of a string of numbers and symbols connected by operations.

For example

1 $4 + 7$ is an expression.

2 $7x^2 - y$ is an expression.

See also **equation**, **number**, **operation**, **symbol**

exterior angle

An exterior angle is formed on the outside of a plane figure when one of its sides is extended beyond the vertex. The exterior angle is between the extension and the other side that meets that vertex.

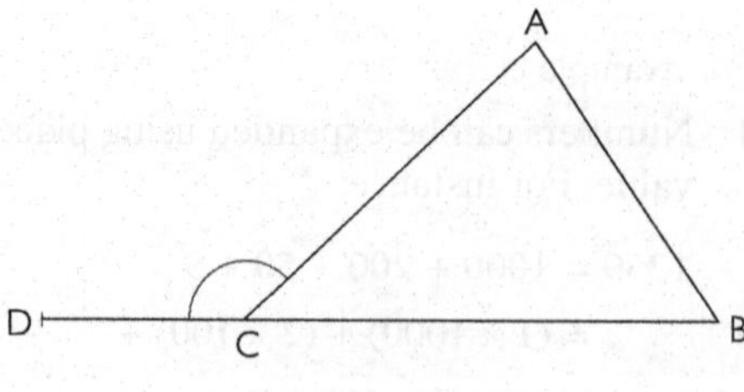

ACD is an exterior angle.

See also **interior angle**, **vertex**

Ff

face

A face is any one of the flat surfaces of a three-dimensional shape. A face is bounded by edges.

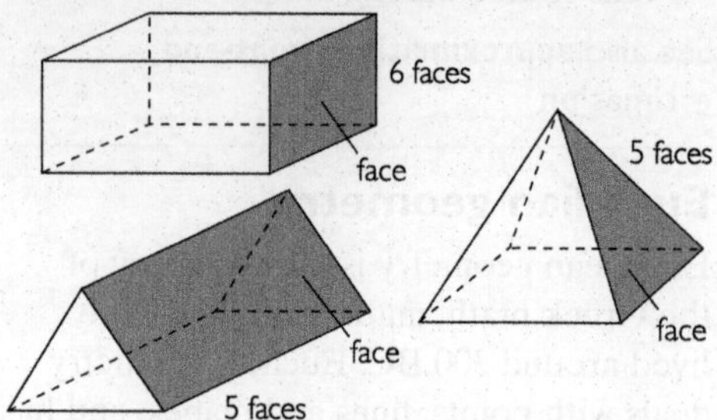

See also **Euler**, **surface**, **three-dimensional**

factorise

To factorise is to break down a given whole number or expression into a product of two or more factors.

For example

Factorise $2x + 2y$.

The number 2 is a factor of $2x + 2y$ since 2 goes into both $2x$ and $2y$.

$$2x + 2y = 2 \times x + 2 \times y$$
$$= 2 \times (x + y) = 2(x + y)$$

See also **distributive**, **factors**, **factor tree**

factors

Factors are the quantities that divide exactly into a given quantity. Factors can be whole numbers, algebraic terms or expressions.

For example

1 The numbers 1, 2, 4 and 8 each divide exactly into 8. The factors of 8 are 1, 2, 4 and 8.

2 The expression $ab + ac$ is equal to $a(b + c)$. The factors of $ab + ac$ are a and $b + c$.

See also **division**, **highest common factor**, **prime number**

factor tree

A factor tree is a diagram that shows all the factors of a given whole number or expression. The diagram is used to find the prime factors of a number.

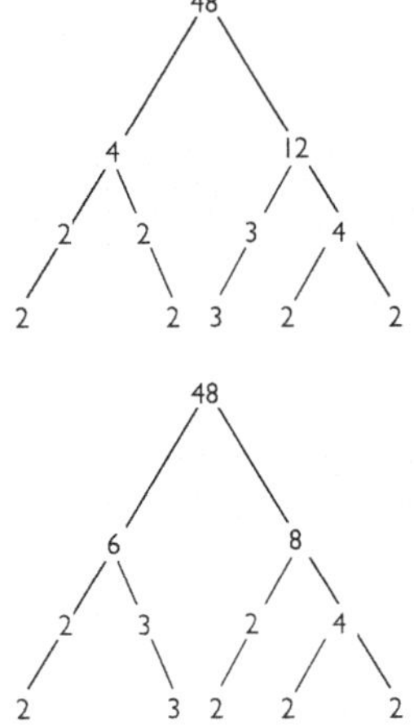

Examples of factor trees for 48.

See also **factors**, **prime factor**

Fibonacci sequence

The Fibonacci sequence of numbers was introduced by Fibonacci, an Italian mathematician who lived in the 12th century. In the Fibonacci sequence, each number after the second is the sum of the two previous numbers. The Fibonacci sequence begins 1, 1, 2, 3, 5, 8, 13, ...

figure

Meaning 1 A figure is a numerical symbol.

For example

Forty-eight written in figures is 48.

See also **symbol**

Meaning 2 A figure is a shape made up of points, lines, curves and surfaces.

For example

1 Circles and squares are plane figures.

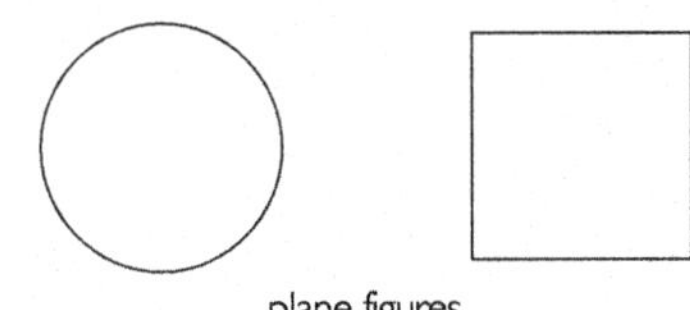

plane figures

2 Spheres and cubes are solid figures.

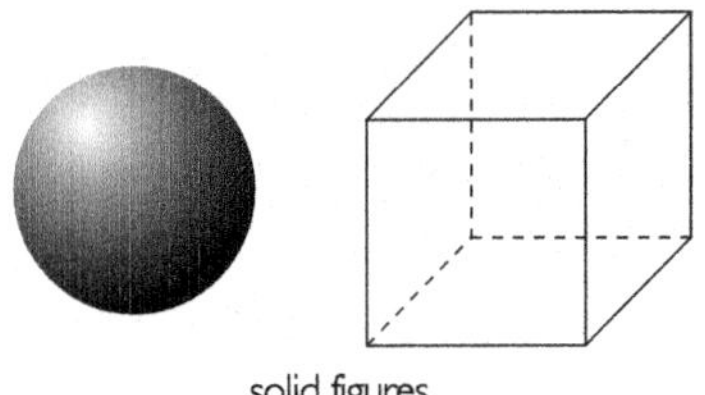

solid figures

See also **Euclidean geometry**, **geometry**

finite

Finite means bounded or limited.

For example

1 A polygon encloses a finite region.

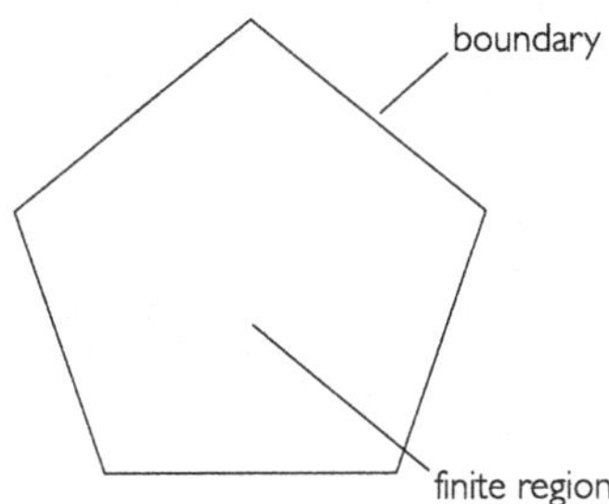

2 A finite set has a limited number of elements. For instance, {1, 2, 3, 4, 5} is a finite set.

See also **infinite**

flat

Flat means only having two dimensions.

For example

The faces of a cube are flat.

See also **plane**, **surface**

flip

A flip is one type of transformation. A flip involves turning over a plane shape about a line, which creates a reflection of the original shape.

☞

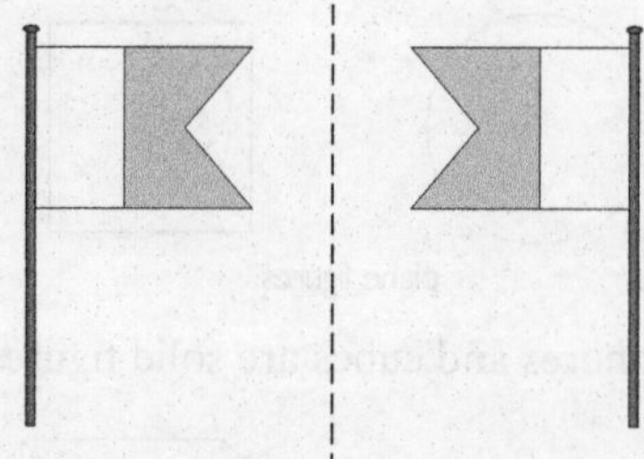

See also **slide, transformation, turn**

flow chart

A flow chart is a diagram of labelled shapes and arrows that shows the sequence of steps in solving a problem or the sequence of steps in a task. A flow chart is also known as a flow diagram.

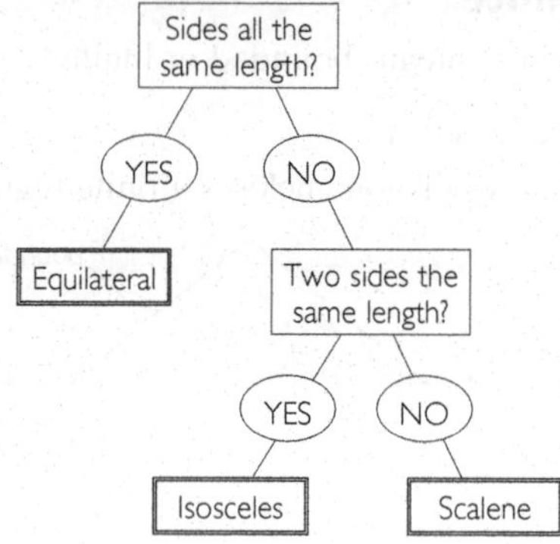

A flow chart for classifying a triangle.

See also **sequence**

formula

A formula is a rule or principle. A formula shows how to work out some unknown value using other known values. A formula is commonly expressed in algebraic symbols. The plural of the word *formula* can be either formulas or formulae.

For example

1 The area of a circle is pi times its radius squared. The formula is $A = \pi r^2$.

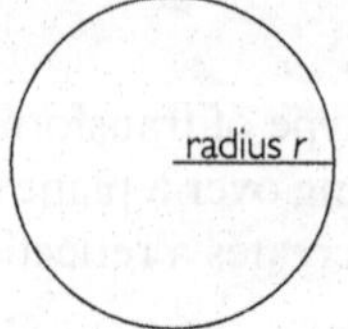

2 The volume of a cube is the length cubed. The formula is $V = l^3$.

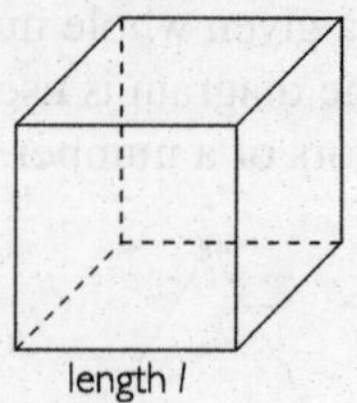

See also **Maths Formulae table** (page 126), **rule**

fortnight

A fortnight is a period of two weeks, which is fourteen days.

See also **day, week**

fractal

Fractal is a geometric term used to describe the irregular shapes in nature that repeat on different scales. Each smaller part of a fractal is a similar version of the larger whole. Fractal shapes include clouds, mountains, trees and coastlines. Computer programs are used to model fractals. The term *fractal* was coined by the mathematician Benoît Mandelbrot in 1975.

The waterfall and the water trickle are similar irregular shapes in nature that repeat on different scales.

See also **fractal geometry, geometry, scale, snowflake curve**

fractal geometry

Fractal geometry is the geometry of nature. It is a new field of geometry that originated in the work of mathematicians between 1875 and 1925.

See also **Euclidean geometry**, **fractal**, **geometry**, **topology**

fraction

A fraction is a part. A fraction can be part of a continuous whole, or part of a group of discrete objects. A fraction can be expressed as a common fraction, a decimal fraction, a percentage or a ratio.

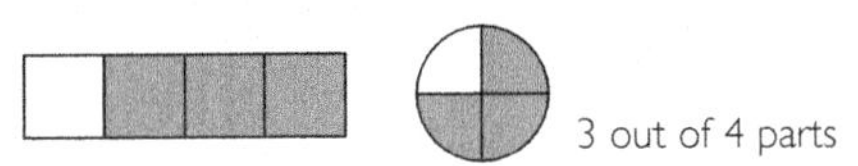

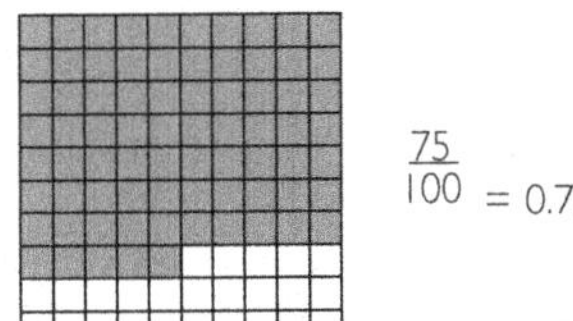

Cream

Milk

Mix 3 parts to 4 parts 3:4

See also **common fraction**, **continuous**, **decimal fraction**, **discrete**, **percentage**, **ratio**

fraction facts

Fraction facts are simple common-fraction addition and subtraction facts, and frequently used common-fraction, decimal and percentage equivalences.

For example

1 $\frac{1}{4} + \frac{1}{2} = \frac{3}{4}$

2 $\frac{1}{2} - \frac{1}{4} = \frac{1}{4}$

3 $0.2 = \frac{1}{5}$

4 $\frac{1}{4} = 25\%$

5 $33\frac{1}{3}\% = \frac{1}{3}$

See also **basic facts**

frequency

Frequency is the number of times that an item occurs in a collection of data.

For example

Among the pets of students in a class, there are 5 rabbits. The frequency of rabbits among the pets is 5.

Pets in 6B		
Dogs	~~||||~~ ~~||||~~ ||	12
Cats	~~||||~~ ~~||||~~ |	11
Birds	|||	3
Mice	||	2
Rabbits	~~||||~~	5

See also **data**, **frequency distribution**

frequency distribution

A frequency distribution is a group of frequencies associated with the different categories of a collection of data.

For example

The pets of the students in a class can be categorised. Each category has a frequency. The group of frequencies forms a frequency distribution.

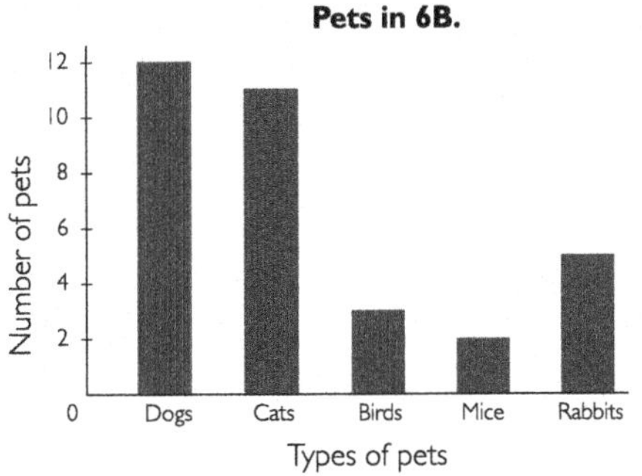

See also **data**, **frequency**

front-end estimation

Front-end estimation is a way to produce approximate answers to addition problems. The strategy is to consider only the leading digits, that is, the largest place value common to the numbers. This strategy is less useful for subtraction, multiplication and division.

472
534
119
+ 860
1800

See also **approximation, estimation strategies, rounding**

front view

A front view is what you see when looking at an object from the front.

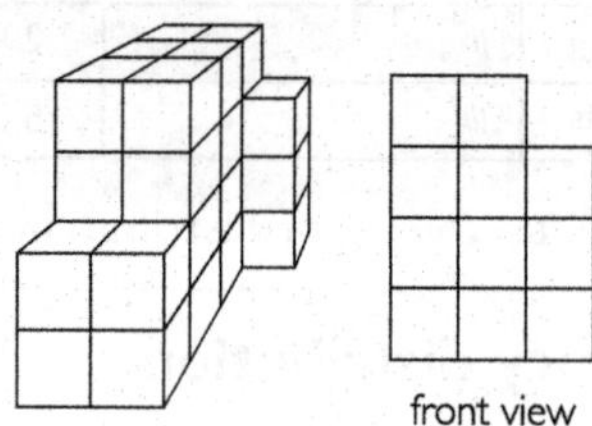

See also **side view, top view**

full-turn

A full-turn is a complete rotation. The angle of a full-turn is 360 degrees, which is the measure of a complete circle.

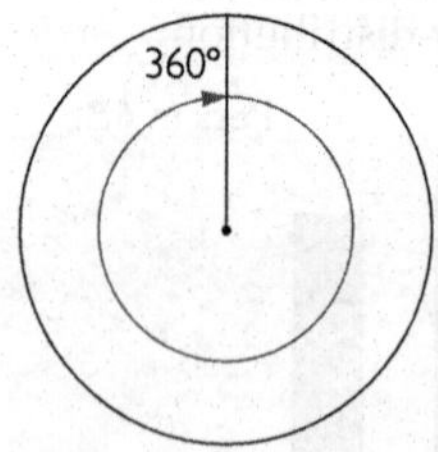

See also **rotation**

function

A function is a type of relation. A function relates each element in a set to exactly one element in another set. Functions can be represented by arrow diagrams, tables or graphs.

For example

1 This arrow diagram and table show the "hair colour" function for a group of students.

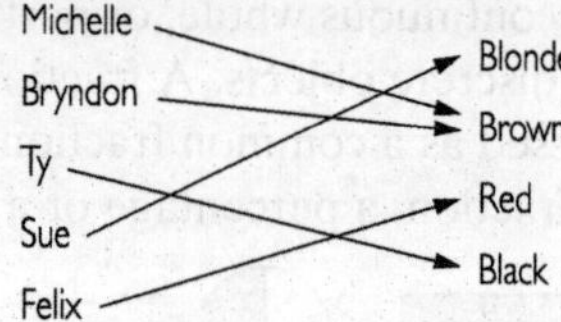

Student	Hair colour
Michelle	Brown
Bryndon	Brown
Ty	Black
Sue	Blonde
Felix	Red

2 This graph shows a function on the set of real numbers.

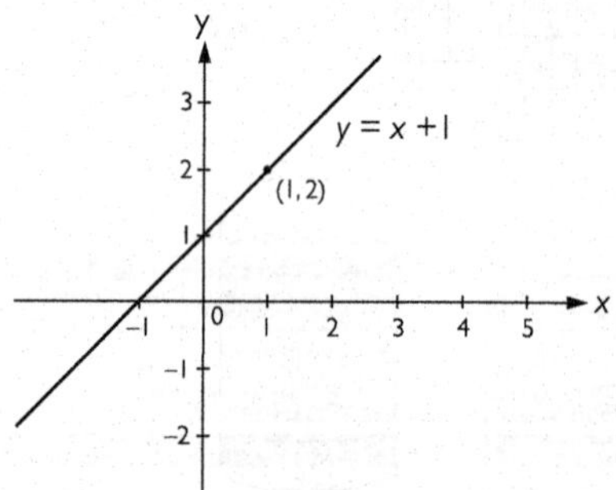

The point (1, 2) is on the graph, so this function relates 1 to 2.

See also **graph, mapping, relation**

Gg

general statement

A general statement expresses a general rule or principle. A general statement applies to more than one example.

For example

"The area of any triangle is half of its base times its height" is a general statement.

See also **rule**

generate

To generate means to bring into existence.

For example

1 We can generate data by collecting information.
2 We can generate a circle by moving a point around a fixed point at a fixed distance.

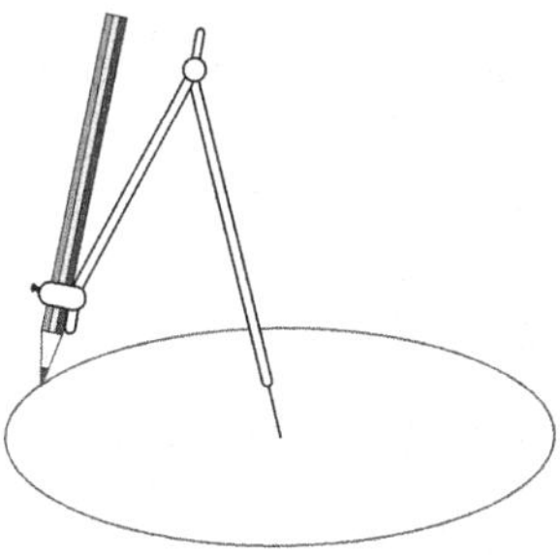

geometrical representation

A geometrical representation is a sketch, diagram, construction or model that communicates geometric concepts and ideas.

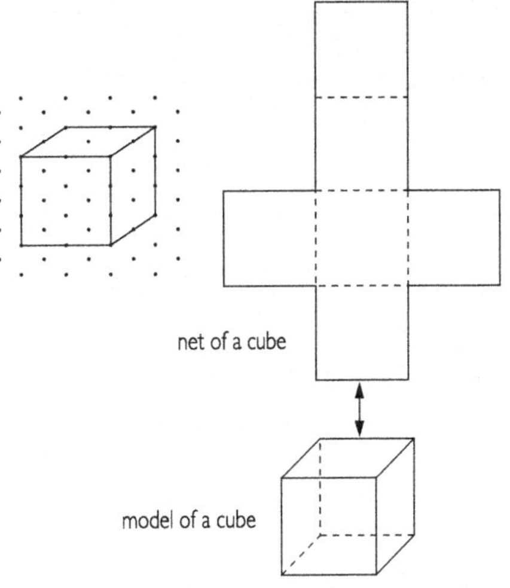

See also **geometry**

geometric progression

A geometric progression is a sequence of numbers in which the ratio between each number and the previous one is always the same.

For example

The sequence 3, 6, 12, 24, 48, ... has a common ratio of 2.

See also **arithmetic progression**, **ratio**, **sequence**

geometry

Geometry is that branch of mathematics that describes shape, size and position in space.

See also **Euclidean geometry**, **fractal geometry**, **topology**

golden ratio

The golden ratio is a special ratio. If a line is divided in the golden ratio, then the ratio of the whole line to the larger part is equal to the ratio of the larger part to the smaller part. This proportion is commonly found in art, architecture and nature, and is pleasing to the eye. The sides of the golden rectangle are in this ratio.

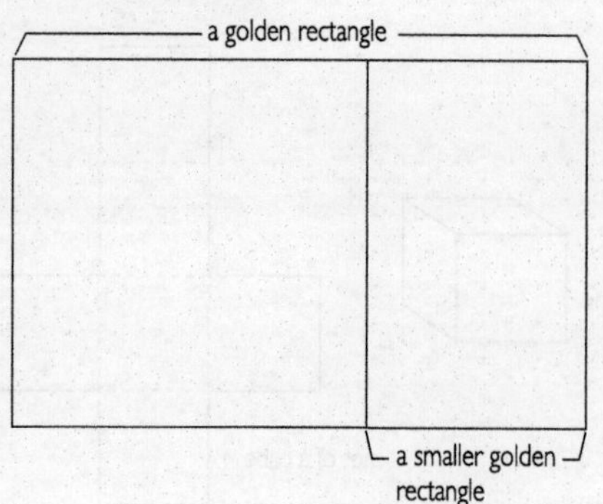

The golden ratio is 1: $\frac{1 + \div 5}{2}$

See also **ratio**

googol

A googol is the number written as 1 followed by 100 zeros, or 10^{100}. The word *googol* was coined by a nine-year-old child.

See also **power, zero**

gradient

Gradient is a measure of slope or steepness. The gradient of a line is the ratio of the distance risen to the horizontal distance travelled.

$$\text{gradient} = \frac{\text{vertical distance}}{\text{horizontal distance}}$$

$$\textbf{gradient} = \frac{\textbf{vertical distance}}{\textbf{horizontal distance}} = \frac{1}{2}$$

See also **ratio**

graduation

A graduation is one of the marks on a measuring jug or measuring instrument that shows the divisions of the measuring scale.

For example

1 The graduations on a measuring jug mark out millilitres.

2 The graduations on a ruler mark out millimetres and centimetres.

See also **calibration, scale**

gram

A gram is a unit of mass. The symbol for gram is g.

1000 grams = 1 kilogram

For example

The minimum mass of each egg in the carton is 61 grams.

See also **kilogram, mass, SI system, standard units of measurement**

graph

A graph is a drawing that displays information. Graphs can make information easier to understand, and show patterns in complicated data. There are many different types of graphs.

Birthdays in Grade 4.

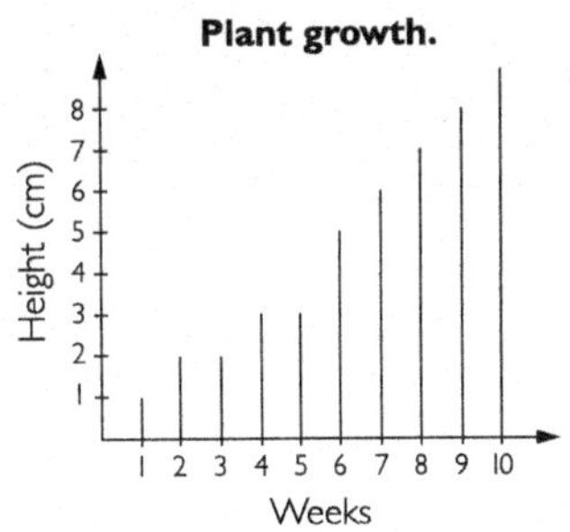

My day.

Other
Eating
Playing &TV
Sleeping
Travel
School

Journey to the shops.

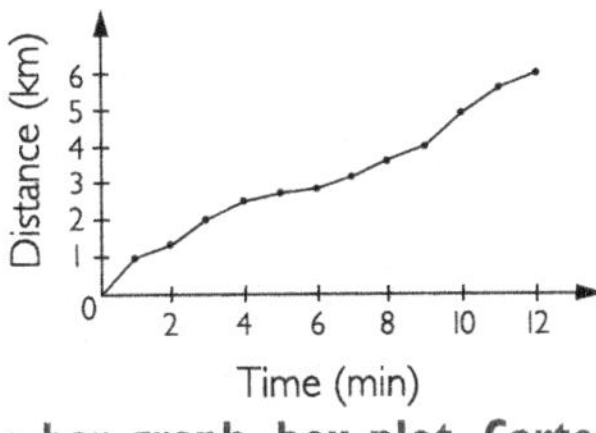

See also **bar graph**, **box plot**, **Cartesian graph**, **dot plot**, **histogram**, **line graph**, **picture graph**, **pie chart**, **stem-and-leaf plot**

greater than

Greater than shows the relationship between two numbers, indicating which is bigger. The symbol for greater than is >.

For example

$9 > 5$ and $5 > 1$

See also **inequality symbols**, **less than**

greatest

The greatest is the largest or the most.

For example

1. The greatest amount of time means the most time.
2. The greatest chance of happening means the most likely occurrence.
3. The greatest mass means the most mass.
4. The greatest number of participants means the largest number.

See also **least**

grid paper

Grid paper is paper marked with a regular pattern of horizontal and vertical lines. Grid paper is often used for graphing or drawing.

See also **rectangular grid**

grid reference

A grid reference is a description of a position on a map. The position description comes from a labelled network of horizontal and vertical lines.

For example

In the street directory on Map 20, a school is located at B3.

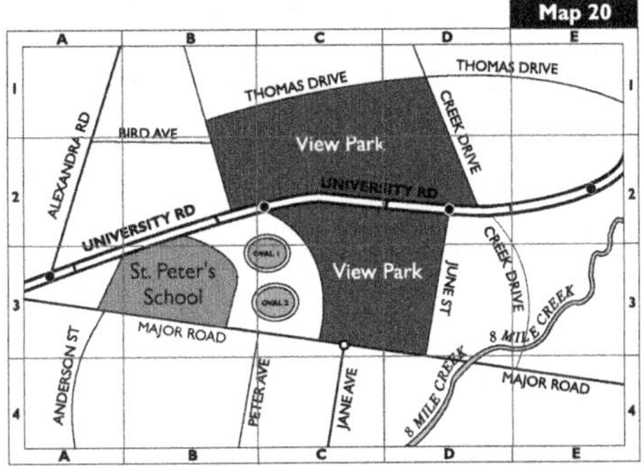

See also **coordinates**

gross

Meaning 1 Gross is the entire or whole amount with no deductions.

For example

1. Gross weight is the weight including packaging.
2. Gross income is income before tax.
3. Gross profit is profit before expenses are deducted.

See also **net**

Meaning 2 A gross is twelve dozen.

G

group

A group is a collection of things.

For example

1 The set of 20 counters was divided into groups of 5.

2 The groups were made according to different criteria.

3 The decimal place-value system is based on groups of 10.

See also **grouping**

grouping

Grouping is sharing objects into equal-sized groups. Objects can be grouped to form an array.

For example

How many groups of 6 can be made from 24?

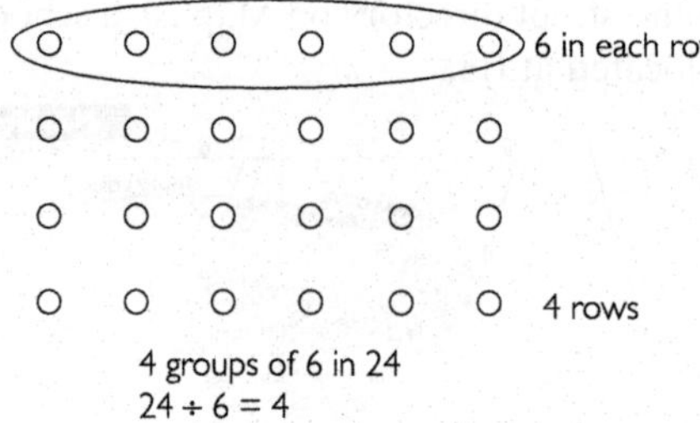

See also **array, division, group, share**

growth pattern

A growth pattern is a number pattern with a constant multiplying factor (right to left) or a constant division factor (left to right). The pattern can be extended forwards and backwards.

For example

The decimal counting system is a based on a growth pattern with a multiplying factor of 10. The pattern is:

$..., 1000, 100, 10, 1, \frac{1}{10}, \frac{1}{100}, \frac{1}{1000}, ...$

See also **constant, geometric progression**

Hh

half

A half is one of two equal parts of a whole.

For example

1 Half a circle is a semicircle.

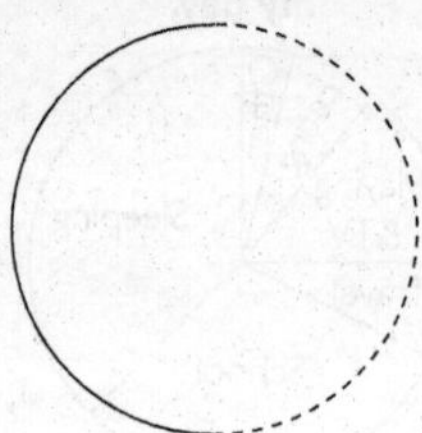

2 Half of 20 is 10.

● ● ○ ○
● ● ○ ○
● ● ○ ○
● ● ○ ○
● ● ○ ○

See also **fraction**

half-turn

A half-turn is half of a complete rotation. The angle of a half-turn is 180 degrees, or half of a full-turn.

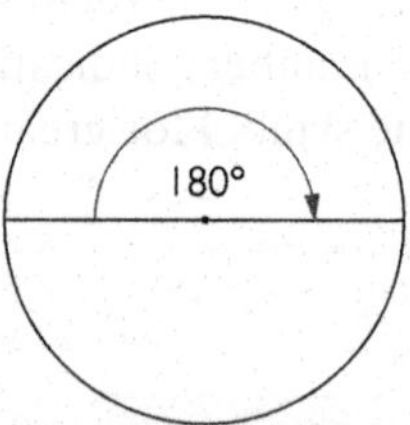

See also **full-turn, rotation**

halve

To halve is to divide into two halves.

For example

To halve an apple is to make fair shares for two people.

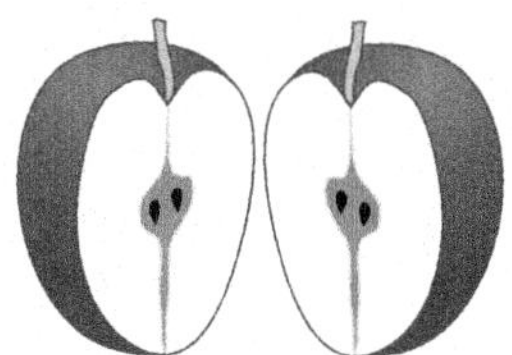

See also **fraction, half, share**

heavy

Heavy means of great weight.

See also **mass, weight**

hectare

A hectare is a unit of area. The symbol for hectare is ha.

1 hectare = 10 000 square metres

For example

The size of the farm is 64 hectares.

See also **area**

heft

To heft is to judge the amount of mass by lifting and comparing the feel of two objects. To heft is to gain a sense of weight by holding.

See also **balance, mass, weight**

height

Height is the vertical distance. Height is usually measured from a base line or plane.

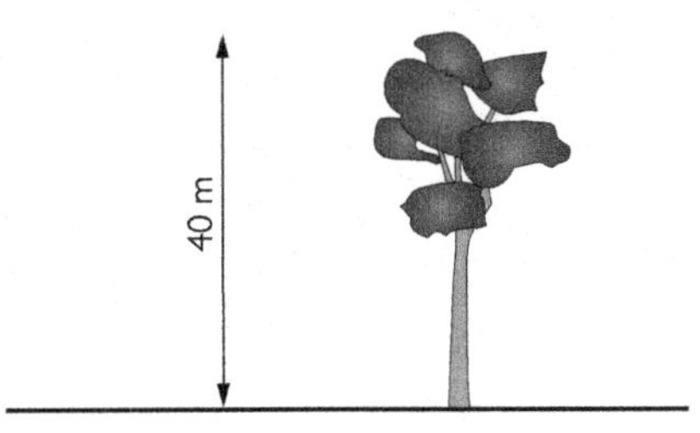

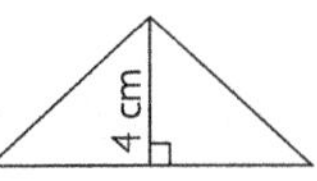

See also **altitude, vertical**

heptagon

A heptagon is a polygon with seven straight sides. In a regular heptagon, all the sides have the same length and all the angles are equal.

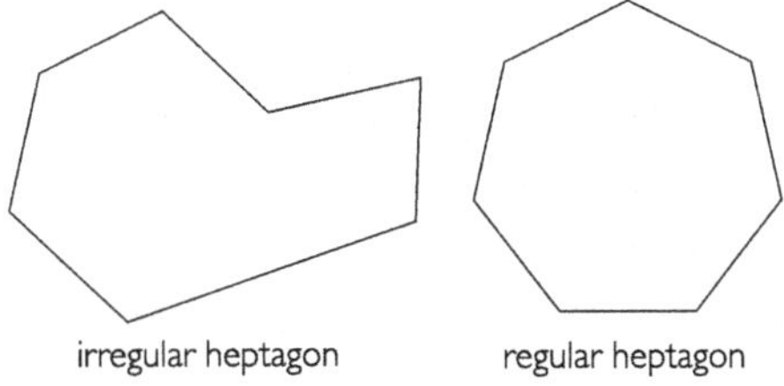

See also **polygon, regular polygon**

heptahedron

A heptahedron is a polyhedron with seven faces.

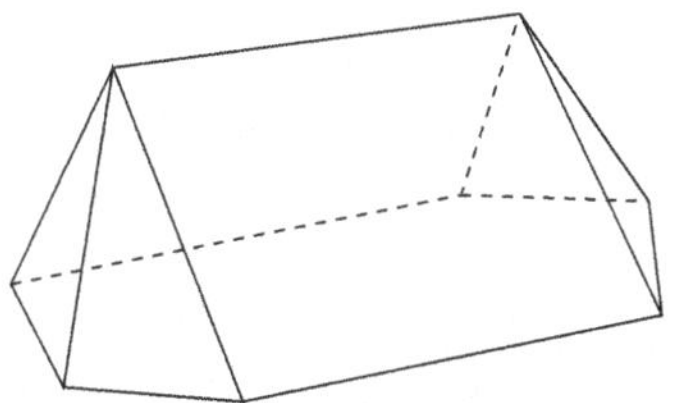

See also **polyhedron**

hexagon

A hexagon is a polygon with six straight sides. In a regular hexagon, all the sides have the same length and all the angles are equal.

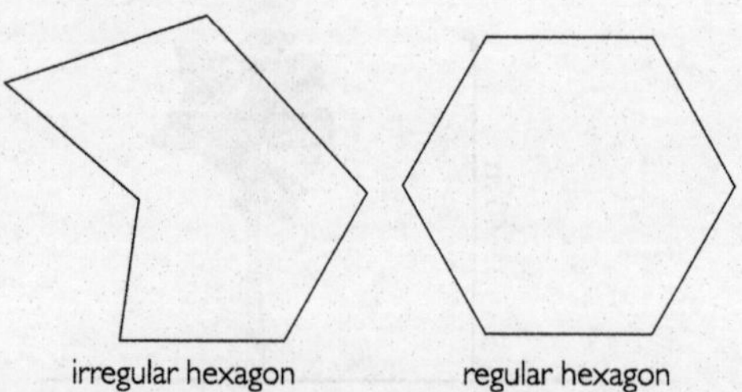

See also **polygon, regular polygon**

hexahedron

A hexahedron is a polyhedron with six faces. A cube is a regular hexahedron and is one of the five Platonic solids.

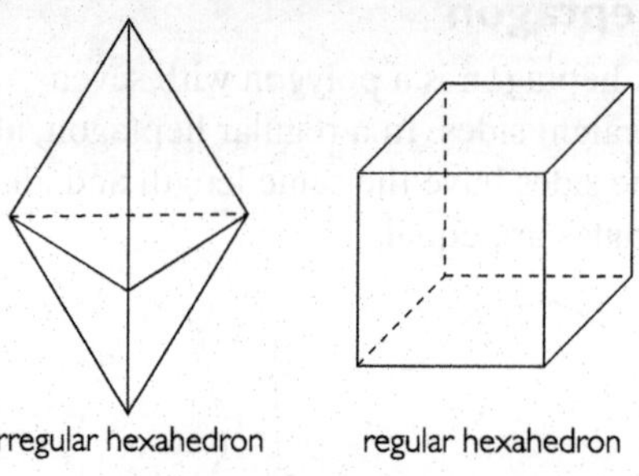

See also **cube, cuboid, Platonic solids, polyhedron, rectangular prism**

highest common factor

The highest common factor of a group of numbers is the largest number that divides exactly into each number in the group.

For example

What is the highest common factor of 15, 18 and 21?

The factors of 15 are 1, 3, 5, 15.
The factors of 18 are 1, 2, 3, 6, 9, 18.
The factors of 21 are 1, 3, 7, 21.
So the highest common factor is 3.

See also **common factor, factors**

histogram

A histogram is a type of bar graph that shows a frequency distribution. The height of the bars shows the number of objects in each category.

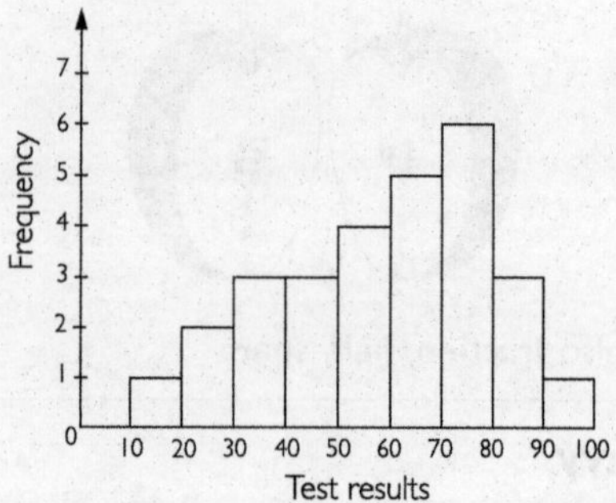

See also **bar graph, frequency distribution**

horizontal

Horizontal means parallel to or level with the horizon.

For example

A horizontal line is a line at right angles to the vertical.

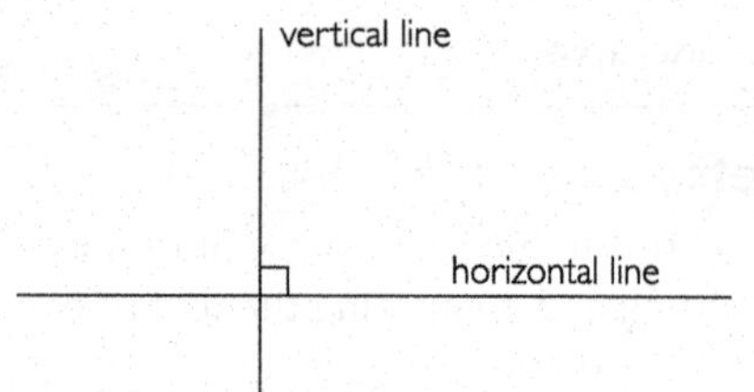

See also **parallel, vertical**

hour

An hour is a unit to measure time.

1 hour = 3600 seconds
1 hour = 60 minutes
24 hours = 1 day

For example

The concert lasted for two hours.

See also **day, minute, second, time**

hundreds

In a base-ten counting system, hundreds is the place value that identifies how many groups of one hundred.

For example

In the number 724, the 7 stands for seven hundreds.

thousands	hundreds	tens	ones
	7	2	4

See also **base, decimal system, place value**

hundredths

In the base-ten counting system, hundredths is a place value in the decimal fraction part of a number. It identifies how many hundredths.

For example

In the number 0.13, the 3 stands for three hundredths.

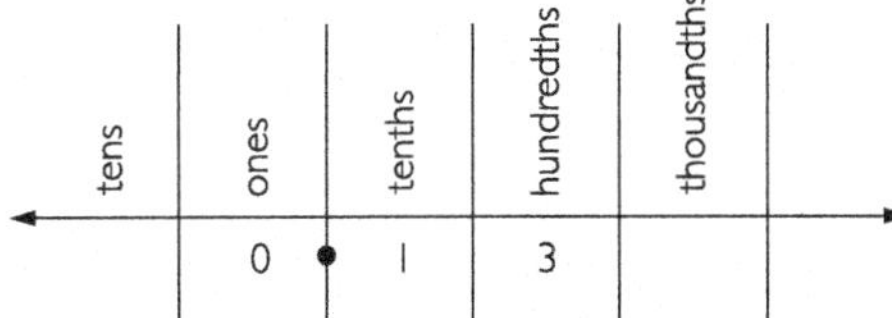

See also **base, decimal place, decimal system, place value**

hypotenuse

The hypotenuse of a right-angled triangle is the side opposite the right angle. The hypotenuse is the longest side. Pythagoras' theorem states that the square of the length of the hypotenuse is equal to the sum of the squares of the lengths of the other two sides.

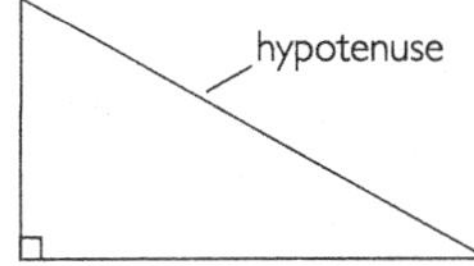

See also **Pythagoras, right-angled triangle**

Ii

icosahedron

An icosahedron is a polyhedron with twenty faces. The faces of a regular icosahedron are congruent equilateral triangles. The regular icosahedron is one of the five Platonic solids.

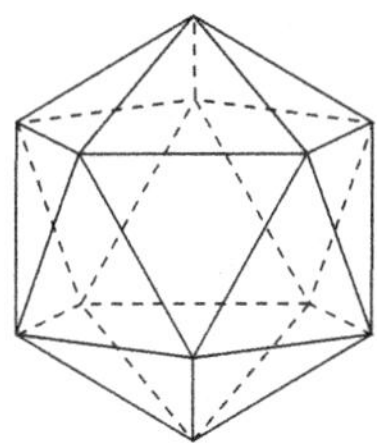

regular icosahedron

See also **Platonic solids, polyhedron**

identity

An identity is an algebraic equation that is true no matter what values are given to the pronumerals.

For example

The equation $2(3x + 4) = 6x + 8$ is an identity. It is true for any value of x.

See also **equation**

improper fraction

An improper fraction is a fraction with a numerator that is greater than or equal to the denominator.

For example

$\frac{7}{5}$, $\frac{16}{3}$ and $\frac{2}{2}$ are improper fractions.

See also **denominator, fraction, numerator**

increase

To increase is to make greater in amount, size or quantity.

For example

1 Increase the amount of water by 50%.
2 Increase the length of the line by 5 centimetres.
3 Increase the number of counters two times.

See also **decrease**

index

The index is the number that indicates the power to which a number is to be raised. The plural of the word *index* is indices.

$$3^2 = 3 \times 3 = 9$$

(2: index; 3: base)

See also **base, exponent, power**

index laws

Index laws are conventions that apply to calculations involving numbers in index form.

- When numbers written in index form (with the same base) are multiplied together, the indices are added: $a^b a^c = a^{b+c}$
- When numbers written in index form (with the same base) are divided, the indices are subtracted: $a^b \div a^c = a^{b-c}$
- When any number is raised to the power of zero, the answer is one: $a^0 = 1$
- When a number written in index form is raised to another power, the indices are multiplied: $(a^b)^c = a^{bc}$
- All numbers that are multiplied together in a bracket are raised to the power outside the bracket: $(ab)^c = a^c b^c$
- All numbers that are divided in a bracket are raised to the power outside the bracket: $\left(\frac{1}{ab}\right)^c = \frac{1}{a^c b^c}$

See also **exponent, index, power**

indirect comparison

To make an indirect comparison is to alter the constitution of a shape or value in order to make a comparison.

For example

To cut up a shape and arrange the parts on top of another shape, in order to compare area, is to make an indirect comparison.

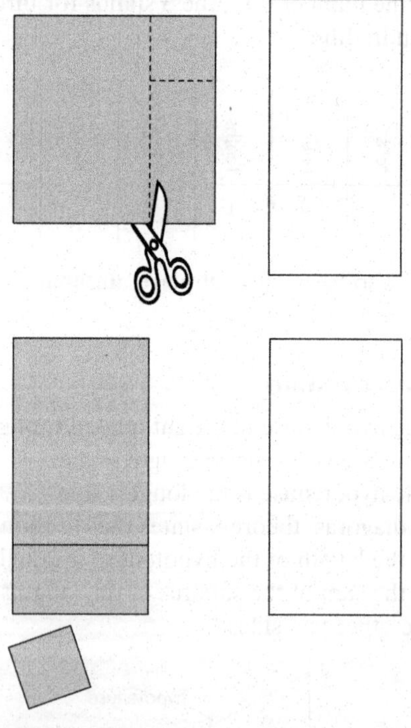

See also **compare**

inequality

Inequality is being not equal. We can compare two different amounts by saying which is bigger or smaller.

For example

Sam ran the race in 42.3 seconds; Kate ran the race in 41.1 seconds. Kate took less time than Sam.

See also **equality**, **inequality symbols**, **inequation**

inequality symbols

The inequality symbols express comparisons between numbers, pronumerals or expressions.

The symbols are:

$>$ is greater than, as in $6 > 3$

$<$ is less than, as in $3.5 < 4$

$\geqslant$ is greater than or equal to, as in $a \geqslant -2$

$\leqslant$ is less than or equal to, as in $-2 \leqslant a$

See also **equality**, **inequality**, **symbol**

inequation

An inequation is a statement of inequality. An inequation is two expressions joined by an inequality symbol.

For example

1. $2.5 + 0.25 < 5.2$
2. $x - 2 > 4$

See also **equation**, **inequality**, **inequality symbols**

inference

Meaning 1 An inference is a conclusion based on reasoning.

For example

If one jug fills three glasses, the inference is that two jugs will fill six glasses.

Meaning 2 An inference is a prediction based on observations.

For example

Based on collected data, we can make an inference on the probability of a team winning its next game.

infinite

Infinite is not finite. Infinite is limitless, not bounded or never-ending.

For example

1. The set of whole numbers is an infinite set.
2. The snowflake curve has infinite length.

See also **finite**

infinity

Infinity is quantity in a never-ending amount that cannot be identified by a real number. The symbol for infinity is ∞.

See also **finite**, **infinite**

informal unit

An informal unit is a non-standard amount that is used for measuring.

For example

1. Hand spans may be used to measure length.
2. Floor tiles may be used to measure area.
3. Bricks may be used to measure mass.
4. Cups full may be used to measure volume.

See also **standard units of measurement**

integer

An integer is a whole number.

The integers consist of:

- the positive whole numbers: 1, 2, 3, 4, 5, ...
- the negative whole numbers: –1, –2, –3, –4, –5, ...
- 0

See also **directed number**, **negative number**, **positive number**, **whole number**, **zero**

intercept

An intercept is the part of a line or a plane cut by another line or plane.

For example

The point at which the curve of a graph intersects a coordinate axis is an intercept.

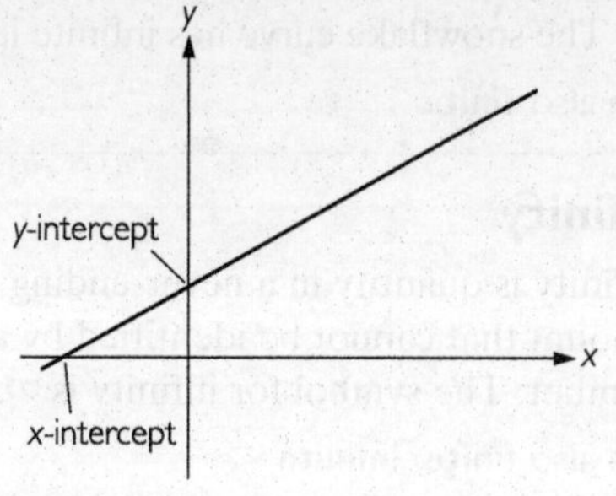

interest

Interest is a fee paid in return for borrowing money. A person can borrow money from a bank. This is called a loan. Or a person can lend money to a bank. This is called an investment. The borrower pays the lender interest, usually once every year. The interest is an agreed percentage of the amount that the borrower owes the lender.

For example

1 Clare borrowed $100 from a bank at the interest rate of 12% per annum. In the first year, she had to pay the bank $12 interest.
2 (Simple interest) Jane invested $1000 at 3% per annum. Every year the bank pays her $30 interest.
3 (Compound interest) Tony invested $1000 at 5% per annum. Each year, the interest earned is added to the investment. After the first year, Tony had $1000 plus $50 interest. After the second year, Tony had $1050 plus $52.50 interest.

See also **percentage**

interior angle

An interior angle is any angle formed inside a figure by two adjacent sides of the figure.

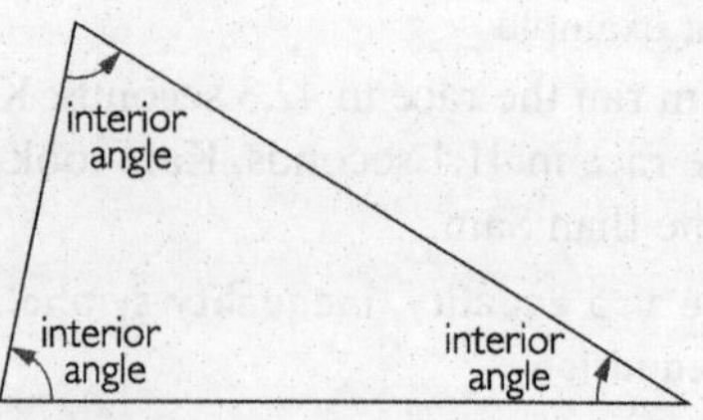

See also **adjacent**

intermediate compass points

The intermediate compass points are the compass points that name the directions at 45-degree angles to the four major compass points.

NE is 45 degrees east of north.
SE is 45 degrees east of south.
SW is 45 degrees west of south.
NW is 45 degrees west of north.

These directions can be divided further, with directions at $22\frac{1}{2}$-degree angles.

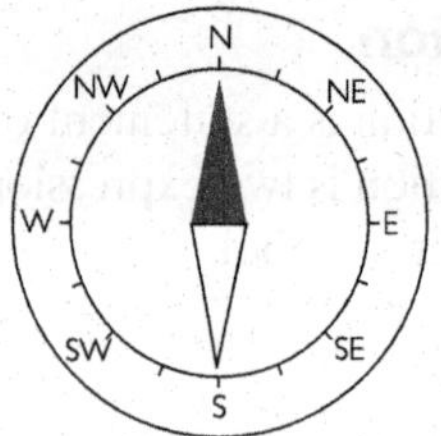

See also **compass points, direction**

interpret

To interpret is to explain the meaning of.

For example

We interpret a graph or a problem.

intersection

The intersection of two objects is where they overlap.

For example

1 The point at which lines cross is an intersection.

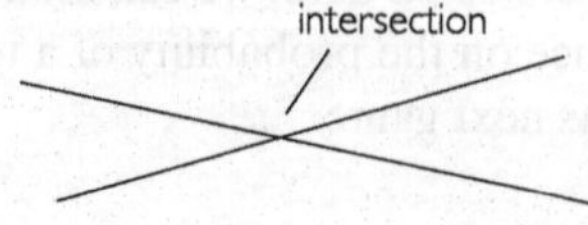

2 The region where shapes overlap is an intersection.

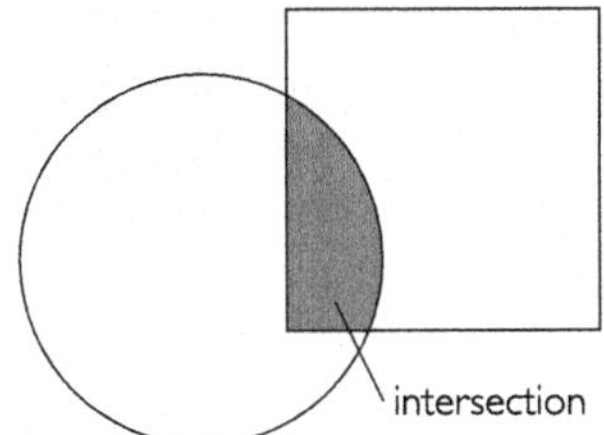

3 The intersection of two sets is the set of elements that they have in common.

Set A = {3, 5, 15, 20}
Set B = {15, 20, 35, 40}

The intersection of Set A and Set B is the set {15, 20}.

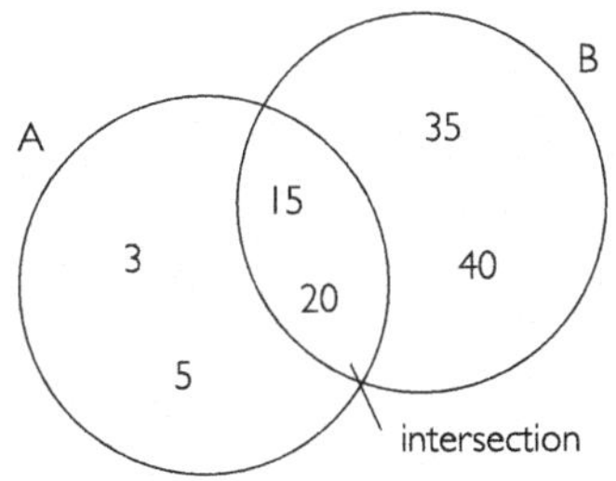

interval

An interval is everything between a beginning and an end.

For example

1 An interval on a number line is the part of the line between two points. This number line is marked in intervals of 2 units.

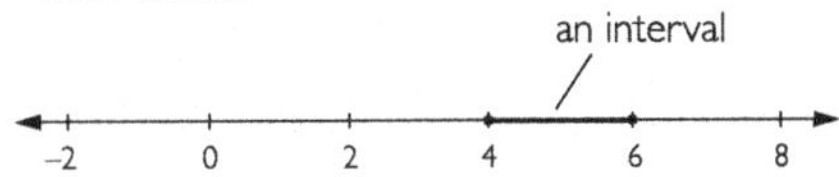

2 A time interval is the period of time between two events. For instance, the temperature readings were made at 5-minute intervals.

3 A line interval is the set of all points between two points on a line. The two end points may or may not be included. For instance, the set of all points between 1 and 2 on the x-axis form an interval.

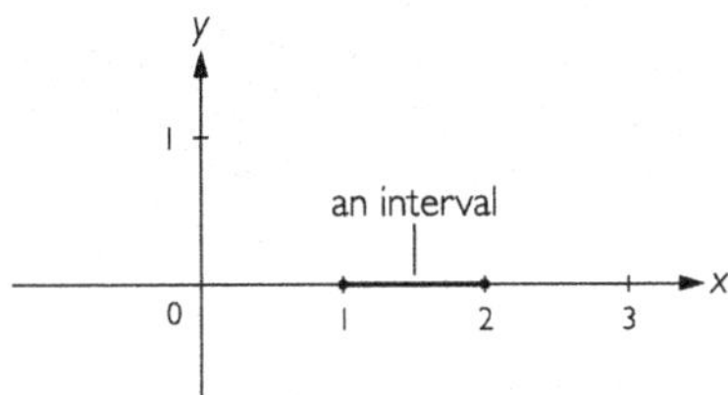

4 A number interval is the set of all numbers between two numbers. For instance, all the numbers between 5 and 10 form an interval.

invariance

Invariance means staying the same after a transformation.

For example

The size and shape of a triangle remain the same as the triangle is rotated.

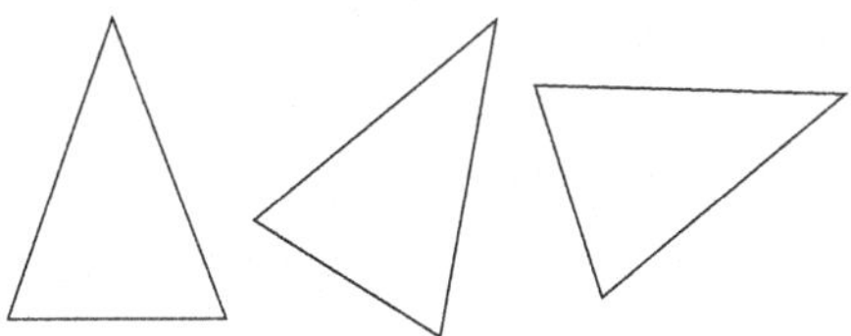

See also **transformation**

inverse

Inverse means in reverse.

For example

1 The inverse of counting backwards from 10 to 1 is counting forwards from 1 to 10.

2 The operation of division is the inverse of multiplication; subtraction is the inverse of addition.

3 The faster you walk to the station, the less time it takes to get there. This is an example of inverse proportion.

4 The inverse of the fraction $\frac{2}{3}$ is $\frac{3}{2}$.

See also **order**

irrational number

An irrational number is a real number that cannot be written as a ratio of integers. In other words, an irrational number cannot be written as a common fraction. Decimal numbers that are never-ending and non-repeating are irrational numbers.

For example

0.1010010001 ... is an irrational number.
Other examples include $\sqrt{2}$ and π.

See also **integer, pi, rational number, real number, root**

irregular

Irregular means not regular. Irregular is without symmetry or rule.

For example

1 An irregular polygon is a plane shape with sides that are not all the same length, or with angles that are not all equal. Compare this with a regular polygon.

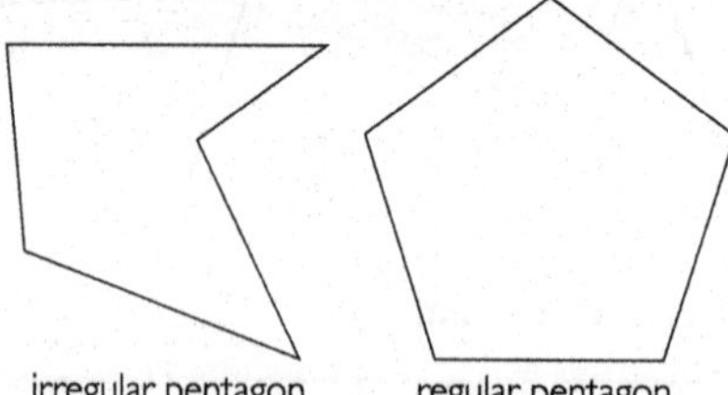

irregular pentagon regular pentagon

2 An irregular polyhedron is a solid shape having faces or vertices that are not equivalent. Compare this with the five Platonic solids.

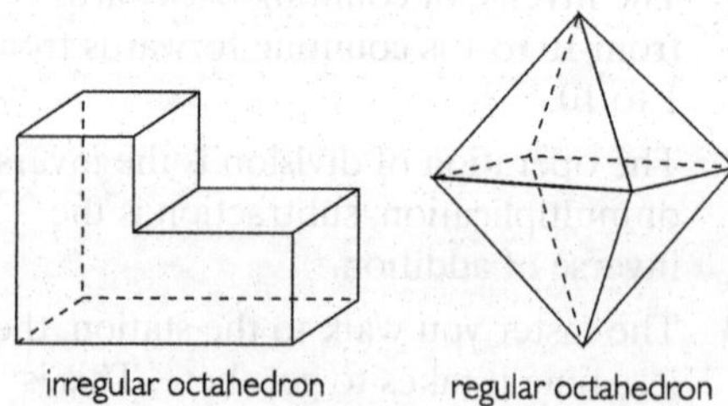

irregular octahedron regular octahedron

3 Intervals of time or space that are not all equal are called irregular intervals.

See also **regular**

isometric drawing

An isometric drawing is a drawing representing three dimensions. The drawing is made using three axes spaced at 120 degrees to each other. Isometric drawing is aided by the use of isometric graph paper, on which printed lines or dots forming equilateral triangles reflect the three axes and are used as a guide to accuracy.

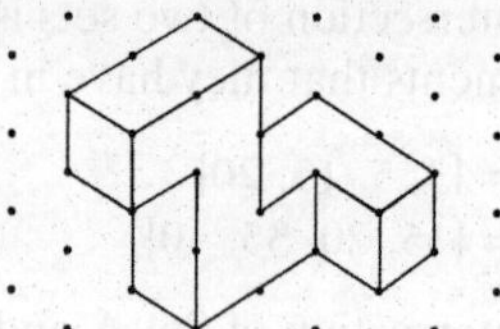

See also **axis, three-dimensional**

isosceles triangle

An isosceles triangle is a triangle with two sides of equal length and two equal angles.

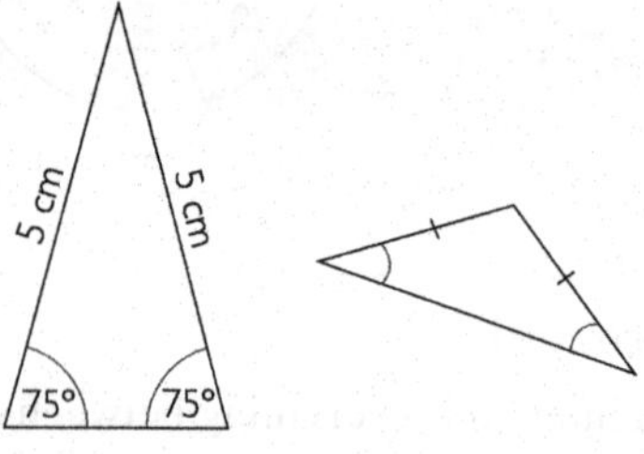

See also **triangle**

iteration

Iteration means repeating an action over and over again.

For example

1 Repeating identical shapes on different scales to form a snowflake curve is an iteration.

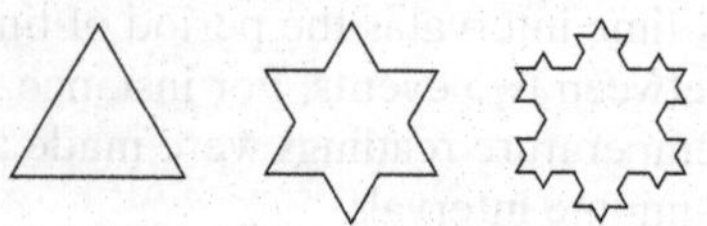

2 The sequence 1, 2, 4, 8, ... is made by iterating the "doubling" operation.

Jj

judgment

To make a judgment is to decide whether or not the method of working out a problem and its result are reasonable.

See also **check**, **estimate**, **reasonable**

Kk

kilogram

A kilogram is the base unit of mass in the International System of Units. The symbol for kilogram is kg.

1 kilogram = 1000 grams

For example

The mass of the bag of rice is 1 kilogram.

See also **gram**, **mass**, **SI system**, **standard units of measurement**

kilolitre

A kilolitre is a unit of volume (or capacity). A kilolitre is used to measure the amount of a liquid. The symbol for kilolitre is kL.

1 kilolitre = 1000 litres

For example

The swimming pool holds 150 kilolitres of water.

See also **capacity**, **litre**, **standard units of measurement**, **volume**

kilometre

A kilometre is a unit of length. A kilometre is used to measure distance. The symbol for kilometre is km.

1 kilometre = 1000 metres

For example

The distance from Melbourne to Port Campbell is 250 kilometres.

See also **length**, **metre**, **standard units of measurement**

kite

A kite is a special quadrilateral. The four sides of a kite come in two pairs. The two sides in a pair are adjacent and have the same length. The diagonals of a kite cross each other at right angles.

☞

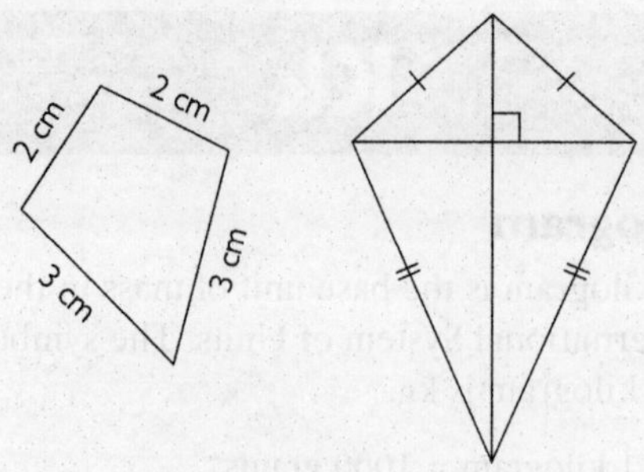

See also **quadrilateral**

Klein bottle

A Klein bottle is a special bottle that has only one surface. It has an outside but not an inside. It passes through itself. The object was named after its creator, the German topologist Felix Klein, who lived 1849–1925.

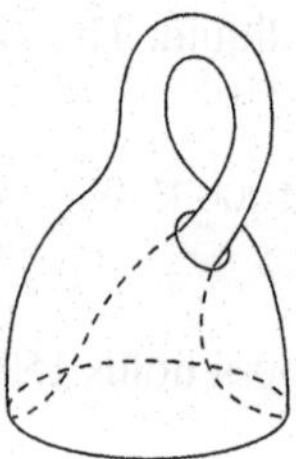

See also **Möbius strip**, **topology**

knot theory

Knot theory is a field of mathematics. A mathematical knot has no ends. It is a closed loop that cannot be rearranged into a circle. Knot theory is a branch of topology.

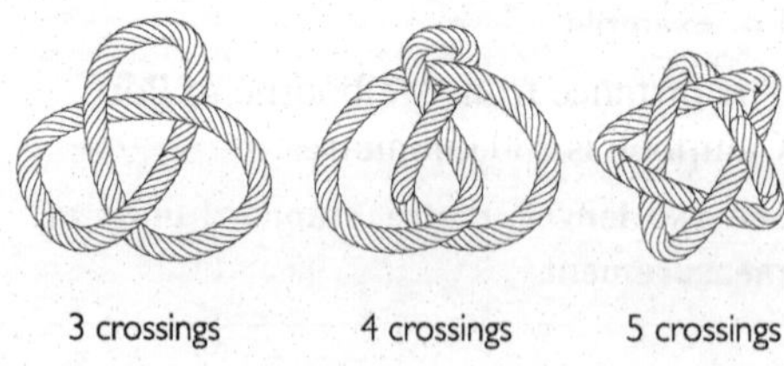

See also **topology**

Ll

label

A label can name something, or it can explain something.

For example

1 In geometry, points are labelled to make them easy to talk about.

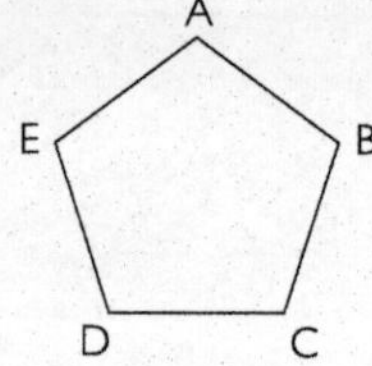

Vertices A and E are adjacent.

2 The axes of a graph are labelled to explain what they represent.

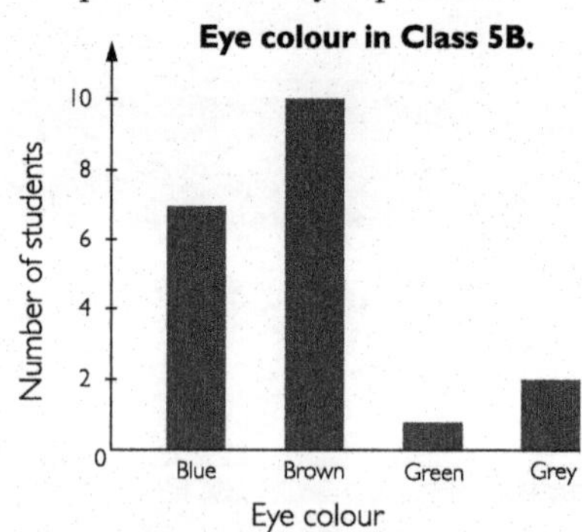

least

The least is the smallest or lowest.

For example

1 The least amount of time means the smallest amount of time.

2 The least likely outcome means the outcome with smallest chance of happening.

3 The least mass means the smallest mass.

4 The least number of participants means the lowest number of participants.

See also **greatest**

length

Meaning 1 Length is the distance from point to point.

For example

The length of the line was 6 centimetres.

See also **distance**

Meaning 2 Length is the measure of a period of time.

For example

The length of time taken was 30 seconds.

See also **time**

less than

Less than shows the relationship between two numbers, indicating which is smaller. The symbol for less than is <.

For example

1 $0.5 < 5$

2 $-5 < 0.5$

See also **greater than, inequality symbols**

likelihood

Likelihood is the probability or chance of an event. An event can be certain, likely, unlikely or impossible.

For example

What is the likelihood of selecting a red marble from a bag containing 25 red marbles, 7 blue marbles and 3 green marbles?

See also **chance, probability**

like terms

Like terms are algebraic terms with identical pronumeral parts. Only like terms can be combined using addition and subtraction.

For example

1 The terms $2xy$ and $-3xy$ are like terms.

2 The terms $5a$ and $4a^2$ are not like terms.

3 To simplify $7a + 4a + 2b$, first note that $7a$ and $4a$ are like terms:

$7a + 4a + 2b = 11a + 2b$.

See also **pronumeral, term, unlike terms**

line

A line is a mark that has length but not width. It has only one dimension. A line can have a beginning and an end, or it can extend infinitely. A line can be straight or curved. A straight line is the shortest line between two points.

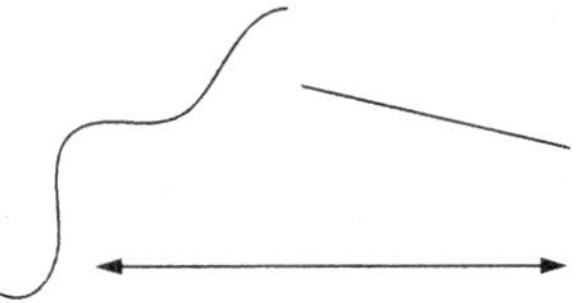

See also **infinite, length, line segment, one-dimensional**

linear

Linear means in a straight line.

For example

1 The set of points $\{(-3,-3), (-2,-2), (-1,-1), (0,0), (1,1)\}$ forms a linear pattern on a Cartesian graph.

2 A linear measure is a unit used to measure length. For instance, the metre is a linear measure.

See also **linear equation**

linear equation

A linear equation is a particular type of algebraic equation. Each term in the equation must be either a number, a pronumeral or the product of a number and a pronumeral. The simplest type of linear equation is the equation of a straight line on a Cartesian graph.

For example

1 The equation $y = \frac{1}{2}x + 1$ is linear. The graph of this equation is a straight line.

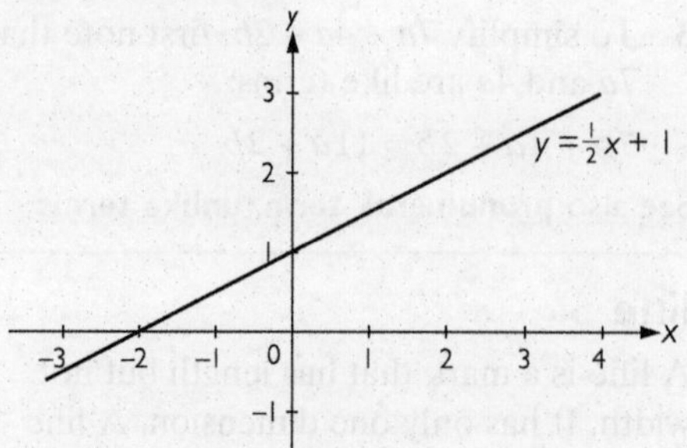

2 The general equation of a straight line on a Cartesian graph is $y = mx + c$, where m and c can be any numbers.

3 The equation $x - 2y + 3z = 4$ is linear.

See also **equation**

line graph

A line graph is a diagram that shows the relationship between two varying quantities. The form of a line graph is a line that joins plotted points. A line graph is used to represent continuous data, that is, data that is measured, not counted. A line graph shows trends or tendencies in the data.

For example

A line graph can show the trend of the temperature during a day, when the temperature has been measured each hour.

The temperature in Sale on 13 March.

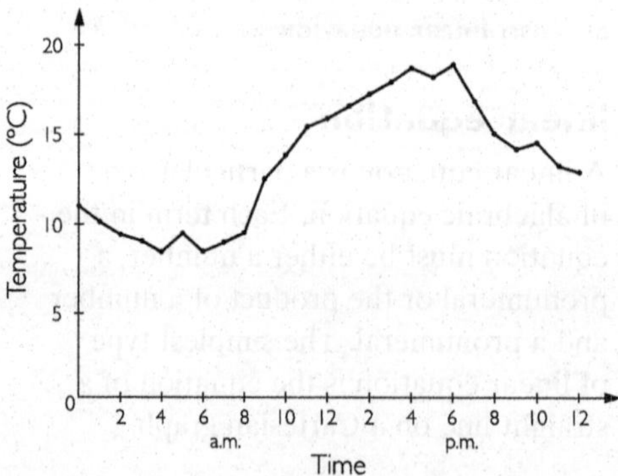

See also **continuous, graph, trend**

line of symmetry

A line of symmetry separates a shape into halves so that each half exactly mirrors the other half. A line of symmetry is also called an axis of symmetry or a mirror line. A shape can have more than one line of symmetry.

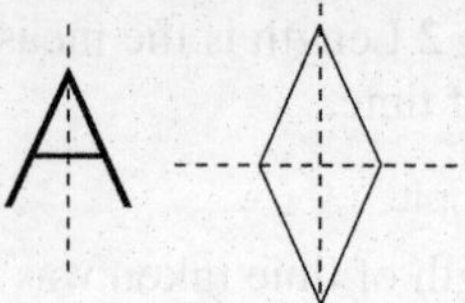

See also **axis, bilateral symmetry, mirror line, symmetry**

line segment

A line segment is the part of a line between two given points.

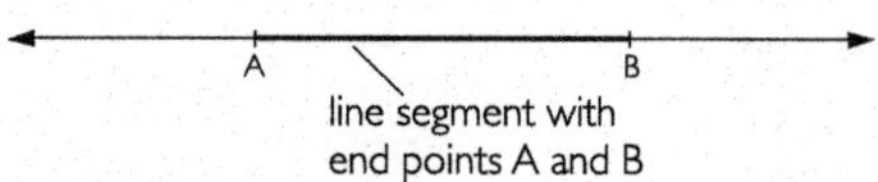

See also **line, point**

litre

A litre is a unit of volume (or capacity). It is used to measure the amount of a liquid. The symbol for litre is L.

1 litre = 1000 cubic centimetres

1 litre = 1000 millilitres

For example

The jug holds 1 litre of water.

See also **capacity, millilitre, SI system, volume**

location

Location means place. Location can be specified using distance and direction, or using a coordinate system.

For example

1 The location of the school is 2 kilometres east of the lake.

2 The location of the school can be found in the street directory on Map 42, G9.

See also **coordinates, compass points**

lowest common denominator

The lowest common denominator is the number that is the lowest common multiple of the denominators of two or more common fractions. Put a different way, it is the smallest number that is divisible by each of the denominators exactly. The name *lowest common denominator* is commonly shortened to LCD.

For example

The lowest common denominator of $\frac{1}{3}$ and $\frac{1}{4}$ is 12.

The number 12 is the smallest number that is a multiple of both 3 and 4.

12 is the smallest number that is divisible by both 3 and 4 exactly.

The multiples of 3 are:

3, 6, 9, (12), 15, 18, 21, (24), 27, 30, …

The multiples of 4 are:

4, 8, (12), 16, 20, (24), 28, 32, …

See also **common denominator, common fraction, division, multiple**

lowest common multiple

The lowest common multiple of a group of numbers is the smallest number that is a multiple of each number in the group. The name *lowest common multiple* is commonly shortened to LCM.

For example

The number 12 is the lowest common multiple of 4 and 6.

12 is the smallest number that is a multiple of both 4 and 6.

12 is the smallest number that has 4 and 6 as factors.

The multiples of 4 are:

4, 8, (12), 16, 20, (24), 28, 32, (36), 40, …

The multiples of 6 are:

6, (12), 18, (24), 30, (36), 42, …

See also **multiple**

Mm

magic square

A magic square is a square array of numbers. The numbers in each row, column and diagonal add up to the same number.

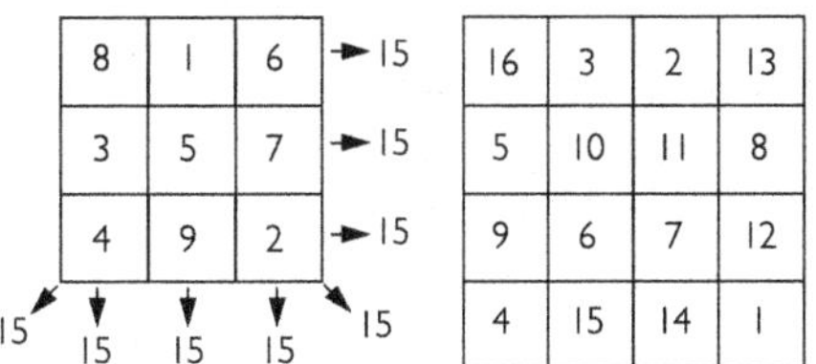

See also **column, diagonal, row**

magnitude

Magnitude means size.

For example

1 The magnitude of the angle is 45 degrees.

2 The magnitude of a whole number is its size, not counting its direction. For instance, the magnitude of the number –2 is 2.

See also **size**

many-to-one correspondence

Many-to-one correspondence is a connection of the elements of two sets. Each element of the first set is connected to an element of the second set. Two different elements in the first set are connected to the same element in the second set.

For example

An arrow diagram can be used to show a many-to-one correspondence.

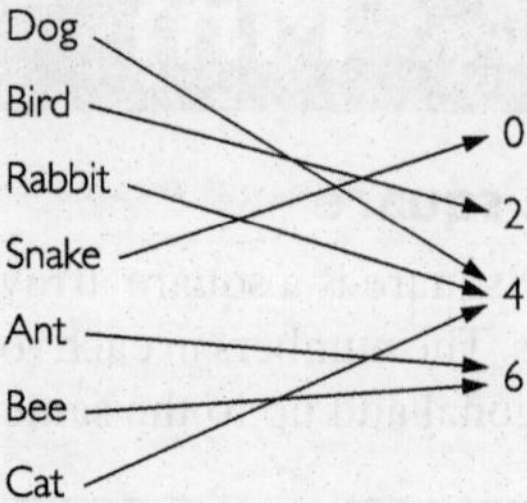

See also **element, one-to-one correspondence, set**

map

A map represents a region of the Earth's surface. A map is made to scale on a convenient-sized flat surface. A grid reference system can be used to describe position on a map.

See also **bird's-eye view, grid reference, plan, scale**

mapping

Mapping is connecting elements in one set to elements in another set. A transformation of a shape is a type of mapping. A mapping is also called a function.

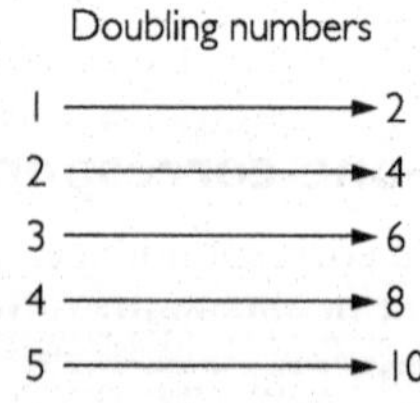

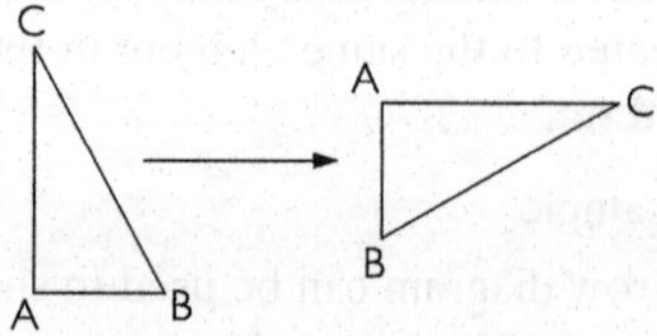

See also **element, function, one-to-one correspondence, set, transformation**

mass

Mass is the amount of matter in an object. Mass is independent of gravity. Mass and weight are commonly confused. The standard metric measures of mass are the gram (g), kilogram (kg) and tonne (t).

1000 grams = 1 kilogram

1000 kilograms = 1 tonne

For example

1. Using the beam balance, the book has the same mass as the standard 1 kilogram mass.

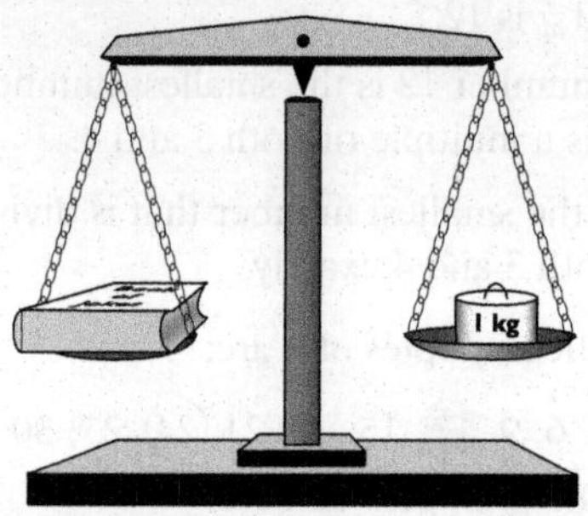

2. An astronaut has the same mass on Earth as on the Moon. But the astronaut weighs more on the Earth because there the pull of gravity is stronger.

See also **metric system, SI system, weight**

matching

Matching means corresponding.

For example

1. Matching patterns are the same pattern.
2. Matching the number of counters is putting out an equal number of counters.
3. Matching shapes are identical shapes.

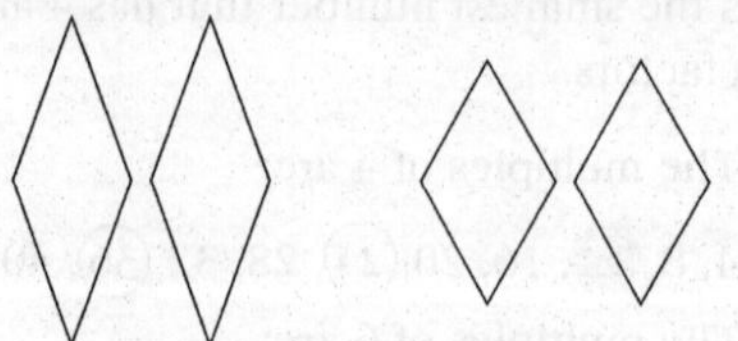

See also **attribute, one-to-one correspondence**

mathematics

Mathematics is a way of thinking about and describing the world we live in and the world of ideas. It uses logic and reasoning. Mathematicians study all kinds of patterns. The patterns can be of shapes, numbers or arrangements.

Mathematical ideas are commonly expressed using symbols. These symbols can be numbers, pronumerals or special symbols that express relationships. Mathematics includes counting, measuring, geometry, arithmetic and algebra.

See also **algebra**, **arithmetic**, **geometry**, **pattern**

maximum

The maximum is the greatest value in a set.

For example

1 The maximum temperature today was 25 degrees Celsius.

2 The maximum element of the set {2, 4, 9} is 9.

See also **minimum**

maze

A maze is a network of confusing paths. The challenge of a maze is to negotiate the paths to reach a place or an object.

See also **network**

mean

The mean is a measure of the "middle" of a group of numbers. The mean is commonly called the average. It is found by adding the numbers together and then dividing by how many numbers there are.

For example

A student got marks of 8, 4, 8, 9 and 7 on five different tests.

$$\text{average mark} = \frac{\text{sum of the marks}}{\text{number of marks}}$$

$$= \frac{8 + 4 + 8 + 9 + 7}{5}$$

$$= \frac{36}{5} = 7.2$$

The student's average mark was 7.2.

See also **arithmetic mean**, **average**, **median**, **mode**

measure

Meaning 1 To measure is to use units to determine the size or amount of an attribute of an object.

For example

1 The length of the tape was measured in centimetres.

2 The area of the playing field was measured in square metres.

3 The capacity of the jug was measured in litres.

4 The volume of the cube was measured in cubic centimetres.

5 The size of the angle was measured in degrees.

6 The mass of the brick was measured in kilograms.

7 The duration of the journey was measured in hours.

8 The temperature of the water was measured in degrees Celsius.

Meaning 2 A measure is a unit of measurement.

For example

Which measure will you use to find out the length of the bicycle track?

See also **informal unit, standard units of measurement**

measurement

Meaning 1 A measurement is the size or amount of a physical attribute of an object. A measurement is determined by measuring.

For example

The measurements of the table are: height 69 cm, length 159 cm, width 78 cm.

Meaning 2 A measurement system is a collection of different measures, or a way of measuring.

For example

The metric system is a system of measurement.

See also **metric system, standard units of measurement**

median

Median means middle.

For example

1 The median of a group of numbers is the middle number, once the numbers have been ordered by size. In the sequence 2, 3, 6, 7, 20, the median is 6.

  When there is an even number of numbers, the median is the mean of the middle two numbers. In the sequence 4, 8, 12, 13, the median is $\frac{1}{2}(8 + 12) = 10$.

2 A median of a triangle is a line from a vertex to the midpoint of the opposite side.

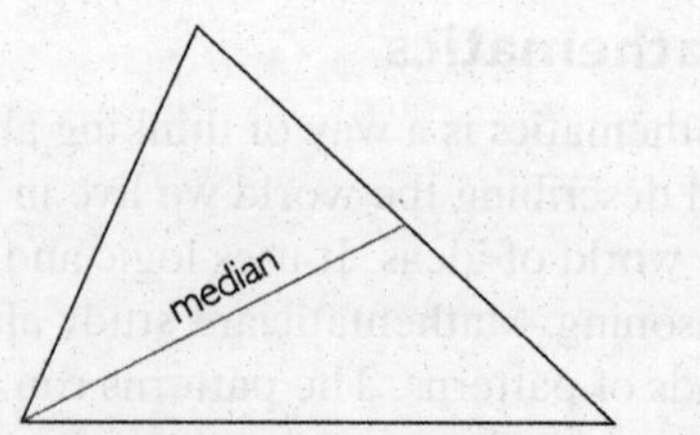

See also **average, mean, midpoint, mode**

mensuration

Meaning 1 Mensuration is the area of mathematics that deals with the measurement of geometric quantities.

For example

Mensuration deals with the measurement of length, area and volume.

Meaning 2 Mensuration is the process of measuring.

For example

Mensuration is used to determine the volume of a room.

See also **geometry, measurement, quantity**

metre

A metre is the base unit for the measurement of length in the International System of Units. The symbol for metre is m.

1000 millimetres = 1 metre

100 centimetres = 1 metre

For example

The measuring tape has an extended length of 25 metres.

See also **length, SI system, standard units of measurement**

metric system

The metric system is a decimal system of measurement. A metric system was introduced in France in 1795. The modern metric system, the International System of Units, was agreed to at a world meeting in 1960, and was adopted in Australia in 1970.

See also **measurement, SI system**

midpoint

The midpoint is the middle point.

For example

1 The midpoint of a line segment is the point that is exactly the same distance from the beginning and end points. The midpoint separates the line segment into two equal parts.

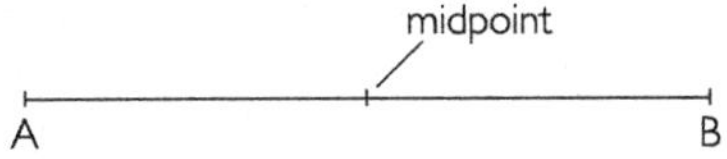

2 The midpoint is the point at the centre of a geometric figure.

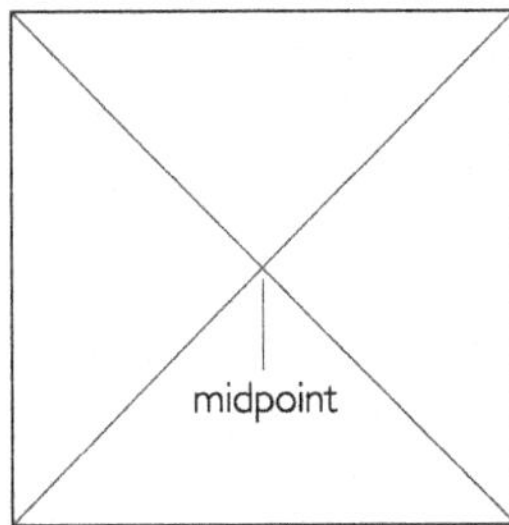

milligram

A milligram is a unit of mass. A milligram is one-thousandth of a gram. The symbol for milligram is mg.

1000 milligrams = 1 gram

For example

The pill contained 425 milligrams of sodium.

See also **gram**, **mass**, **standard units of measurement**

millilitre

A millilitre is a unit of volume (or capacity). A millilitre is one-thousandth of a litre. The symbol for millilitre is mL.

An object with a volume of one cubic centimetre displaces one millilitre of water. One millilitre of water at four degrees Celsius has a mass of one gram.

1000 millilitres = 1 litre

For example

1 A teaspoon holds about 5 millilitres.

2 A cup holds about 200 millilitres.

See also **capacity**, **litre**, **standard units of measurement**

millimetre

A millimetre is a unit of length. A millimetre is one-thousandth of a metre. The symbol for millimetre is mm.

10 millimetres = 1 centimetre

1000 millimetres = 1 metre

For example

The line measured 75 millimetres.

See also **centimetre**, **metre**, **standard units of measurement**

millions

In a base-ten counting system, millions is the place value that identifies how many groups of one million. A million is one thousand thousands.

For example

In the number 6 734 298, the 6 stands for six millions.

millions	hundreds of thousands	tens of thousands	thousands	hundreds	tens	ones
6	7	3	4	2	9	8

See also **base**, **decimal system**, **place value**

minimum

The minimum is the least value in a set.

For example

1 The minimum temperature today was 5 degrees Celsius.

2 The minimum element of the set {2, 4, 9} is 2.

See also **maximum**

minuend

The minuend is the number from which another number is to be subtracted.

For example

In 15 – 7 = 8, the minuend is 15.

See also **subtraction, subtrahend**

minus

Meaning 1 Minus means take away or subtract. The symbol for minus is –.

For example

24 – 9 = 15

See also **subtraction**

Meaning 2 Minus is the symbol of negative numbers.

For example

–1, –2, –3, –4, –5, ...

See also **positive number**

minute

Meaning 1 A minute is a unit of time.

1 minute = 60 seconds

60 minutes = 1 hour

For example

It takes 15 minutes to walk from home to the station.

See also **hour, second, time**

Meaning 2 A minute is a unit for measuring angles. The symbol for minute is ′.

60 seconds = 1 minute

60 minutes = 1 degree

For example

The angle measures 45 degrees 10 minutes.

See also **angle, degree, second**

mirror line

A mirror line separates a shape into halves so that each half reflects the other half exactly. A mirror line is also known as a line of symmetry.

See also **bilateral symmetry, line of symmetry, symmetry**

missing addend

The missing addend is the unknown addend, when the total and one addend are known. The missing addend is found by subtracting the known addend from the total.

For example

Michelle has 6 stamps; she needs 10 stamps altogether. How many more stamps does she need?

See also **addend, addition, subtraction, total**

missing factor

The missing factor is the unknown factor, when the product and one factor are known. The missing factor is found by dividing the product by the known factor.

For example

20 lettuces grow in 5 rows. How many grow in each row?

See also **division, factors, multiplication, product**

mixed number

A mixed number is a number that comprises a whole number and a common fraction.

For example

1. $3\frac{1}{2}$, $8\frac{1}{4}$ and $12\frac{3}{7}$ are mixed numbers.
2. Improper fractions convert to mixed numbers, for instance $\frac{7}{2} = 3\frac{1}{2}$.

Note: A number that comprises a whole number and a decimal fraction is called a decimal number.

See also **common fraction**

Möbius strip

A Möbius strip is a special surface with only one side and only one edge. A Möbius strip is made by taking a strip of paper and twisting it, before gluing its ends together. It is named after the German mathematician Augustus Möbius, who lived 1790–1868.

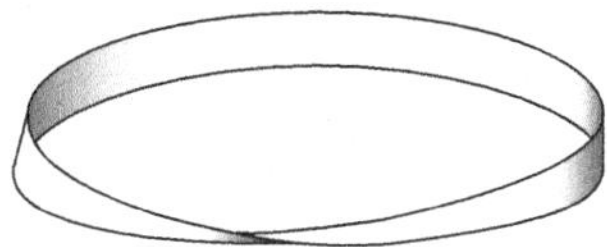

See also **Klein bottle**, **topology**

mode

The mode of a group of numbers is the number that occurs most often.

For example

Over the year, a student's test scores were:

5, 6, 7, 7, 7, 8, 8, 9, 9, 9, 9

The mode of these scores is 9 because 9 occurs the greatest number of times in the scores.

See also **average**, **mean**, **median**

model

Meaning 1 A model is a mathematical pattern that is applied to a real world situation.

For example

A graph of sales figures over time can be used to predict future sales.

Sales at "Snazzy Sock Shop".

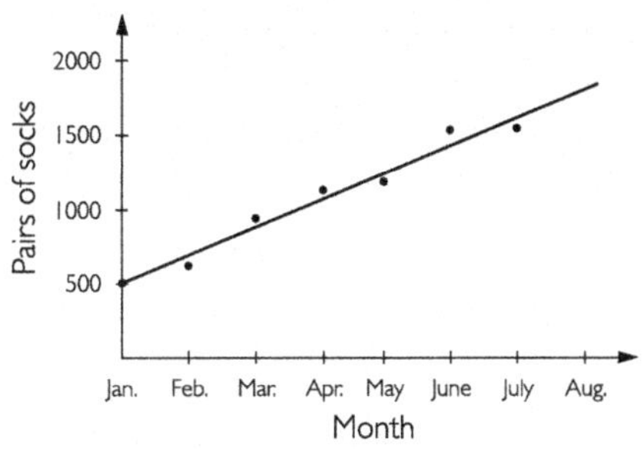

See also **pattern**

Meaning 2 A model is a representation, in particular, of geometric shapes.

For example

A three-dimensional model of a cube can be made from the net of a cube.

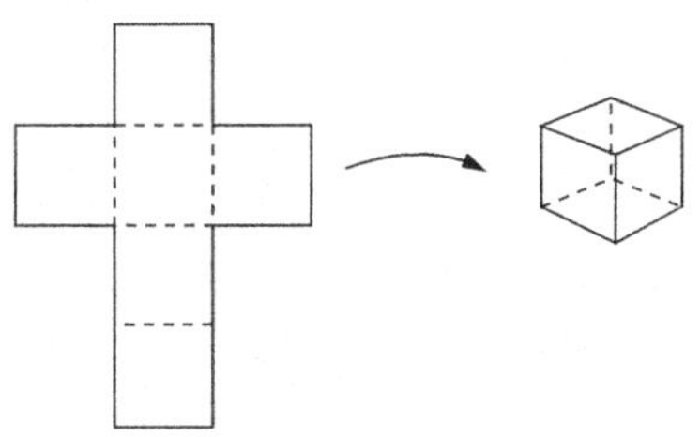

See also **geometrical representation**, **net**

month

A month is one of the twelve divisions of the calendar year. Each month is approximately four weeks. The number of days in a month varies from 28 days to 31 days.

See also **calendar**, **year**

multiple

A multiple is a number that has a given number as a factor. Multiples are found by multiplying the given number by any of the counting numbers.

For example

6, 12 and 18 are all multiples of 6.

See also **factors**, **multiplication**

multiplicand

The multiplicand is the number to be added to itself in the multiplication process. It is the number that is multiplied by another.

For example

```
*  *  *  *   4 × 5 = 20
*  *  *  *   |
*  *  *  *   multiplicand
*  *  *  *
*  *  *  *
```

The 4 is added to itself 5 times to make 20.

The 4 is multiplied by 5.

See also **multiplication, multiplier**

---

multiplication

Multiplication is the operation of adding a number to itself a given number of times. The symbol for multiplication is ×.

For example

1 $4 + 4 + 4 + 4 + 4 = 4 \times 5 = 20$

2 $a + a + a = a \times 3 = 3a$

See also **addition, operation**

---

multiplier

The multiplier is the number that says how many times the multiplicand is to be added in the multiplication process.

For example

```
*  *  *  *   4 × 5 = 20
*  *  *  *   |
*  *  *  *   multiplier
*  *  *  *
*  *  *  *
```

The 5 says that 4 is to be added 5 times.

The 4 is multiplied by 5.

See also **multiplicand, multiplication**

Nn

natural number

A natural number is a counting number. This is a positive whole number.

The natural numbers are 1, 2, 3, 4, 5, 6, …

See also **positive number, whole number**

---

negative number

A negative number is a number less than zero. On a number line, the negative numbers are in the opposite direction to the positive numbers. A negative number is written with the minus symbol (–) in front.

For example

–1 and –2.5 are negative numbers.

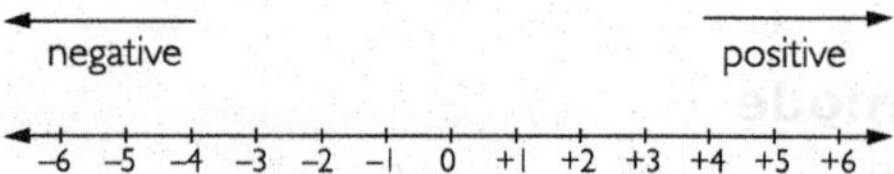

See also **positive number**

---

net

Meaning 1 Net is the final amount, after all necessary additions and subtractions have been taken into account.

For example

1 Net weight is the weight of the contents after subtracting the weight of the packing materials.

2 Net income is the amount of income left, after tax and levies have been deducted.

3 Net profit is the amount remaining from a sale, after all taxes and costs are deducted.

See also **gross**

Meaning 2 A net is a flat shape that will fold up to make a three-dimensional shape.

For example

1 A net of a cube folds into a model of a solid cube.

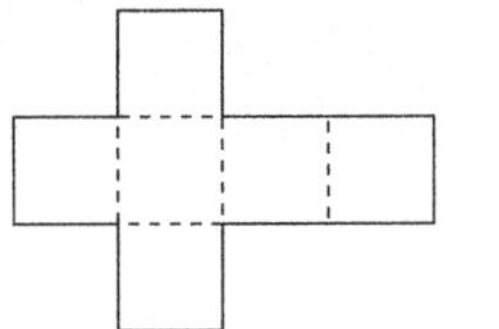

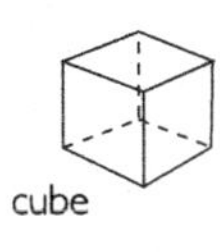

2 A net of a tetrahedron folds into a model of a solid tetrahedron.

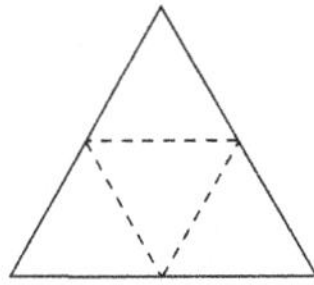

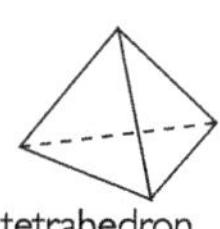

See also **model**

network

A network is a group of connecting lines. The lines are called arcs, and the points of connection are called nodes. Some networks can be traced without lifting the pen from the paper, or going over the same arc twice. Can these?

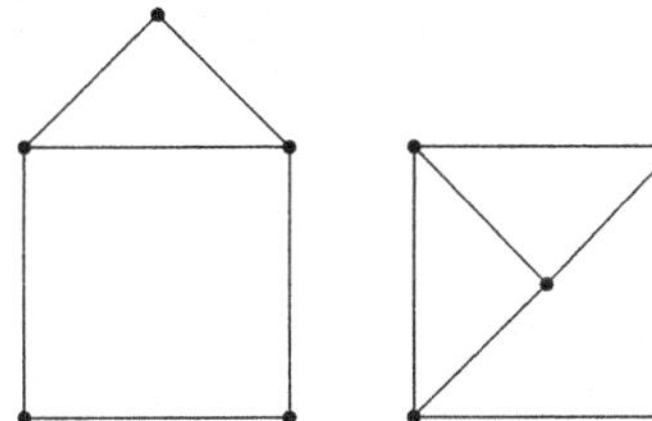

See also **Euler**

node

A node is a point at which arcs in a network meet.

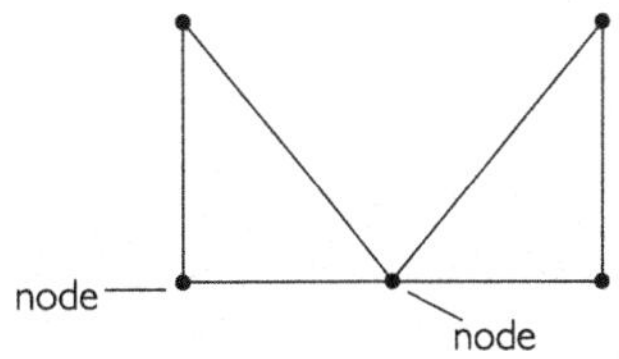

This network has five nodes.

See also **network**

nonagon

A nonagon is a polygon with nine straight sides. In a regular nonagon, all the sides have the same length and all the angles are equal.

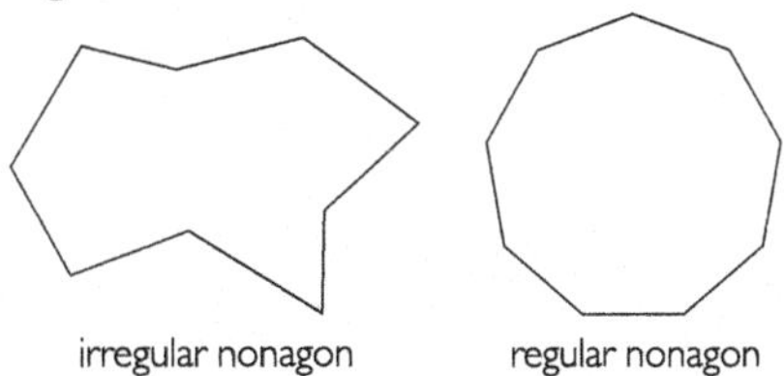

See also **polygon**, **regular polygon**

non-Euclidean geometry

Non-Euclidean geometry is a geometry that is not based on all of Euclid's assumptions about points, lines and planes.

For example

Geometry on the surface of a sphere is non-Euclidean geometry. In Euclidean geometry, the angles of a triangle add up to 180 degrees. But on the surface of a sphere, the angles of a triangle add up to more than 180 degrees.

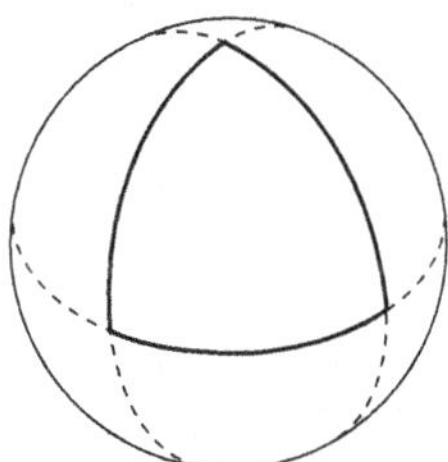

See also **Euclidean geometry**, **topology**

normal distribution

A normal distribution is a standard distribution of data. The highest frequency is at the mean. The mean, median and mode are equal. A normal distribution shows up clearly when the data is displayed as a line graph. The graph forms a bell curve.

☞

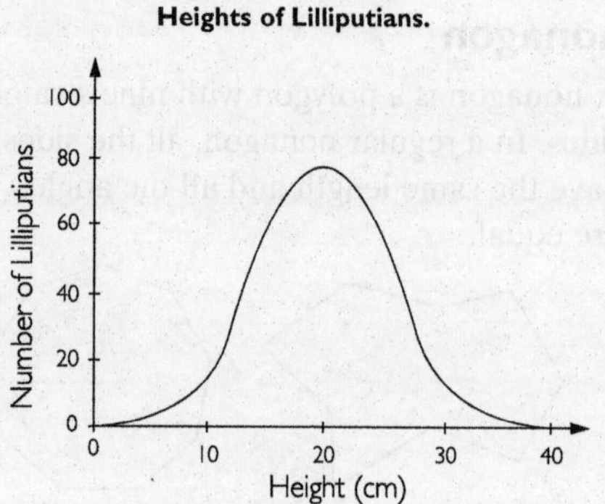

See also **data, frequency**

notation

Notation is a particular way of writing numbers and mathematical statements.

For example

1 In index notation, the expression $2 \times 2 \times 2$ is written 2^3.

2 Position notation is used to write numbers. In a decimal number, the position of the decimal point gives value to the digits. For instance, in the number 123.4, the 1 stands for one hundred, the 2 stands for two tens, the 3 stands for three ones and the 4 stands for four tenths.

nought

Nought is the symbol for zero.

For example

0.5 means 0 units and 5 tenths. It is read as "nought point five".

See also **symbol, zero**

number

Meaning 1 A number is a measure of how many or how much.

For example

★★★★ There are four stars.

Meaning 2 A number is the symbol that shows how many. This is also called a numeral.

For example

The number four is represented by the symbol 4.

See also **cardinal number, composite number, counting number, even number, fraction, integer, irrational number, natural number, odd number, ordinal number, prime number, rational number, real number, whole number**

number expander

A number expander is a visual aid to the understanding of the place value of digits in any given number.

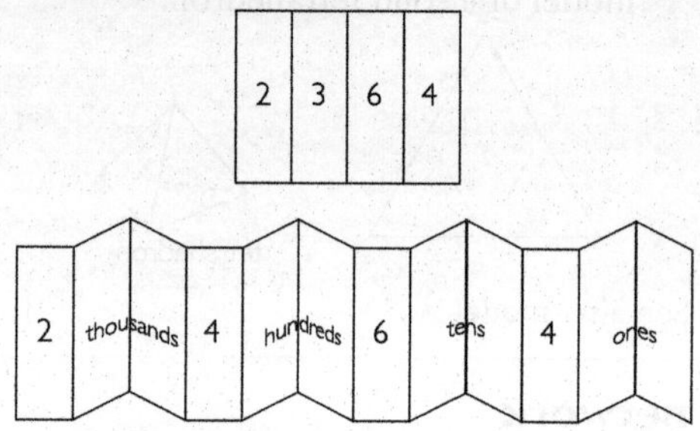

See also **place value**

number line

A number line is used to show the position of a number in terms of both distance and direction from zero. A number line extends indefinitely in both the positive and the negative directions.

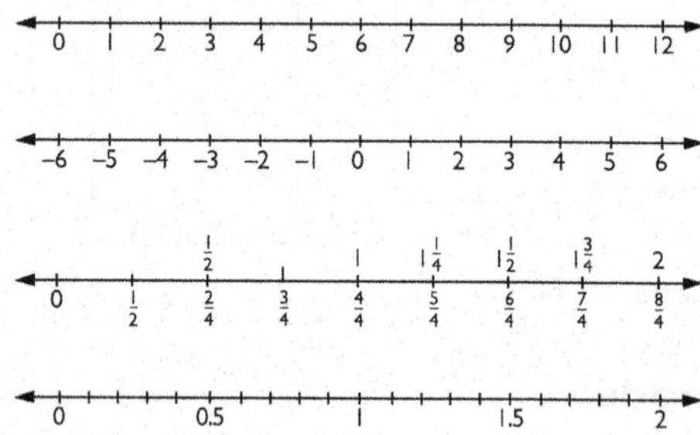

See also **negative number, positive number, zero**

number pair

A number pair is a group of two numbers. A number pair describes the position of a point on a plane. The two numbers determine horizontal and vertical distance and positive and negative direction from the origin on a Cartesian graph.

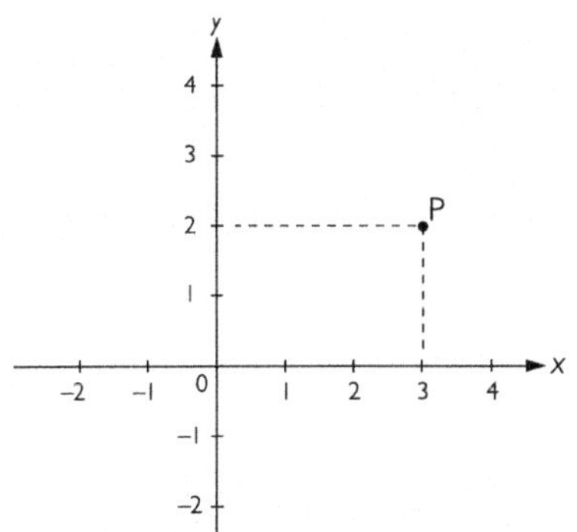

The number pair for the point P is (3,2). Its horizontal position is 3. Its vertical position is 2.

See also **coordinates**, **ordered pair**

number pattern

A number pattern is a sequence or arrangement of numbers based on a rule.

For example

1. The number pattern 22, 24, 26, 28, ... is counting by twos.
2. The number pattern 4, 8, 16, 32, ... is doubling.
3. The pattern 1, 1, 2, 3, 5, 8, 13, 21, 34, ... is the Fibonacci sequence.
4. Below is Pascal's triangle.

```
         1
      1     1
   1     2     1
1     3     3     1
```

number sentence

A number sentence is a statement of a relationship between numbers.

For example

1. $5 + 3 = 8$
2. $24.3 \neq 24.5$
3. $5 \times 10 > 50 \div 10$

See also **relation**

number system

A number system is a way of representing numbers.

For example

Today (Arabic)	1	2	3	4	5	6	7	8	9	10	11	12	13
Babylonian (1500 BC)	𒁹	𒈫	𒐈	𒐉	𒐊	𒐋	𒐌	𒐍	𒐎	𒌋	𒌋𒁹	𒌋𒈫	𒌋𒐈
Chinese (500 BC)	一	二	三	四	五	六	七	八	九	十	十一	十二	十三
Egyptian (300 BC)	I	II	III	IIII	IIIII	IIIIII	IIIIIII	IIIIIIII	IIIIIIIII	∩	∩I	∩II	∩III
Roman (200 BC)	I	II	III	IV	V	VI	VII	VIII	IX	X	XI	XII	XIII
Hindu (11th century)	१	२	३	४	५	६	७	८	९	१०	११	१२	१३
Base-two (computers)	1	10	11	100	101	110	111	1000	1001	1010	1011	1100	1101

See also **Arabic numerals**, **binary system**, **decimal system**, **Roman numerals**

numeral

A numeral is a symbol or a group of symbols that stand for a number.

For example

The number three can be represented by the numerals 3, $\frac{6}{2}$, III (Roman) and 11 (binary).

See also **symbol**

numerator

The numerator is the part of a common fraction that is above the fraction line. It indicates how many of the equal parts have been chosen.

For example

$\frac{3}{4}$ names three out of four equal parts of a whole. The numerator of $\frac{3}{4}$ is 3.

See also **common fraction**, **denominator**

numerical

Numerical means of numbers.

For example

1. Numerical data is information in numbers, as in the maximum temperature every day for a month.
2. A numerical statement is a number sentence.

See also **number**

Oo

oblique

Oblique means slanting or sloping.

For example

1 An oblique line is neither horizontal nor vertical.

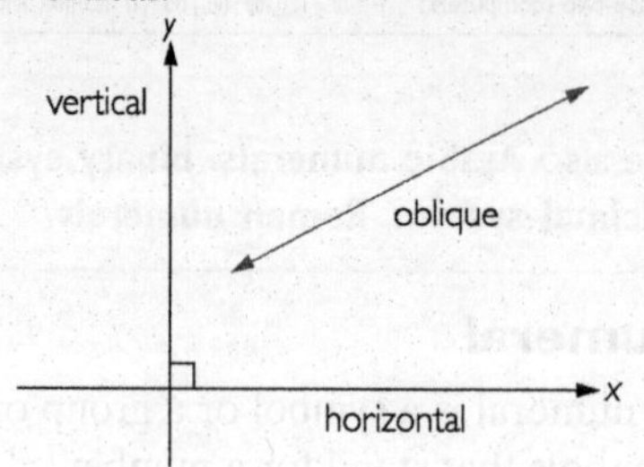

2 Cones, prisms and pyramids can be oblique.

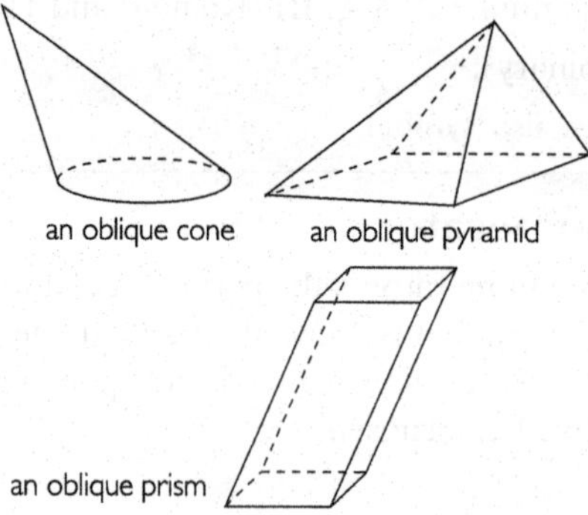

3 An oblique triangle does not have a right angle.

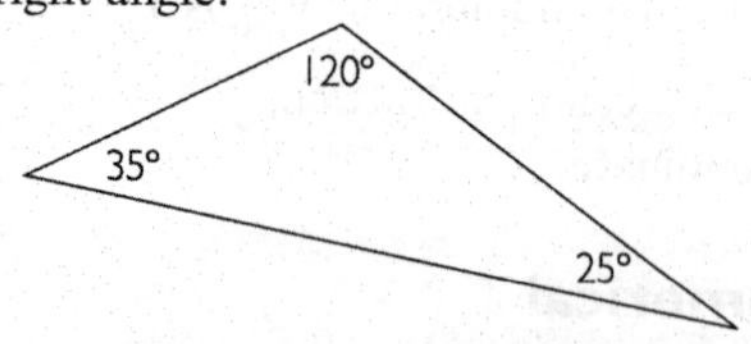

oblong

An oblong is a rectangle that is not a square.

See also **rectangle, square**

obtuse angle

An obtuse angle is an angle that measures more than 90 degrees (a right angle) but less than 180 degrees (a straight angle).

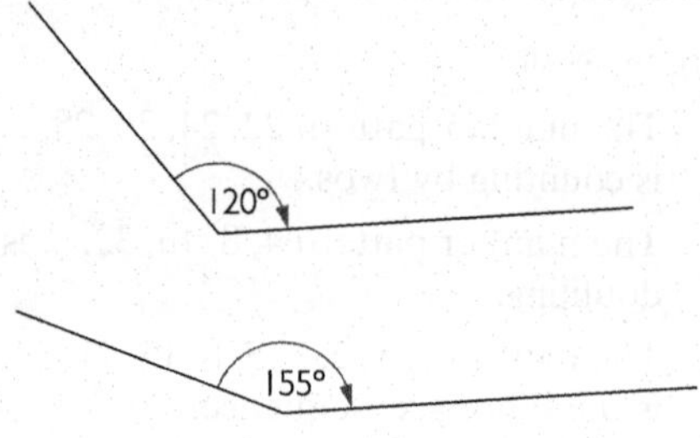

See also **angle**

obtuse-angled triangle

An obtuse-angled triangle is a triangle with one obtuse angle.

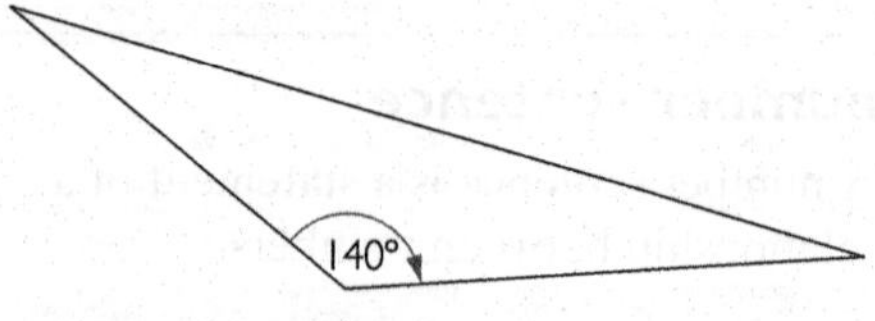

See also **obtuse angle, triangle**

octagon

An octagon is a polygon with eight straight sides. In a regular octagon, all the sides have the same length and all the angles are equal.

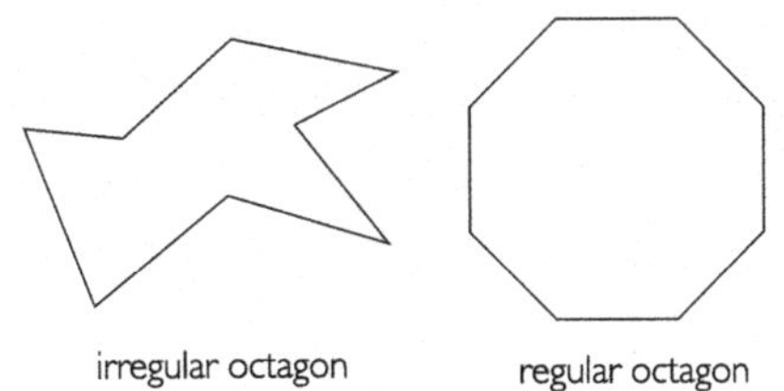

irregular octagon regular octagon

See also **polygon**, **regular polygon**

octahedron

An octahedron is a polyhedron with eight faces. The faces of a regular octahedron are equilateral triangles. The regular octahedron is one of the five Platonic solids.

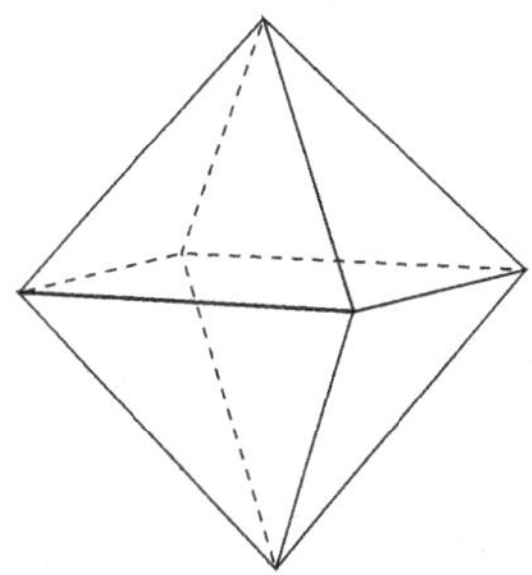

regular octahedron

See also **Platonic solids**, **polyhedron**

odd number

An odd number is a counting number that does not divide equally by two.

For example

3, 17, 99, 181, 1625 are all odd numbers. When divided by two, each leaves a remainder of one.

See also **even number**

one-dimensional

One-dimensional means able to be measured in one direction only.

For example

A line is one-dimensional. Only its length can be measured.

See also **dimension**, **three-dimensional**, **two-dimensional**, **zero-dimensional**

ones

In a base-ten counting system, ones is the place value that identifies the number of leftover ones. A group of ten ones becomes one group of ten in the tens place. Ones is sometimes known as units.

For example

In the number 467, the 7 stands for seven ones, that is, just seven.

hundreds	tens	ones
4	6	7

See also **base**, **decimal system**, **place value**, **tens**

one-to-one correspondence

One-to-one correspondence is the matching of two sets. Each element of one set is paired with an element of the other set. There is one-to-one correspondence between two sets if they have the same number of elements.

For example

1 This arrow diagram shows a one-to-one correspondence between Kylie's T-shirts and shorts.

2 When counting a group of objects, each object is tagged with exactly one number word.

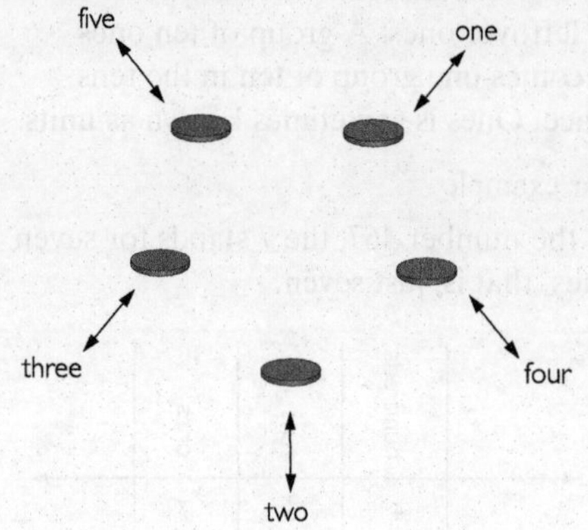

See also **many-to-one correspondence**, **mapping**

open curve

An open curve is a curved line with different beginning and end points.

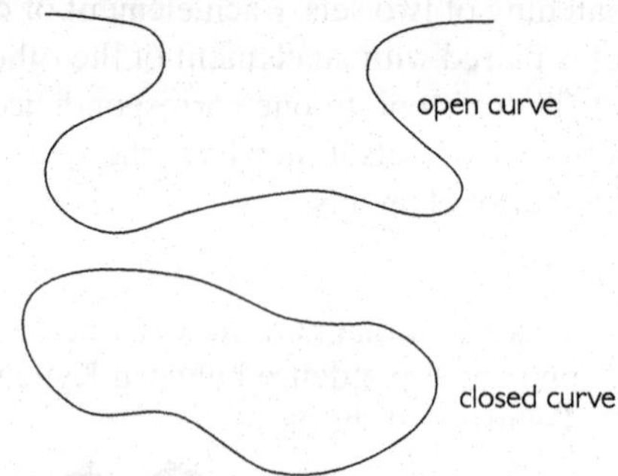

See also **closed curve**, **line**

open number sentence

An open number sentence is a number sentence with a number missing. The place for the missing number is marked with a box or a letter.

For example

1 6 + □ = 15

2 9 – □ < 5

3 $x + 10 > 12$

See also **equation**, **inequation**, **number sentence**

operation

Meaning 1 An operation is any one of the four methods of solving the problems of arithmetic: addition, subtraction, multiplication and division. These operations are also known as the four processes.

For example

1 3 + 9 = □ is solved by addition.

2 9 – 4 = □ is solved by subtraction.

3 7 × 6 = □ is solved by multiplication.

4 40 ÷ 10 = □ is solved by division.

See also **addition**, **division**, **multiplication**, **subtraction**

Meaning 2 An operation is the process of taking a number, or group of numbers, and obtaining a new number.

For example

Taking square roots is an operation on positive numbers.

operator

An operator is the symbol, also known as sign, that indicates the operation to be carried out.

For example

+, –, × and ÷ are operators.

See also **operation**, **symbol**

order

Meaning 1 To order is to arrange objects or numbers according to some rule.

For example

1 The spheres are arranged in order of size from largest to smallest.

2 The groups of counters are in order from least to greatest number.

3 The number sequence 2, 4, 6, 8, 10 is arranged in ascending order.

See also **pattern**, **sequence**

Meaning 2 In a network, the order of a node is the number of lines that meet at the node.

For example

In this network, the order of the node N is 4.

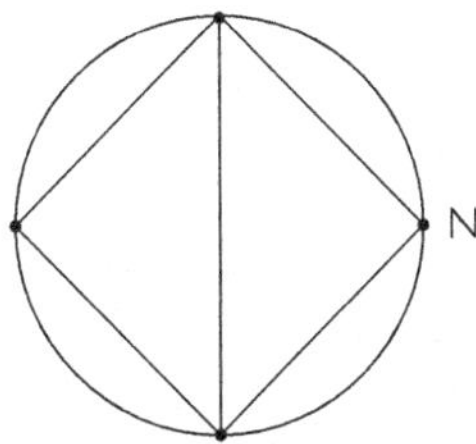

See also **network**

Meaning 3 The order of a shape is the number of times it appears to be the same during a complete rotation. This is called the order of rotational symmetry. A shape with order of symmetry 1 has no rotational symmetry.

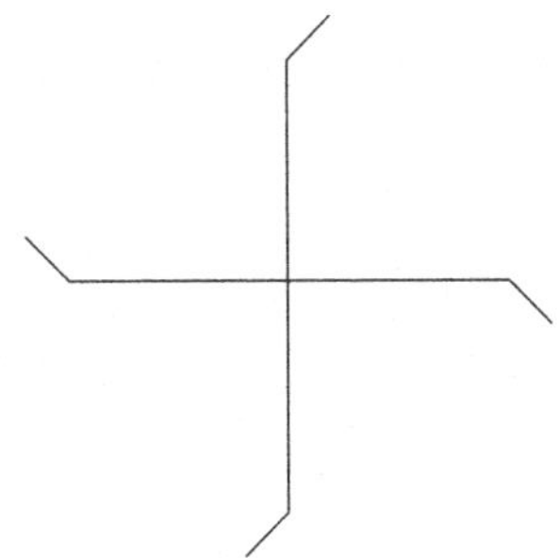

This shape has rotational symmetry of order 4.

See also **symmetry**

Meaning 4 The order of an array is its number of rows and columns.

For example

The order of this array is 3×4.

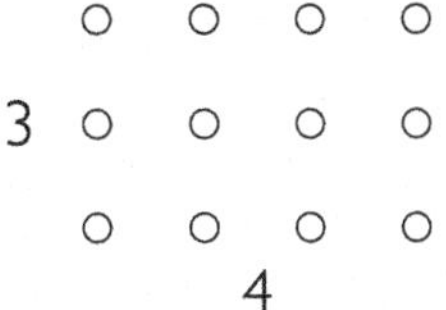

See also **array**

ordered pair

An ordered pair is a group of two numbers in a particular order. Ordered pairs are used in the Cartesian coordinate system to locate a point on the plane.

For example

(2,3) is an ordered pair. The order of the numbers is important. The ordered pairs (2,3) and (3,2) are different.

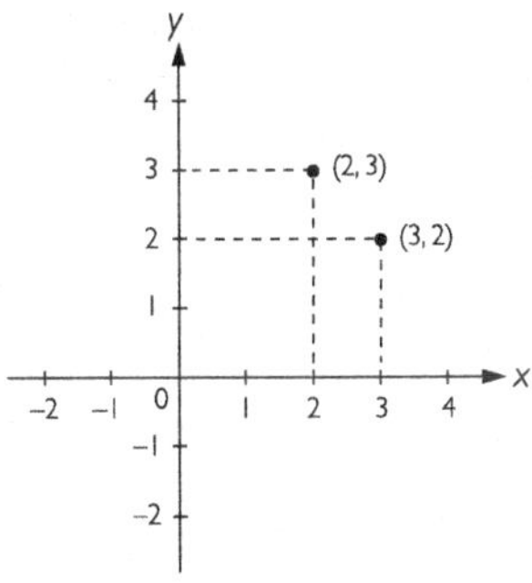

See also **Cartesian coordinates**, **coordinates**, **order**

order of operations

The order of operations is the set of conventions to be followed when solving an arithmetic problem involving several operations. Without these conventions, different solutions to the problem can be found, causing confusion.

☞

The conventions are to:

- carry out calculations within **brackets** first
- work a fraction **of** a number as a multiplication
- complete **multiplication** and **division** before **addition** and **subtraction**
- work from left to right

These conventions have the status of rules. The acronym BOMDAS is a reminder of the order of operations.

For example

$64 \div 8 + (4 - 2) \times 5 - \frac{1}{2}$ of 6

$= 64 \div 8 + 2 \times 5 - \frac{1}{2}$ of 6	B
$= 64 \div 8 + 2 \times 5 - 3$	O
$= 8 + 10 - 3$	MD
$= 18 - 3$	AS
$= 15$	left to right

See also **addition, brackets, division, fraction, multiplication, subtraction**

ordinal number

An ordinal number is a number that names a place in a sequence. First, second, third, fourth and so on are ordinal numbers.

For example

1. Susan was first in the line.
2. Of the numbers 5, 6, 8 and 10, the second highest number is 8.
3. The results are recorded in the third column to the right.

Mathematics Class 7B			
Family name	*Given name*	*Result*	*Comments*
Abbot	Sue	B+	Been ill
Allan	Chris	C	Tries hard
Au	Li	A	Excellent
Boukas	Joe	A	Excellent

See also **cardinal number, sequence**

origin

An origin is a source or a beginning.

For example

1. The point where the axes of a coordinate system cross is called the origin.

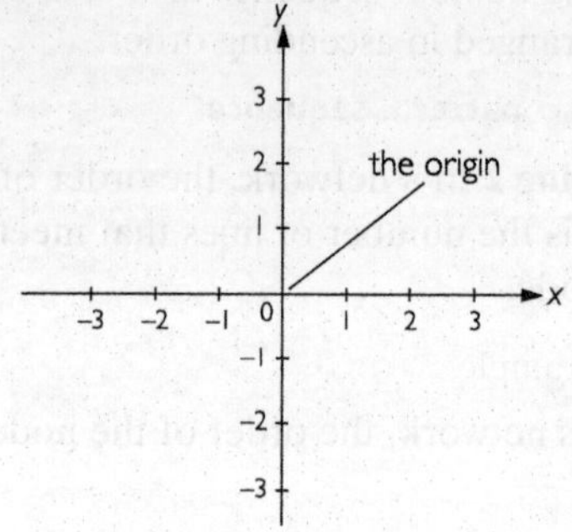

2. The origin is the point from which a measurement is taken.

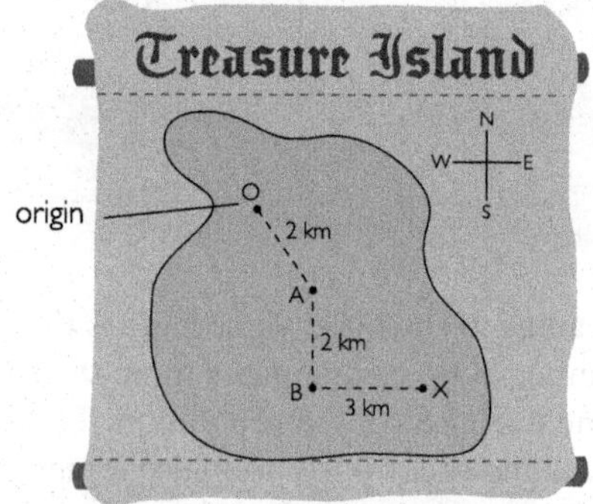

oval

An oval is a plane curve that is egg-shaped. An ellipse is a special type of oval.

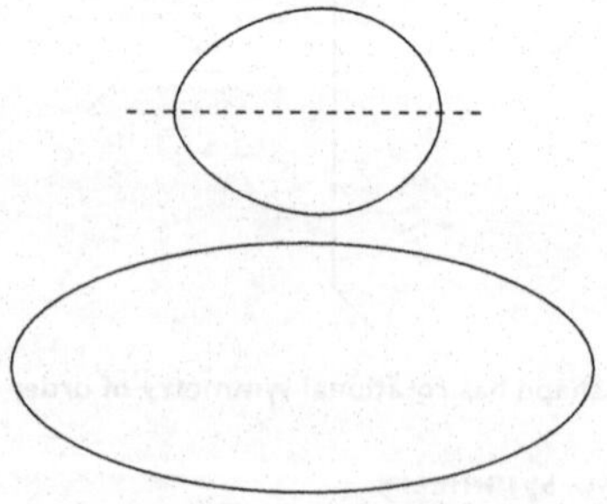

See also **closed curve, ellipse**

Pp

pair

Meaning 1 A pair is a group of two. It is two things for use together.

For example

Pairs of numbers describe a point on a plane, such as:

(-1, +6), (-2, +8), (-3, +10)

See also **coordinates**, **ordered pair**

Meaning 2 To pair is to link or to arrange in pairs.

For example

To pair one member of one set with one member of a different set is to show correspondence.

See also **arrow diagram**, **one-to-one correspondence**

pair of compasses

A pair of compasses is an instrument used to draw circles and to measure and mark out distances. The instrument consists of a pair of hinged, movable arms. The name *pair of compasses* is sometimes shortened to compass, but it should not to be confused with the instrument that shows direction.

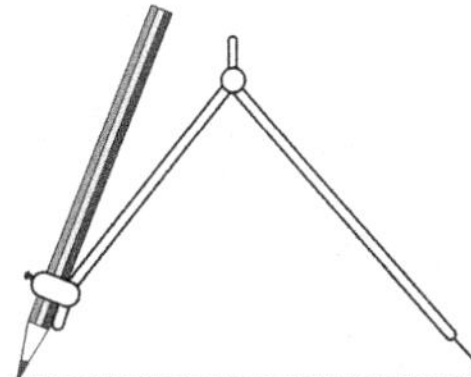

See also **compass**

palindromic number

A palindromic number is a number that reads the same both forwards and backwards.

For example

33, 121 and 87 978 are palindromic numbers.

parallel

Parallel lines are two or more lines that are always the same distance apart. They never meet, no matter how far they extend. Similarly, parallel planes are planes that are always the same distance apart.

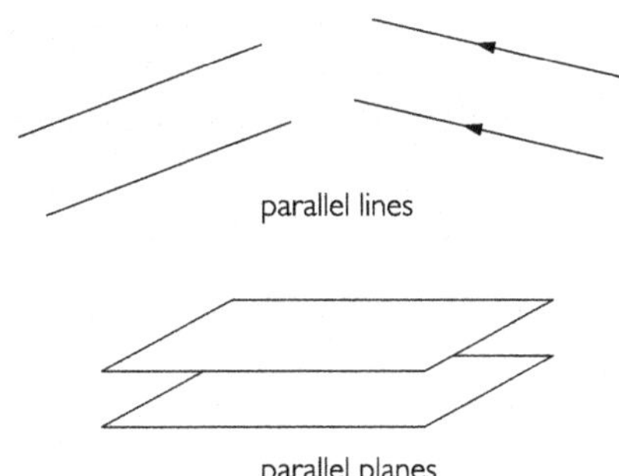

See also **line**, **plane**

parallelogram

A parallelogram is a polygon with four straight sides, and with opposite sides parallel and equal in length. The opposite angles of a parallelogram are equal.

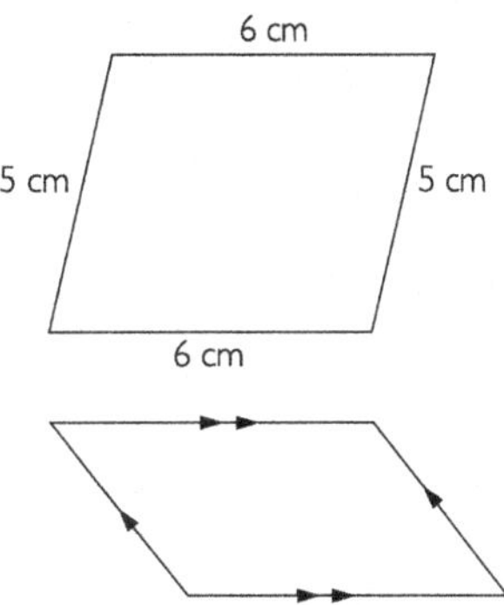

See also **parallel**, **polygon**

part

A part is a piece of a whole.

For example

The common fraction $\frac{1}{3}$ names one part out of three equal parts of a whole pie, or one part out of three equal parts of a bag of nine apples.

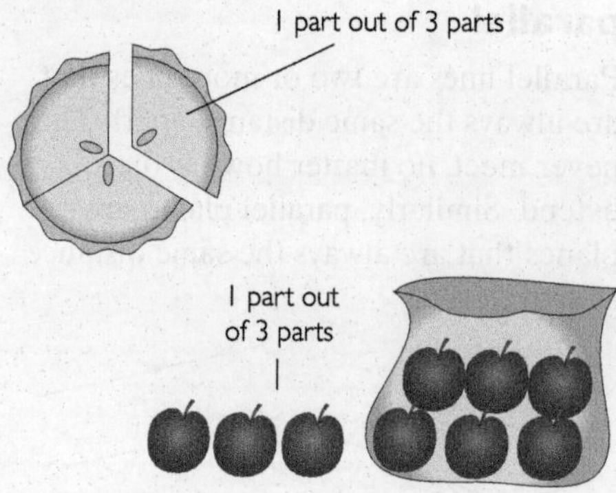

See also **continuous**, **discrete**, **whole**

partition

Partition means division, when the number of groups is known. The problem is to find how many are in each group.

For example

There are 20 golf balls to be shared among 5 players. How many balls will each player get?

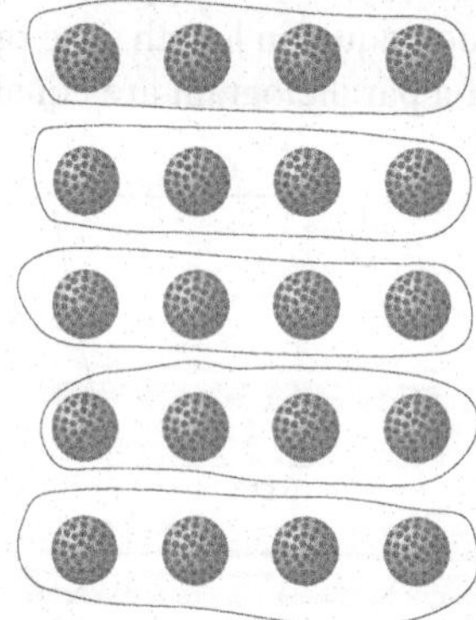

$20 \div 5 = 4$

See also **division**, **group**, **quotition**

Pascal's triangle

Pascal's triangle is a triangular pattern of numbers, named after the French mathematician Blaise Pascal, who lived 1623–1662. Each number in the triangle is the sum of the number above it to the right and the number above it to the left. Pascal's triangle can be used to find probabilities because its rows model possible outcomes.

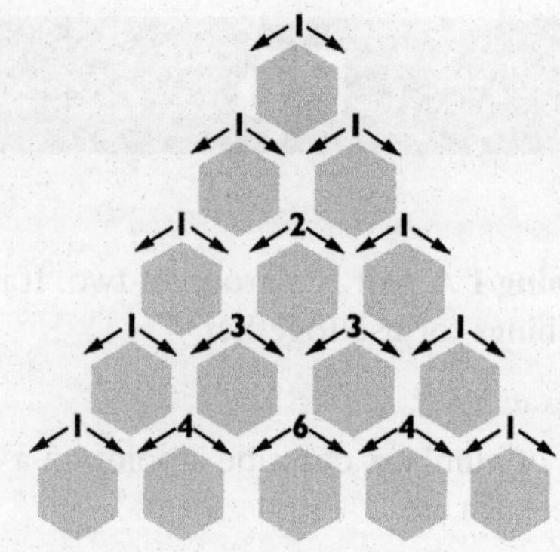

See also **pattern**, **probability**, **tree diagram**

path

A path is a route or a line that connects points.

For example

1 On a map, trace the shortest path from one location to another.

2 On a network, trace a path that passes through each arc only once.

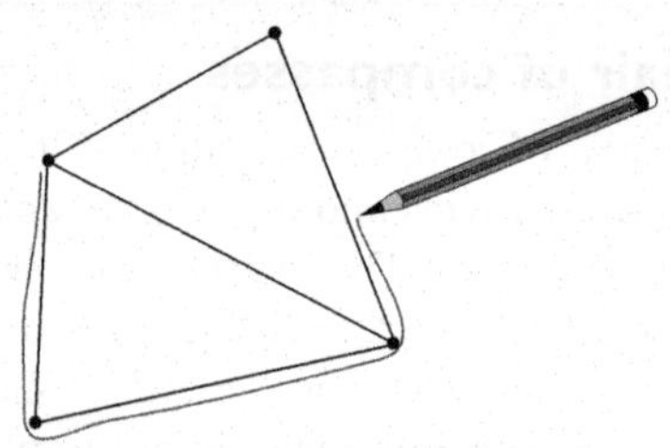

See also **line**, **point**

pattern

A pattern is an arrangement of objects in keeping with a rule.

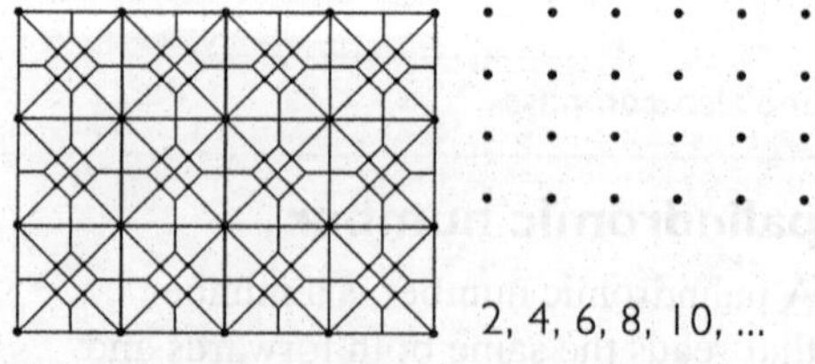

See also **number pattern**, **regular**, **rule**, **sequence**, **spatial pattern**

pentagon

A pentagon is a polygon with five straight sides. In a regular pentagon, all the sides have the same length and all the angles are equal.

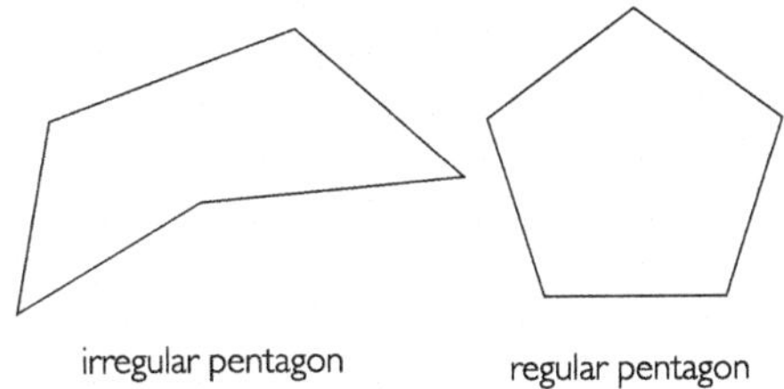

irregular pentagon   regular pentagon

See also **polygon**, **regular polygon**

pentahedron

A pentahedron is a polyhedron with five faces.

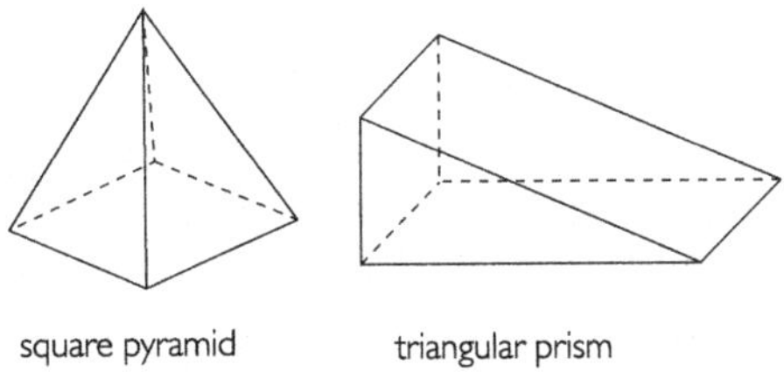

square pyramid   triangular prism

See also **polyhedron**

per cent

Per cent means out of each hundred. The symbol for per cent is %.

For example

25 per cent means 25 parts out of every 100 parts. Written as a common fraction, this is $\frac{25}{100}$. Written as a decimal fraction, this is 0.25.

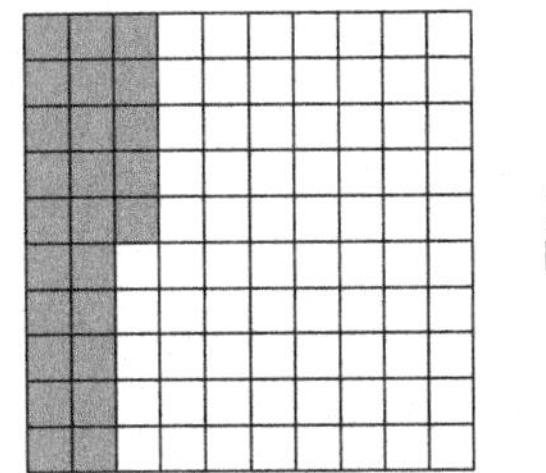

$\frac{25}{100} = 25\%$

See also **percentage**

percentage

A percentage is a fraction expressed in hundredths, using the per cent symbol (%). Fractions used in everyday life are often written as percentages, such as in sale discounts, interest rates and opinion polls.

For example

1 The fraction $\frac{9}{100}$ written as a percentage is 9%.

2 In a shop sale, "20% off the marked price" involves a calculation of the percentage discount. The discount is $\frac{20}{100}$, or $\frac{1}{5}$, of the marked price.

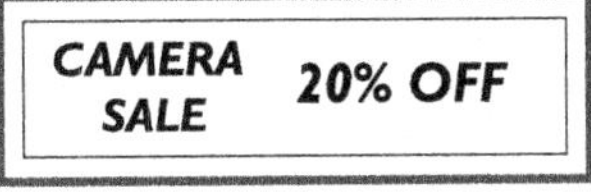

See also **per cent**

perfect number

A perfect number is a counting number that is equal to the sum of all its factors, excluding itself.

For example

6 is a perfect number. This is because the factors of 6, excluding itself, are 1, 2 and 3, and 6 = 1 + 2 + 3.

See also **factors**, **sum**

perfect square

A perfect square is a number or expression that is the product of two equal factors.

For example

1 The number 16 is a perfect square since $16 = 4 \times 4 = 4^2$.

2 The expression $a^2 + 6a + 9$ is a perfect square since:

$$\begin{aligned} a^2 + 6a + 9 &= (a + 3)(a + 3) \\ &= (a + 3)^2. \end{aligned}$$

See also **factors**, **product**, **square number**

perimeter

The perimeter of a shape is the distance around its boundary.

For example

1 The perimeter of a rectangle is twice its length plus twice its width.

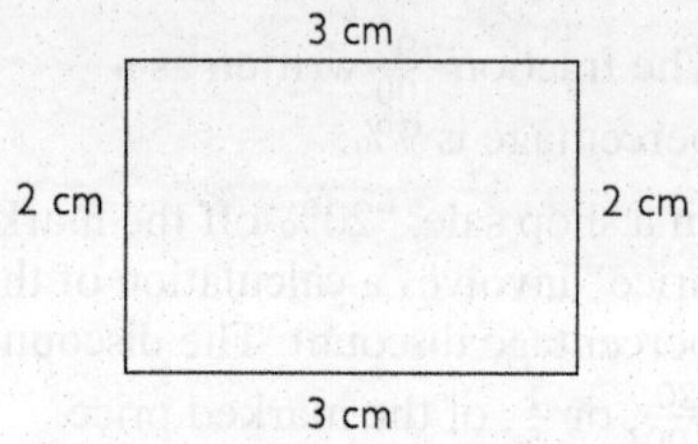

2 x 3 + 2 x 2 = 6 + 4 = 10
The perimeter of this rectangle is 10 cm.

2 The perimeter of a circle is called its circumference. It is equal to the diameter times the number π.

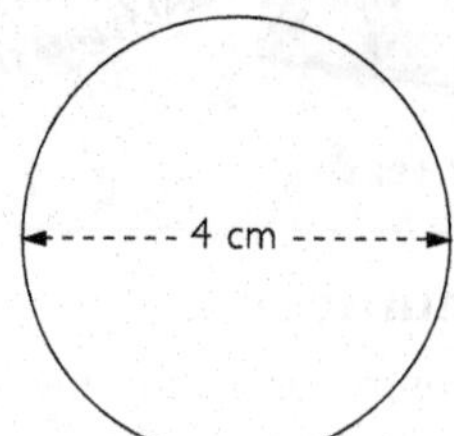

4 x π ≈ 4 x 3.14 = 12.56
The perimeter of this circle is approximately 12.56 cm.

See also **distance**

perpendicular

Perpendicular means at right angles. Lines or planes can be perpendicular to each other.

For example

1 Two lines are perpendicular if they are at right angles.

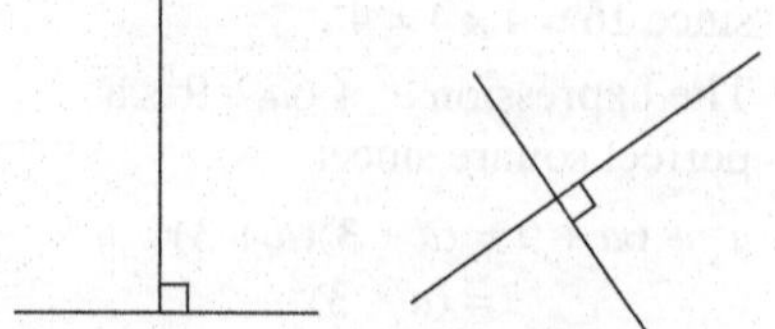

2 The height of a pyramid is the perpendicular distance from its base to its highest point.

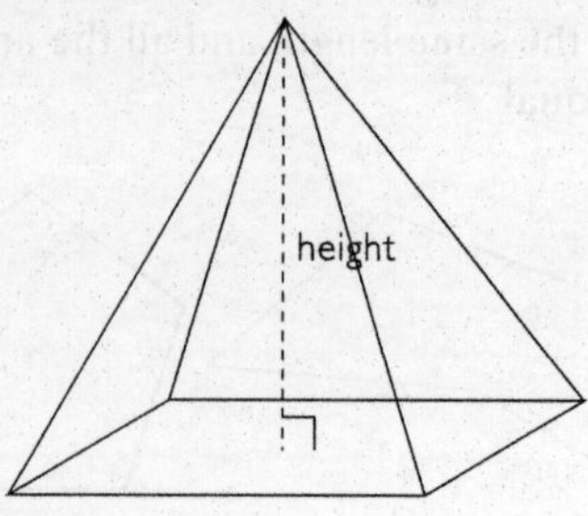

See also **right angle**

perspective

Perspective is the appearance of objects, which is affected by their relative positions and sizes.

For example

In a drawing that shows perspective, objects are smaller as they are positioned into the distance, and parallel lines converge at the horizon.

See also **converge, relative**

pi

The number pi is the ratio of the circumference of any circle to its diameter. Pi is approximately 3.1416. The symbol for pi is π.

The number π cannot be written out in full as a decimal number. Its string of digits goes on forever without repeating.

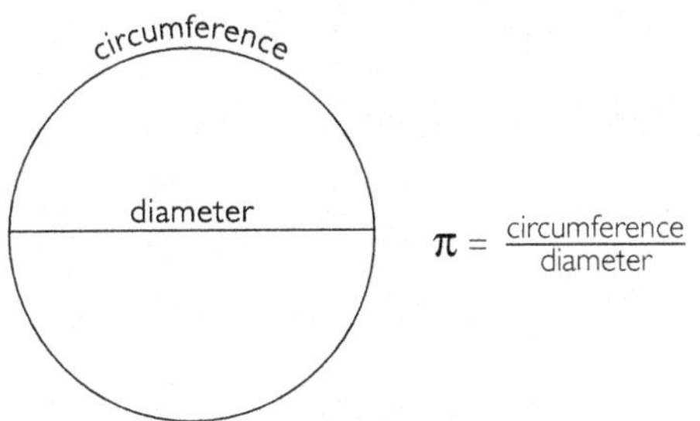

See also **irrational number**, **ratio**

picture graph

A picture graph is a graph that uses pictures to represent information. The number of pictures shows how many objects there are in each category. A picture graph is also known as pictograph, pictogram or ideograph.

Time I spent watching TV this week.

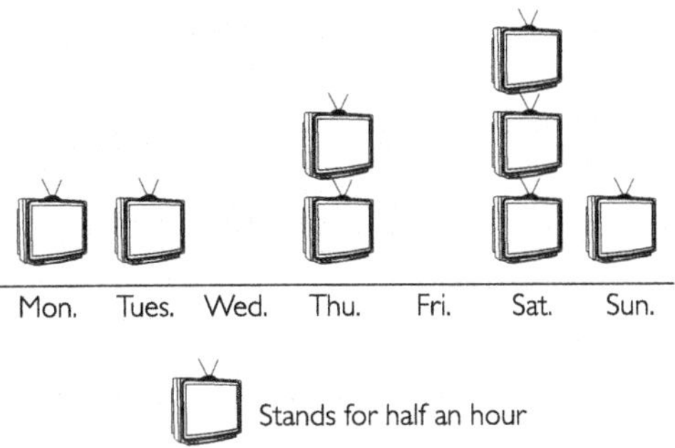

See also **bar graph**, **data**, **graph**

pie chart

A pie chart is a graph that represents information inside a circle. The circle is divided into sectors. Each sector represents a category of the whole group. The angle of each sector corresponds to the size of each category as a fraction of the whole group. A pie chart is also known as a pie graph.

Favourite sports in Year 7.

See also **data**, **fraction**, **graph**, **sector**

place value

Place value is the position that determines the value of any digit in a number. In a base-ten system, just ten digits are used to write all whole numbers and decimal fractions by means of place value.

For example

346 is 3 hundreds, 4 tens and 6 ones.

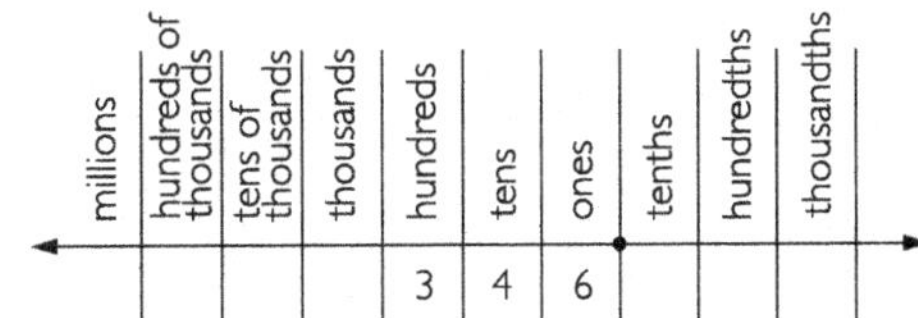

See also **base**, **decimal system**, **digit**

plan

A plan is a diagram that shows a bird's-eye view of a structure or arrangement.

For example

The floor plan of a room is a diagram of the room, as seen from above.

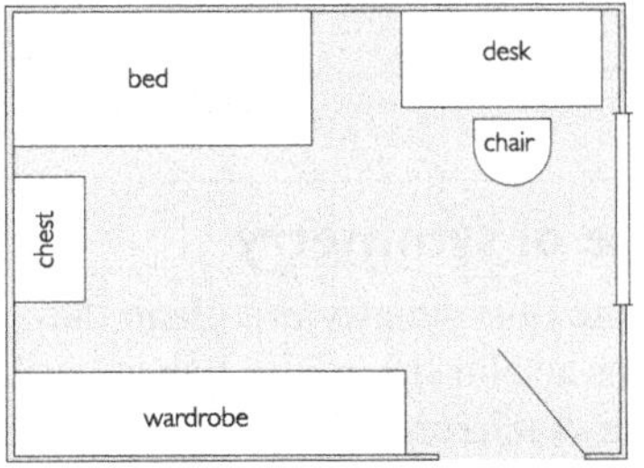

Scale: 1 cm = 1 m

See also **bird's-eye view**, **map**

P

plane

A plane is a flat surface that extends infinitely in all directions.

For example

1 A two-dimensional shape is called a plane shape because it is drawn in one plane.

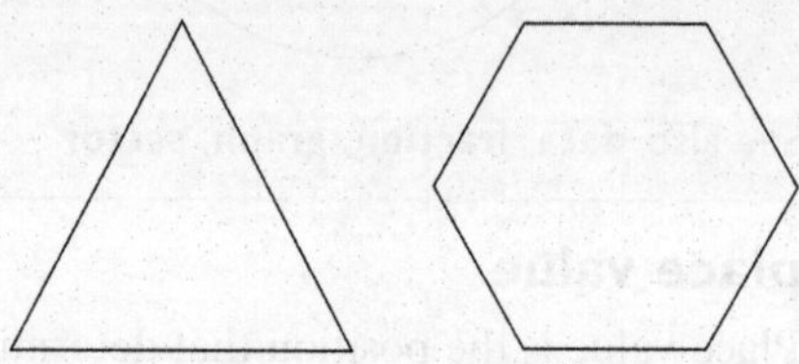

2 A horizontal plane is a plane that is parallel to the horizon.

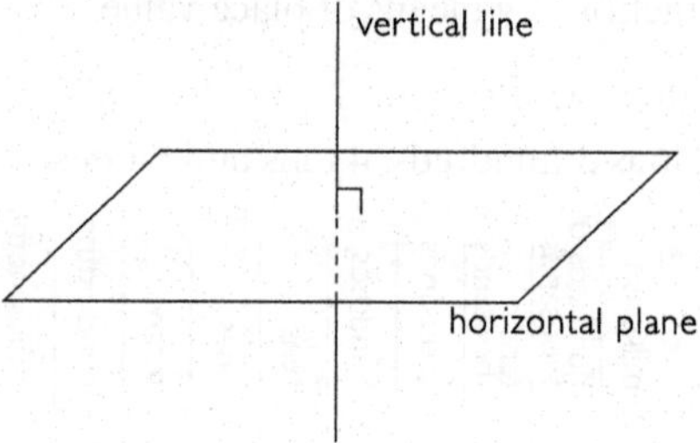

3 Cartesian coordinates locate points on a plane, with reference to two axes.

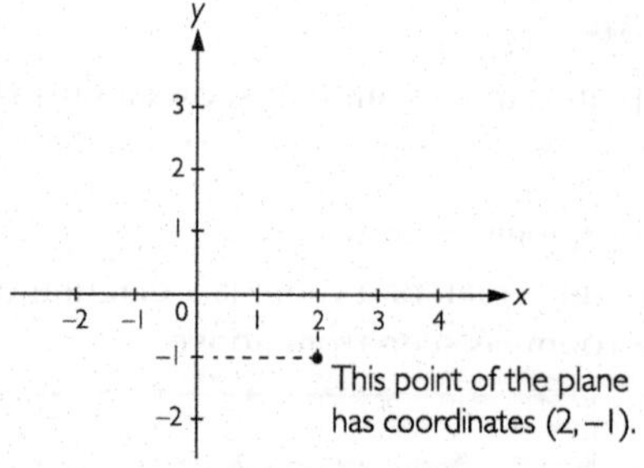

plane of symmetry

A plane of symmetry is a plane that divides a solid shape into halves so that each half mirrors the other half exactly. A plane of symmetry is also called a mirror plane. A solid shape can have more than one plane of symmetry.

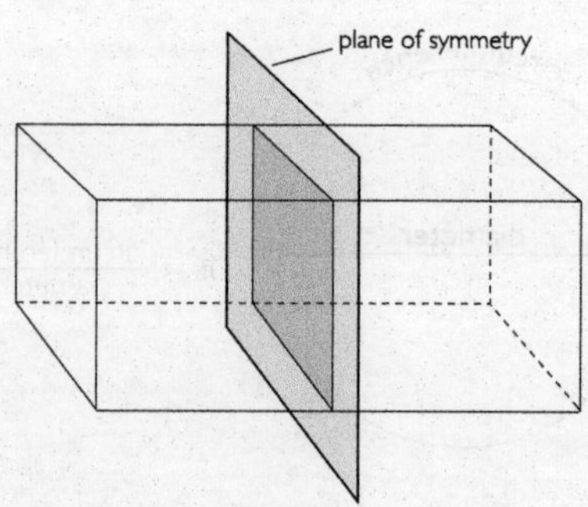

See also **line of symmetry, symmetry**

Platonic solids

The Platonic solids are the regular polyhedra. They are named after the Greek philosopher Plato, who lived around 400 BC. There are only five types of Platonic solids. They are the tetrahedron, hexahedron (cube), octahedron, dodecahedron and icosahedron. The faces of each Platonic solid are identical polygons, and all their vertices are equivalent.

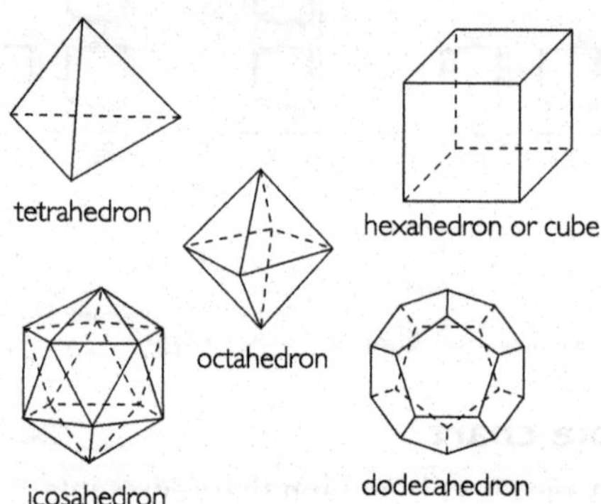

See also **congruent, dodecahedron, hexahedron, icosahedron, octahedron, polygon, polyhedron, tetrahedron**

plot

To plot is to locate and mark points, as on grid paper, using quantities or coordinates.

For example

You can plot a graph to show the relationship between time and distance in a journey.

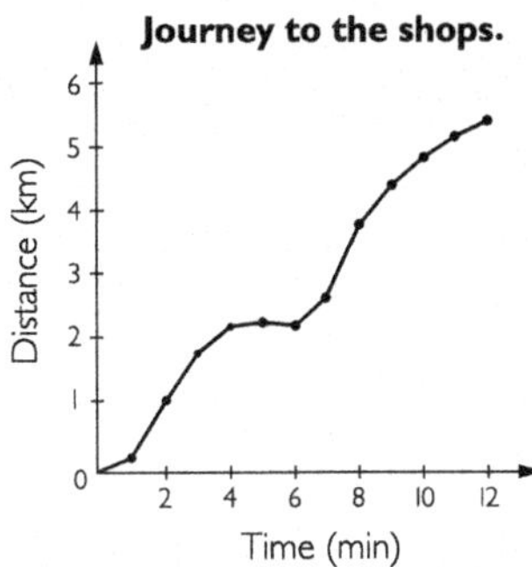

See also **Cartesian coordinates**, **grid paper**, **point**

plus

Meaning 1 Plus means to be added. The symbol for plus is +.

For example

5 + 9 = 14

See also **addition**

Meaning 2 Plus is the symbol of positive numbers.

For example

..., –3, –2, –1, 0, +1, +2, +3, ...

See also **negative number**

p.m.

p.m. is the period between 12 o'clock in the middle of the day and 12 o'clock at night that is used with 12-hour time. The letters *p.m.* are an abbreviation of the Latin words *post meridiem*.

For example

The television news is broadcast at 7 p.m.

See also **a.m.**, **twelve-hour time**

point

Meaning 1 A point has a location but it has no size. A point has no dimension: no length, no height, no width.

For example

1 A point is a position in space. For instance, the points that make up a circle are all the same distance from its centre.

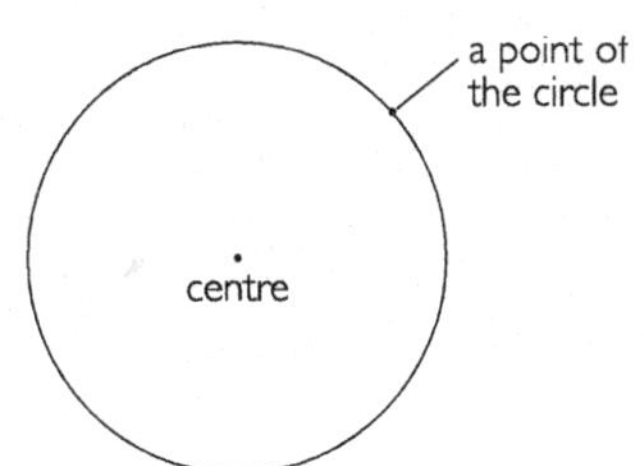

2 A point is a position plotted on a graph with reference to two axes. Cartesian coordinates locate points on a plane.

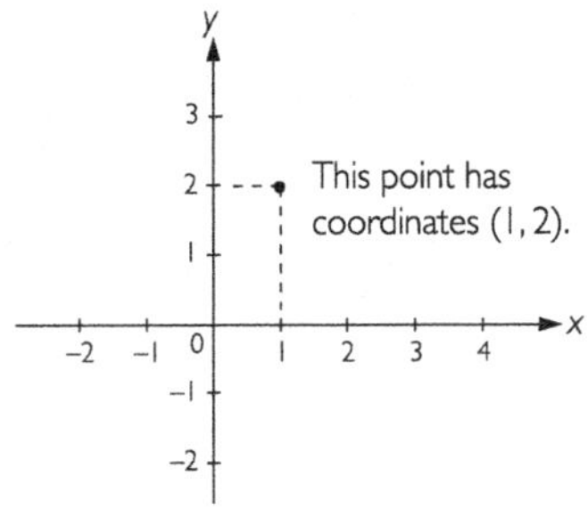

3 Two points on a line mark a line segment.

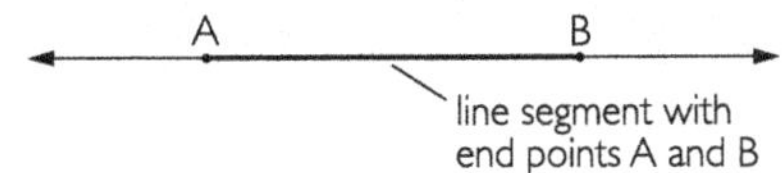

See also **zero-dimensional**

Meaning 2 A decimal point separates a number into a whole-number part and a decimal-fraction part.

For example

In the number 3.426, the 3 is separated by a decimal point from the decimal-fraction part of the number.

See also **decimal point**

Meaning 3 The points of a compass indicate direction.

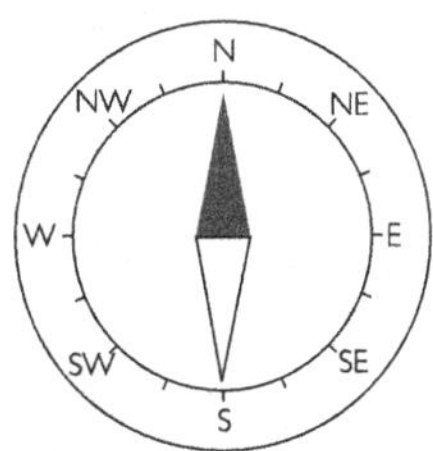

See also **compass points**

P

polygon

A polygon is a straight-sided plane shape. Polygons are classified according to their number of sides. In a regular polygon, all the sides have equal length.

For example

A hexagon is a six-sided polygon.

Figure	Number of sides	Name
	3	Triangle
	4	Quadrilateral
	5	Pentagon
	6	Hexagon
	7	Heptagon
	8	Octagon
	9	Nonagon
	10	Decagon
	12	Dodecagon

See also **irregular, regular polygon, two-dimensional**

polyhedron

A polyhedron is a solid shape with polygons as faces. Polyhedrons are classified according to their number of faces. In a regular polyhedron, all the faces are congruent, regular polygons and all the vertices are equivalent. Types of polyhedrons include prisms and pyramids.

For example

An octahedron is a polyhedron with eight faces.

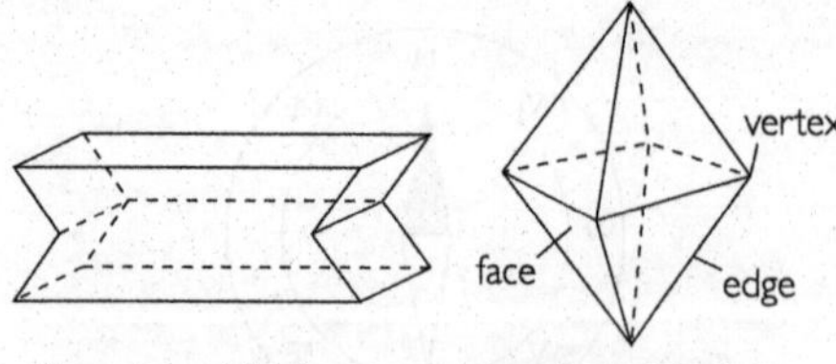

See also **irregular, Platonic solids, regular polyhedron, three-dimensional**

positive number

A positive number is a number greater than zero.

For example

1 The counting numbers, 1, 2, 3, 4, 5, ... are positive numbers.

2 2.5 is a positive number.

3 $17\frac{2}{3}$ is a positive number.

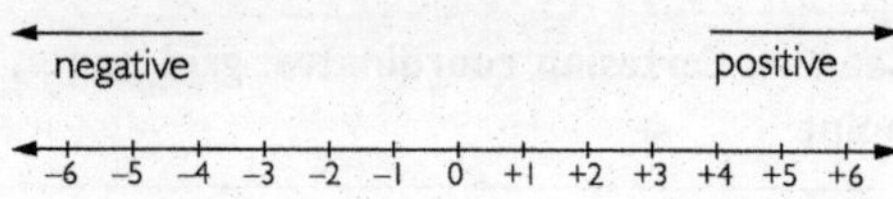

See also **integer, natural number, negative number, zero**

power

Taking a power of a number means multiplying that number by itself several times.

For example

1 "2 to the power of 4" is the number you get by multiplying 2 by itself 4 times. A short way of writing this is 2^4.

$2^4 = 2 \times 2 \times 2 \times 2 = 16$

The original number is called the base. The number of times the base is repeated is called the exponent or index.

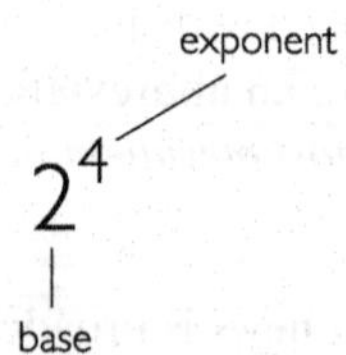

2 Powers of 10 are often used to simplify the writing of very large numbers.

For example

$10^5 = 10 \times 10 \times 10 \times 10 \times 10 = 100\,000$

See also **cube, square**

prime factor

A prime factor is a factor of a given number that is a prime number.

For example

1 The prime factors of 21 are 3 and 7.

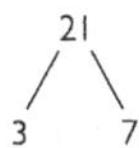

2 The prime factors of 18 are 2 and 3.

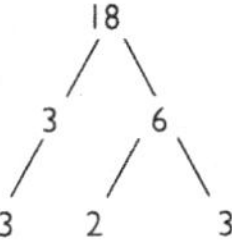

See also **factors**, **prime number**

prime number

A prime number is a counting number with only two different factors: the number itself and one. The number 1 is not counted as a prime number.

For example

7 is a prime number. Its only factors are 7 and 1.

See also **composite number**, **factors**, **sieve of Eratosthenes**

prism

A prism is a solid shape with two parallel, congruent polygons as faces. These two faces are called bases. All the other faces of the prism are parallelograms. These faces are called lateral faces. If the lateral faces are rectangles, the prism is called a right prism. A prism is named according to the shape of the bases.

For example

1 A triangular prism has two triangular bases and three lateral faces.

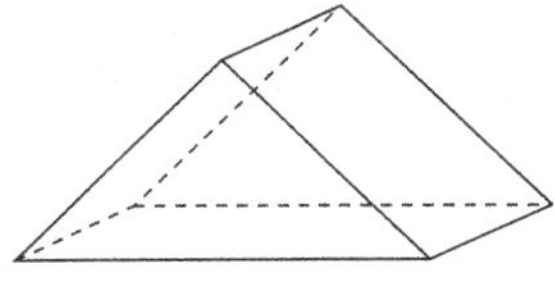

triangular prism

2 A pentagonal prism has two pentagonal bases and five lateral faces.

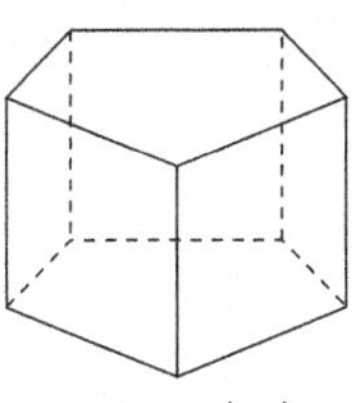

pentagonal prism

See also **congruent**, **face**, **parallel**, **polygon**, **polyhedron**, **solid**

probability

Meaning 1 Probability is a consideration of the likelihood or chance of an event occurring in everyday situations. The probability of an event occurring can be expressed using statements such as *certain*, *likely*, *equal chance*, *unlikely* and *impossible*.

For example

It is likely that it will rain tomorrow.

See also **chance**, **likelihood**

Meaning 2 Probability is an evaluation of the likelihood or chance of an event occurring. A probability is expressed as a number between 0 and 1, and can be written as a fraction, decimal or ratio.

For example

1 When a drawing pin is tossed on a table, it can land point up or point down. To estimate the probability of the drawing pin landing point up, we toss the drawing pin 100 times and record whether or not it lands point up. Say the pin lands point up 32 times, then the experimental probability that the pin will land point up is $\frac{32}{100}$.

2 In one throw of a fair die, the probability of throwing a five is $\frac{1}{6}$. This is because there is one chosen outcome among six possible outcomes. This probability is a theoretical probability because it can be worked out without an experiment.

P

product

A product is the answer when two or more numbers are multiplied together.

For example

48 is the product of 6 and 8.

See also **multiplication**

pronumeral

A pronumeral is a symbol representing an unknown value. The unknown value may be a constant or a variable.

For example

a and □ are pronumerals.

$a = 5$

$6a - 2a = \square$

See also **algebra, constant, variable**

proper fraction

A proper fraction is a common fraction with the numerator less than the denominator.

For example

$\frac{1}{2}, \frac{1}{3}, \frac{5}{8}$ and $\frac{75}{100}$ are all proper fractions.

See also **common fraction, denominator, improper fraction, numerator**

proportional

Proportional means having the same ratio.

For example

1 Similar triangles have corresponding sides proportional.

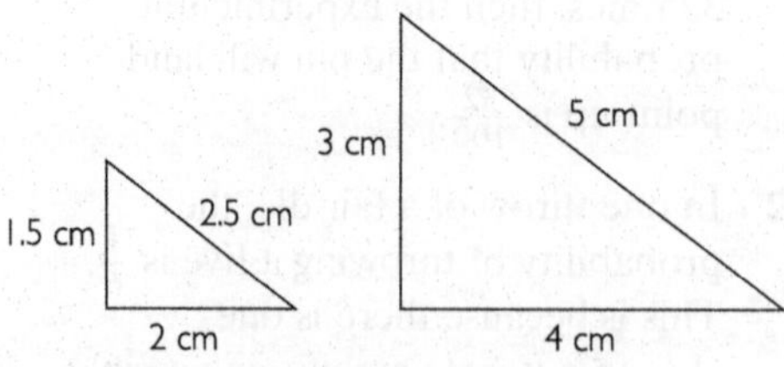

Corresponding sides are in the ratio 1:2.

2 When making different amounts of cordial, the amount of concentrate is proportional to the amount of water.

Concentrate (mL)	20	40	60
Water (mL)	100	200	300

The ratio of concentrate to water is 1:5.

See also **direct proportion, inverse, ratio**

protractor

A protractor is a flat, circular or semicircular instrument with a graduated scale in degrees. A protractor is used to measure or mark off angles.

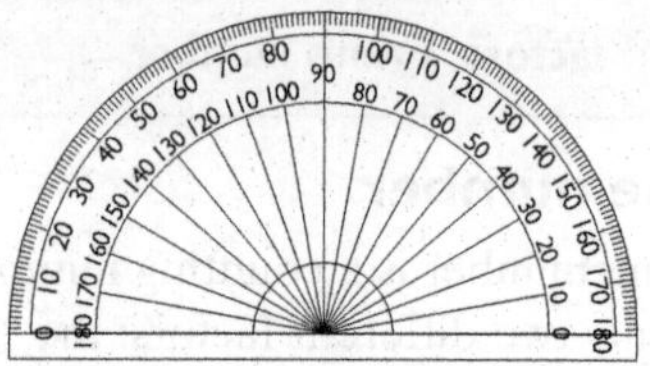

See also **degree, graduation, scale**

pyramid

A pyramid is a solid shape with a polygon as its base. The other faces are all triangles with a common vertex. A pyramid is called a right pyramid if the line from the centre of the base of the pyramid to its vertex is perpendicular. A pyramid is named according to the polygon that is its base.

For example

1 A pyramid with a square as its base is called a square pyramid.

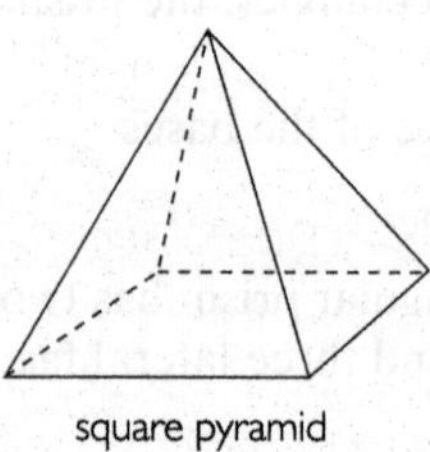

square pyramid

2 A pyramid with a hexagon as its base is called a hexagonal pyramid.

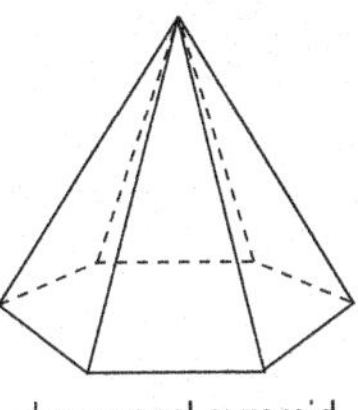

hexagonal pyramid

See also **polygon**, **polyhedron**, **three-dimensional**

Pythagoras

Pythagoras was a Greek mathematician who lived around 540 BC. He founded a school, the members of which were among the first to study theoretical mathematics. The Pythagoreans were the first to record written proofs of the theorem that is now called Pythagoras' theorem. This theorem states that, in a right-angled triangle, the square on the hypotenuse is equal to the sum of the squares on the other two sides.

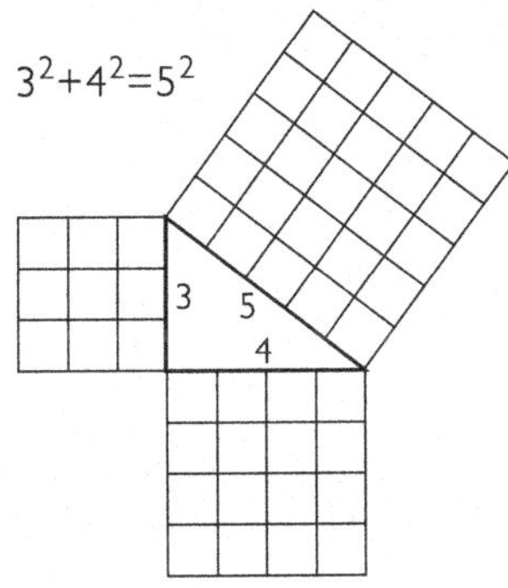

See also **hypotenuse**, **right-angled triangle**, **square**

Qq

quadrant

Meaning 1 A quadrant is a quarter of a circle. A quadrant is an arc that makes an angle of 90 degrees.

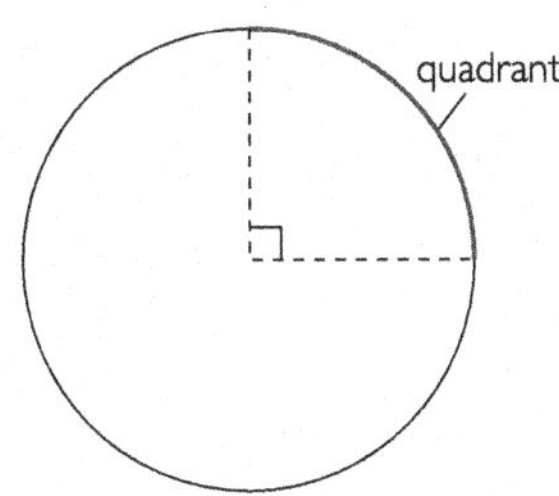

See also **arc**

Meaning 2 A quadrant is a quarter of the region inside a circle. A quadrant is a sector formed by two radii at right angles to each other.

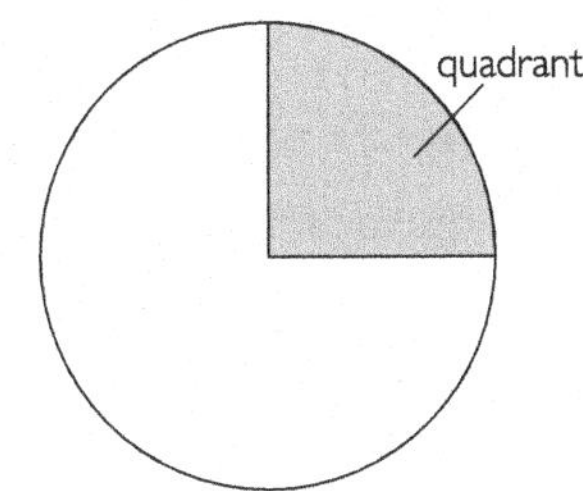

See also **radius**, **sector**

Meaning 3 A quadrant is one of the four parts of the plane formed by a pair of axes crossing at right angles.

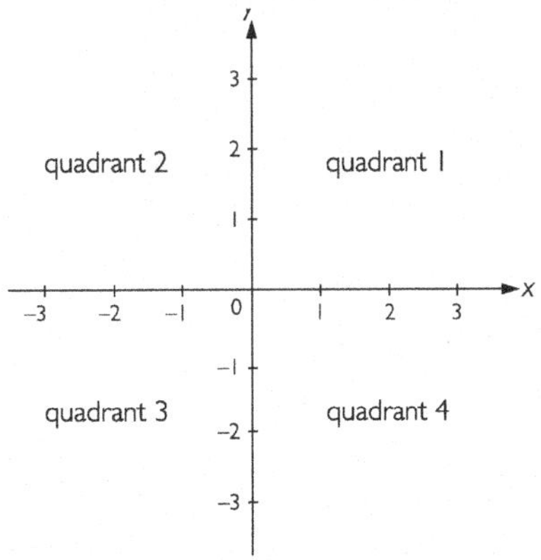

See also **Cartesian graph**, **plane**

Q

quadrilateral

A quadrilateral is a polygon with four straight sides. Rectangles, kites and parallelograms are all quadrilaterals. In a regular quadrilateral, all the sides have the same length and all the angles are equal. A regular quadrilateral is usually called a square.

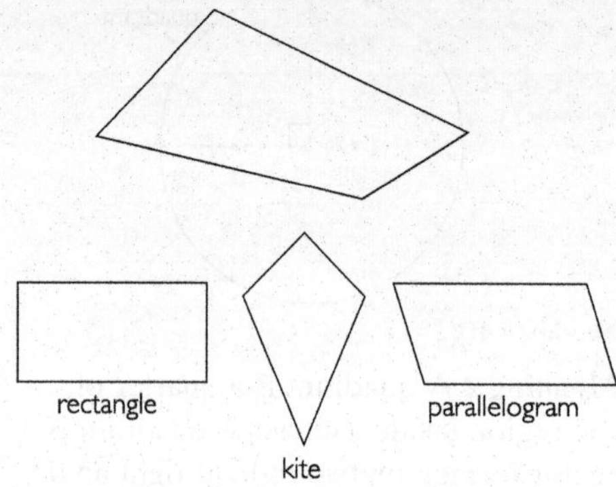

Irregular quadrilaterals

See also **polygon, regular polygon**

quadruple

Quadruple means four times as much or four times as many. The quantity is multiplied by four.

For example

1 Quadruple 3 is 12.

2 To quadruple the amount of cake batter, multiply the quantities in the recipe by four.

See also **double, treble**

quantitative data

Quantitative data is numerical data. Quantitative data is information given as numbers.

For example

1 The maximum temperature every day for a month is quantitative data.

2 The rating of pop groups on a scale of numbers is quantitative data.

See also **data, quantity**

quantity

Quantity is how much or how many. Quantity is an amount or a number.

For example

1 The quantity of juice in the jug is 1 litre.

2 Each guest at the party will eat 3 cup cakes. What quantity of cup cakes will we need if there are 10 guests?

See also **amount, magnitude, measure**

quarter

A quarter is one part out of four equal parts of a whole.

For example

1 $\frac{1}{4}$ of one apple is one part out of four parts. $\frac{1}{4}$ is a piece of the apple.

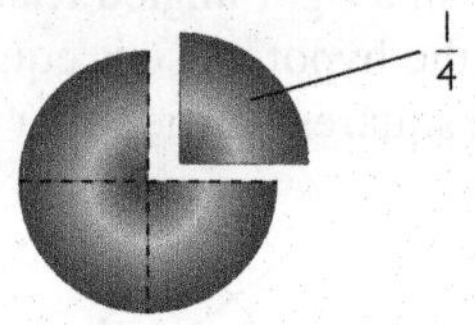

2 $\frac{1}{4}$ of 8 apples is one part out of four parts. $\frac{1}{4}$ is 2 whole apples.

See also **continuous, discrete, fraction, part, whole**

quarter-turn

A quarter-turn is one quarter of a complete rotation. The angle of a quarter-turn is 90 degrees, or a quarter of a full-turn.

For example

To see the sign, face the building and make a quarter-turn to the right.

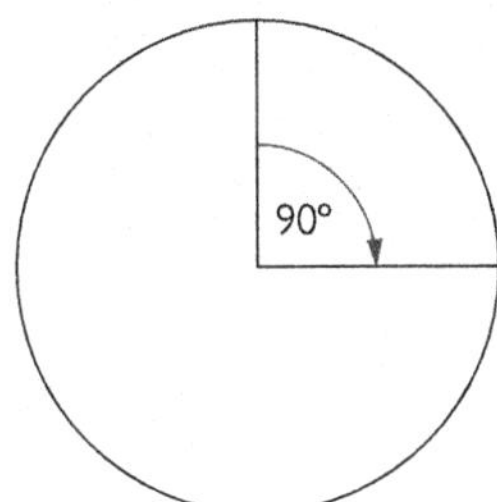

See also **full-turn**, **quarter**, **rotation**

---

quotient

The quotient is the answer after dividing one number by another.

For example

In 18 ÷ 9 = 2, the quotient is 2.

See also **division**

---

quotition

Quotition means division, when the size of each group is known. The problem is to find the number of groups.

For example

There are 20 golf balls. Four golf balls are to be placed in each can. How many cans are needed?

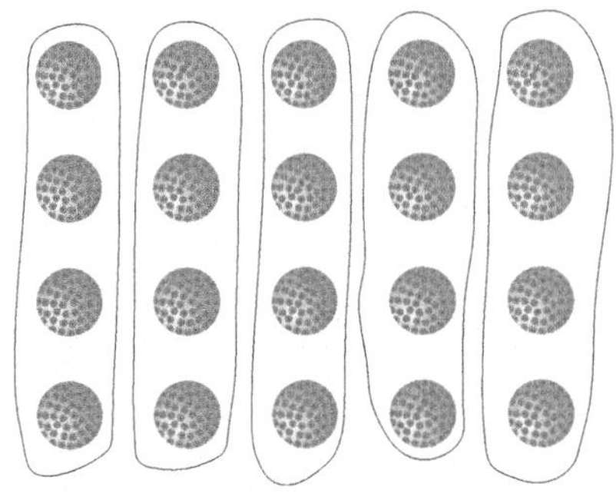

20 ∏ 4 = 5

See also **division**, **group**

Rr

radius

Meaning 1 The radius of a circle is the distance from its centre to any point on the circle. The radius of a sphere is the distance from its centre to any point on the sphere. The plural form of the word *radius* is radii.

Meaning 2 A radius is a straight line from the centre of a circle or sphere to any point on the circle or sphere.

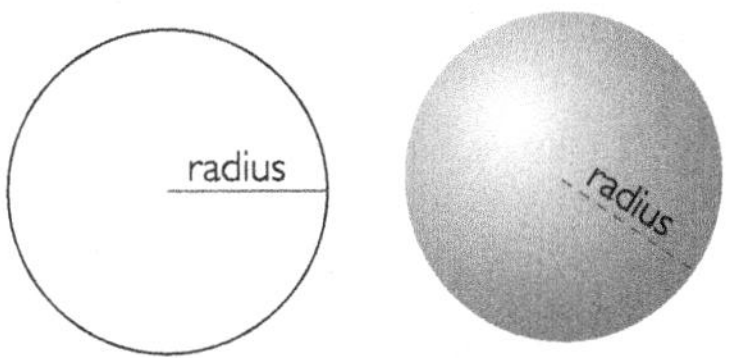

See also **centre**, **circle**, **sphere**

---

random

Random means not according to a pattern. Random means not methodical.

For example

1. Tossing a coin many times gives a random sequence of heads and tails.
2. In a random sample of the population, every member of the population has an equal chance of being included.

See also **pattern**

---

random device

A random device is a piece of equipment that is designed to turn out objects or numbers according to chance.

For example

A die, a spinner, a bag of coloured counters and a random-number generator are random devices.

See also **random**

R

range

Meaning 1 The range of a set of data is the difference between the highest and lowest values.

For example

At a shop, the cheapest watch costs $14 and the most expensive costs $880. The price range is $866.

See also **spread, value**

Meaning 2 The range of a variable is all the values that it can have.

For example

The table below gives values that satisfy the equation $y = 2x$.

The range of x is $\{0, 1, 2, 3\}$.
The range of y is $\{0, 2, 4, 6\}$.

x	0	1	2	3
y	0	2	4	6

rank

To rank is to give a relative position.

For example

1 We may rank colours in order of preference.
2 We may rank outcomes according to their likelihood.

See also **relative**

rate

A rate is a comparison of two quantities of different types.

For example

1 The rate of water flow from the tap was 3 litres per minute. This rate compares volume with time.
2 The rate for advertising in a newspaper is $1 per word. This rate compares cost with length.
3 An exchange rate compares two different currencies. For instance, at the time of writing one Australian dollar was worth 36 British pence.

See also **ratio, relation**

ratio

A ratio is a comparison of two quantities. The quantities are usually of a similar type.

For example

1 To make a cordial drink, Betsy used 1 part of cordial for every 5 parts of water. The ratio of cordial to water was 1:5.

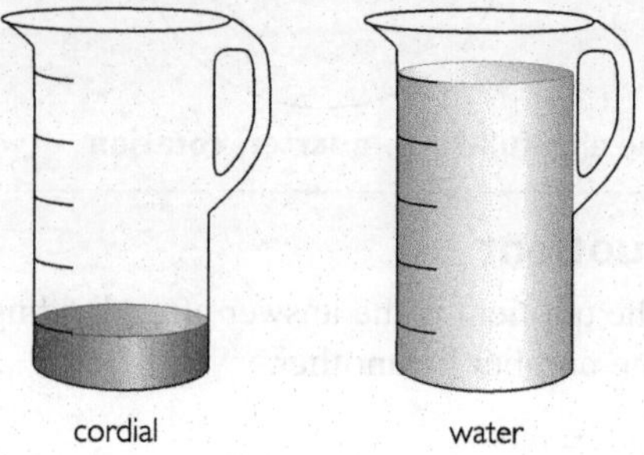

2 The class has 12 boys and 14 girls. The ratio of boys to girls is 12:14, which is the same as 6:7.
3 A ratio of numbers is a fraction. $\frac{1}{2}$, 0.5, 50% and 1:2 are equivalent fractions.

See also **compare, quantity**

rational number

A rational number is a number that can be written as a ratio of integers. In other words, a rational number can be written as a common fraction.

For example

1 $\frac{3}{4}$ and $-\frac{17}{5}$ are rational numbers.

2 7 is a rational number since $7 = \frac{7}{1}$.

Every terminating or recurring decimal number is a rational number.

For example

1 $0.5 = \frac{1}{2}$

2 $0.333333... = \frac{1}{3}$

See also **fraction, integer, ratio, recurring decimal**

ray

A ray is a straight line that has a starting point but no end point. A ray extends in one direction only.

For example

Two rays with a common starting point form an angle.

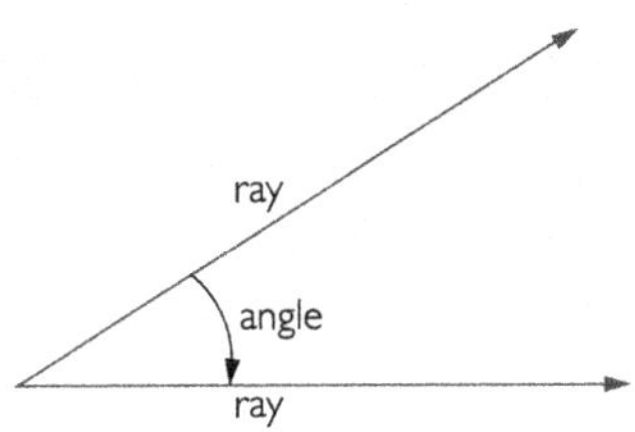

See also **direction**, **line**

real number

A real number is a number that can be written as a decimal number. Real numbers can be shown on a number line. They are the numbers used in everyday counting and measurement. The real numbers include both the rational and the irrational numbers.

For example

5.472 is a real number.

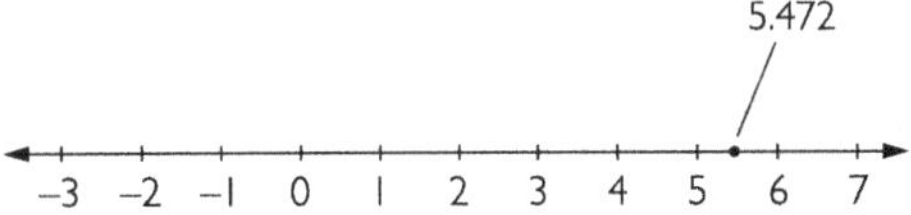

See also **decimal number**, **irrational number**, **number line**, **rational number**

reasonable

A reasonable solution to a problem is a result that seems sensible and that has been found using an appropriate method.

For example

It is not reasonable that the area of your desk is 50 square metres. There must have been a mistake in the working of the problem.

See also **accurate**, **check**, **verify**

rectangle

A rectangle is a polygon with four straight sides and four right angles. Opposite sides have the same length and are parallel. The two diagonals are equal in length. A square is special type of rectangle. A rectangle with unequal side lengths is called an oblong.

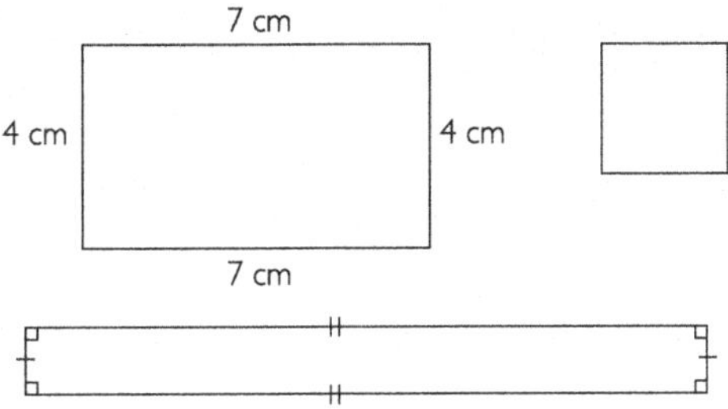

See also **oblong**, **parallelogram**, **polygon**, **quadrilateral**

rectangular grid

A rectangular grid is a pattern of horizontal and vertical lines forming equal squares.

For example

1 The framework for a graph is a rectangular grid.

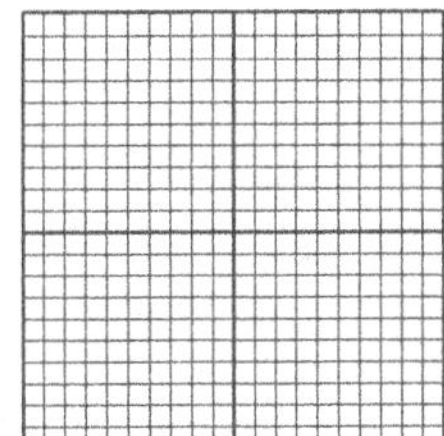

2 The reference system on a map in a street directory is a rectangular grid.

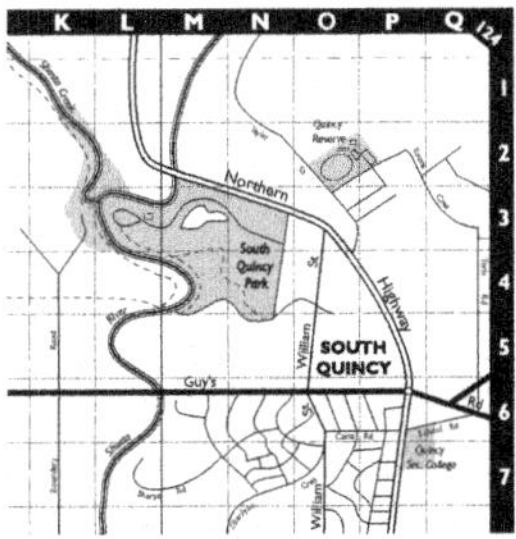

See also **grid paper**, **grid reference**

R

rectangular number

A rectangular number is a number that can be represented as a rectangular dot pattern. The array of dots must have at least two rows and at least two columns.

For example

4, 12, 18 are all rectangular numbers. These numbers can each be shown as a rectangular array.

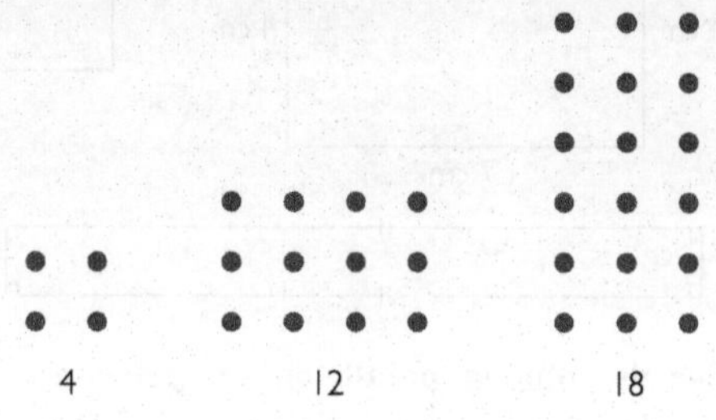

See also **array, composite number, rectangle, triangular number**

rectangular prism

A rectangular prism is a prism with two rectangular bases. The four other faces are parallelograms. Cubes and cuboids are rectangular prisms.

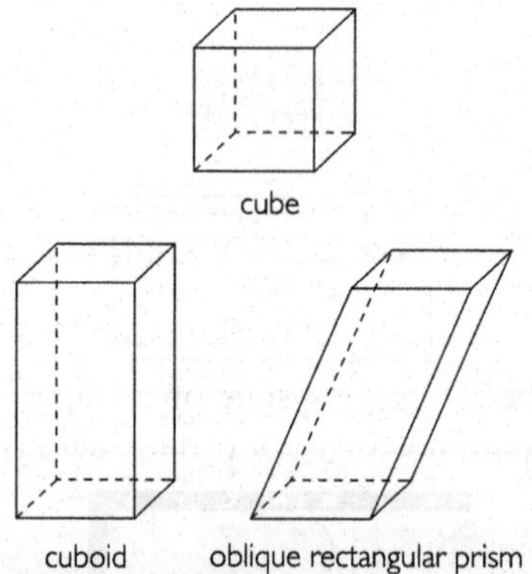

See also **cube, cuboid, polyhedron, prism, rectangle**

recurring decimal

A recurring decimal is a decimal fraction that is a repeating sequence of one or more digits. A recurring decimal is also known as a repeating decimal. A recurring decimal fraction always has a common fraction equivalent.

For example

1 $\frac{1}{3} = 0.3333333...$, which can be written as $0.\dot{3}$ to show that the 3 is recurring.

2 $\frac{23}{99} = 0.23232323...$, which can be written as $0.\dot{2}\dot{3}$ to show that the 23 is recurring.

See also **common fraction, decimal fraction, rational number**

reduce

Meaning 1 Reduce is to make simpler.

For example

We can reduce a fraction to its simplest form, for instance $\frac{12}{8} = \frac{3}{2}$.

See also **simplify**

Meaning 2 Reduce is to make smaller without changing the shape. Ways to reduce a shape include using a photocopier and the method of squares.

For example

To reduce a figure drawn on a grid, draw the figure on a grid with smaller squares.

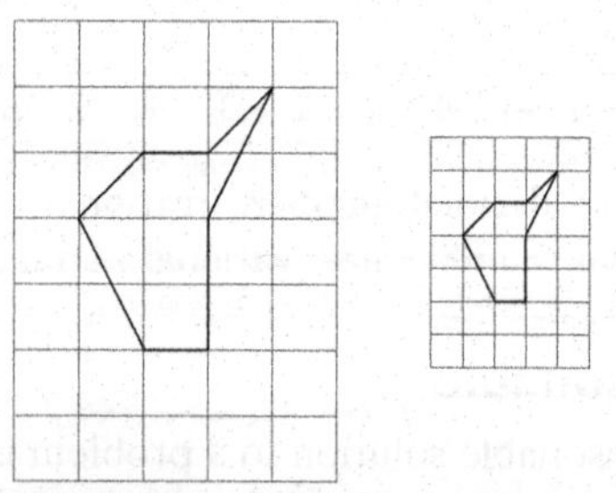

See also **enlargement, ratio, scale, transformation**

reflection

A reflection of a shape is the shape as seen in a mirror, that is, reversed.

For example

If a shape looks the same as its reflection along a mirror line, then the shape has bilateral symmetry.

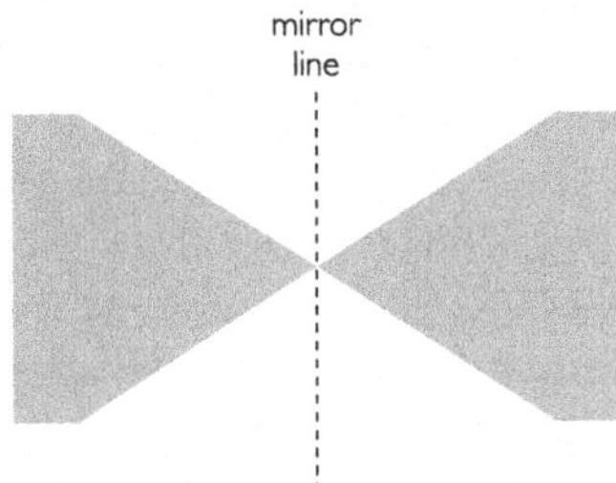

See also **bilateral symmetry**, **flip**, **line of symmetry**, **mirror line**

reflex angle

A reflex angle is an angle that measures more than 180 degrees (a straight angle) and less than 360 degrees (one full revolution).

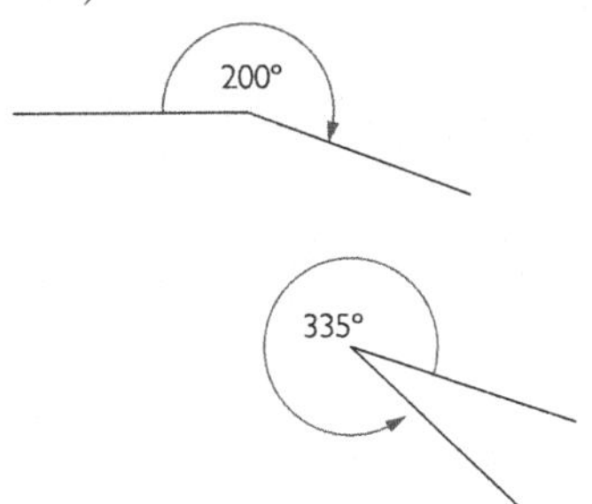

See also **angle**, **revolution**, **straight angle**

region

A region is the part of a surface within a boundary.

For example

There are four regions in the Venn diagram below.

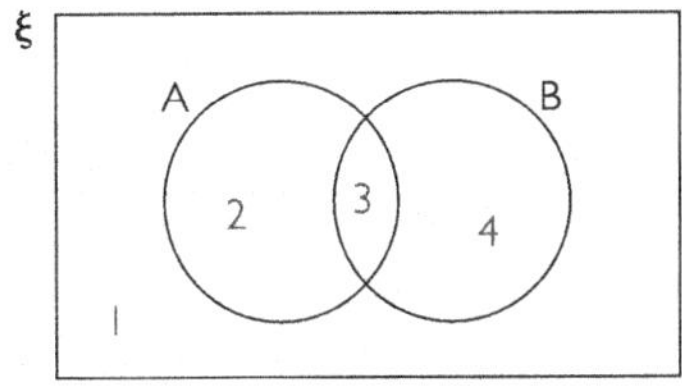

regroup

To regroup is to rearrange a number. To regroup means to exchange units from one place-value position to another. A single ten could be exchanged for 10 ones, or a single hundred could be exchanged for 10 tens. Regrouping usually occurs for the purpose of carrying out an operation.

For example

It helps to regroup when subtracting 17 from 42. The 4 tens and 2 ones can be regrouped as 3 tens and 12 ones.

```
  4 2  →   40 + 2  →    30 +12
- 1 7    - 10 + 7     - 10 + 7
                        20 + 5 =25
```

```
 ³4 ¹2
- 1  7
  2  5
```

See also **group**, **operation**, **place value**

regular

Regular means that the same rule is applied throughout.

For example

1 A tessellation is a regular arrangement of shapes.

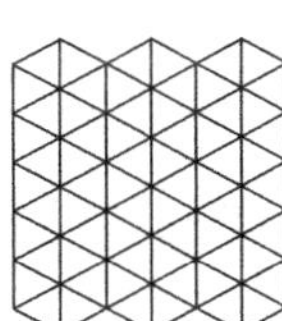

2 An analogue clock face is marked in regular intervals.

See also **irregular, pattern, regular polygon, regular polyhedron.**

R

regular polygon

A regular polygon is a polygon with all sides equal in length and all angles equal in size.

For example

An equilateral triangle and a square are regular polygons.

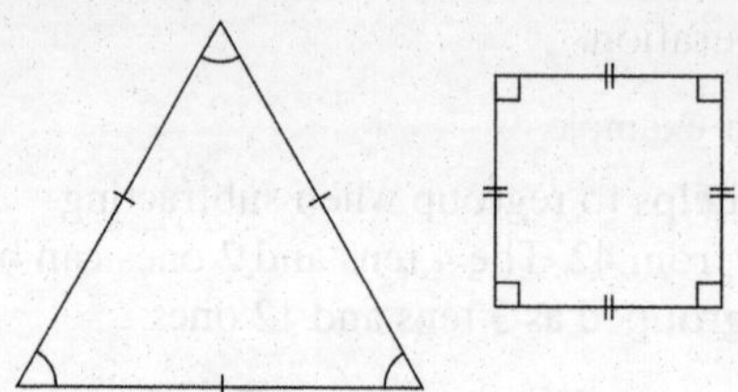

See also **irregular, polygon**

regular polyhedron

A regular polyhedron is a polyhedron with all faces congruent, regular polygons and all vertices equivalent. The plural form of the word *polyhedron* is polyhedra. There are only five regular polyhedra. These are called the five Platonic solids.

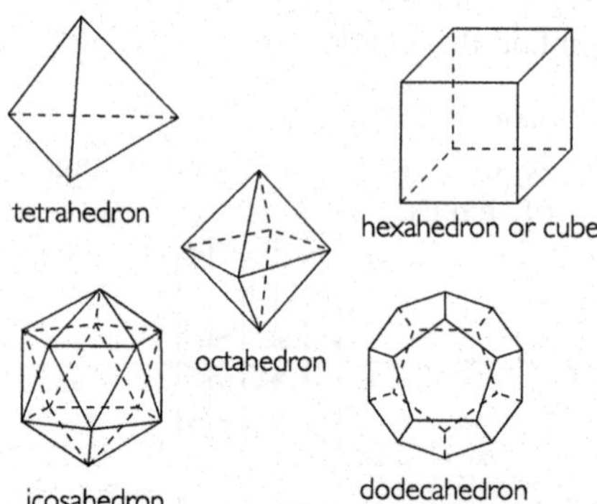

See also **congruent, Platonic solids, polyhedron, regular polygon, vertex**

relation

A relation is a connection that can occur between objects. Not all the objects have to be connected. For instance, "is a brother of" is a relation between people. Sam is a brother of Kate, but Kate is not a brother of Sam.

For example

1. "Is less than" is a relation between numbers, for instance, $5 < 7$ and $-2 < 5$.
2. An arrow diagram shows a relation. The related objects are connected by arrows.
3. Congruence is a relation between shapes. Matching triangles are congruent.
4. An equation can determine a relation. For instance, $y = 2x$ forms a relation. The table below shows pairs of related numbers. The Cartesian graph of all the related pairs is a straight line.

x	0	1	2	3	4
y	0	2	4	6	8

See also **arrow diagram, equality, inequality**

relative

Relative is one attribute or measure with reference to another. This is sometimes expressed as a ratio.

For example

1. We can test the relative strength of a range of shapes constructed from rods with punched holes.
2. We can establish the relative frequency of a drawing pin landing point up by repeated tosses.
3. We need to determine the relative proportion of all measurements when drawing a scaled plan.

See also **ratio**

remainder

The remainder is what is left over.

For example

1. The remainder of the time was spent checking the results.
2. After subtracting 8 from 24, there is a remainder of 16.

3 The solution to a division problem may have a remainder, for instance, $20 \div 3 = 6$ and 2 remainder. The 2 remainder is $\frac{2}{3}$ of a group of 3.

See also **difference**, **division**

rename

To rename is to give an alternative name.

For example

1 We can rename a mixed number as a common fraction: $2\frac{1}{2}$ as $\frac{5}{2}$.

2 A fraction can be renamed as an equivalent fraction: $\frac{1}{2} = \frac{2}{4}$.

3 It is sometimes useful to rename a common fraction as a percentage or a decimal: $\frac{50}{100}$ as 50% or 0.5.

4 We can rename a measure using different units: 2 metres as 200 centimetres.

repeated addition

Repeated addition is the addition of equal-sized groups a number of times. Repeated addition is an alternative to the multiplication process.

● ● ●
● ● ●
● ● ●
● ● ●

$3 + 3 + 3 + 3 = 12$

$3 \times 4 = 12$

$4 + 4 + 4 = 12$ $4 \times 3 = 12$

See also **addition**, **multiplication**

repeated subtraction

Repeated subtraction is the subtraction of equal-sized groups a number of times. Repeated subtraction is an alternative to the division process.

For example

24 divided by 4 can be interpreted as taking groups of 4 away from 24. Six groups of 4 can be subtracted from 24.

See also **division**, **quotition**, **subtraction**

repeating decimal

See **recurring decimal**

repeating factors

Repeating factors are factors of a number that occur more than once.

For example

2 is a repeating factor of 12.

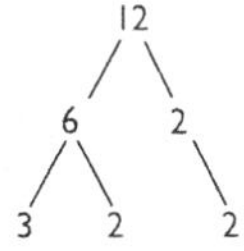

Repeating factors can be written in shorthand using index notation: $12 = 3 \times 2^2$.

See also **factors**

result

A result is an answer or an outcome.

For example

1 Twelve is the result of dividing 96 by 8.

2 The results of the survey showed that most of the children owned a pet.

See also **solution**

reverse

Reverse means the opposite direction or order.

For example

1 Numbers may be arranged in order, either from highest to lowest, or the reverse, from lowest to highest.

2 The process of subtraction is the reverse of the process of addition. The process of division is the reverse of the process of multiplication.

See also **direction**, **inverse**, **order**

R

revolution

A revolution is a complete turn about a central point. There are 360 degrees in a complete turn.

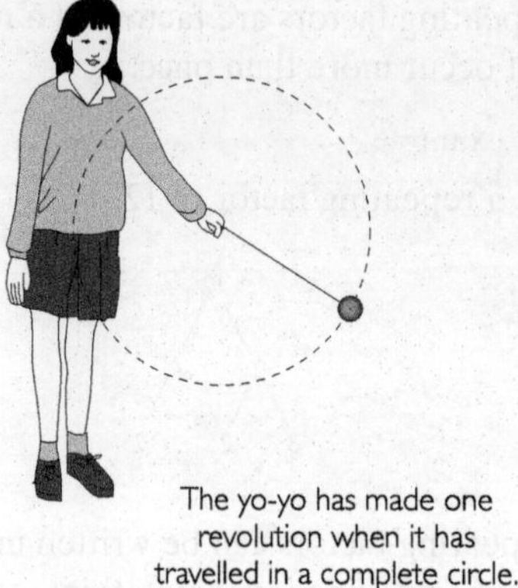

The yo-yo has made one revolution when it has travelled in a complete circle.

See also **full-turn**

rhombus

A rhombus is a polygon with four sides all the same length. Opposite sides are parallel and opposite angles are equal. The diagonals of a rhombus intersect at right angles. A square is a special type of rhombus.

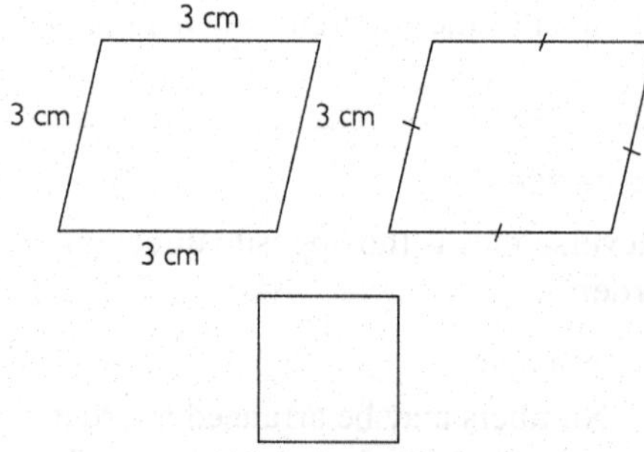

See also **parallelogram**, **polygon**, **quadrilateral**

right angle

A right angle is an angle of 90 degrees. A right angle is a quarter of a complete turn.

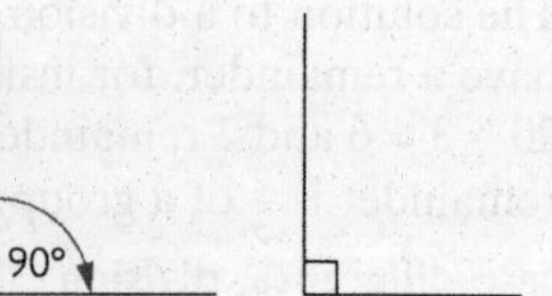

See also **angle**

right-angled triangle

A right-angled triangle is a triangle with one angle of exactly 90 degrees.

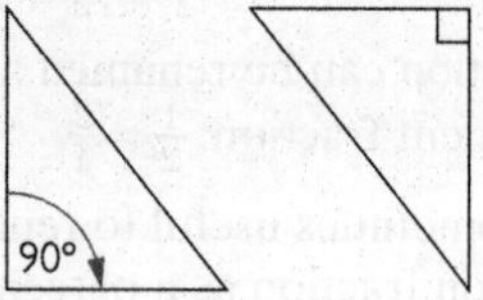

See also **triangle**

rigid structure

A rigid structure is a construction that is non-flexible or fixed. The joints cannot move, so the angles do not change.

For example

A triangle is a rigid structure, but a square is not. The sides of a square can be skewed.

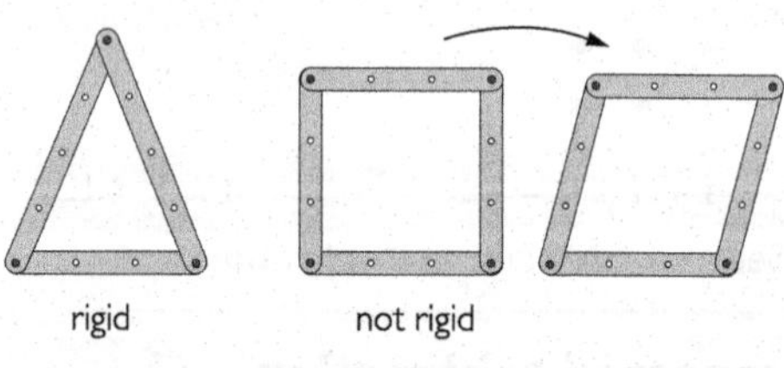

See also **angle**

Roman numerals

Roman numerals are the symbols of the ancient Roman number system. The system is sometimes used today, as in numerals on a clock face. The numerals are commonly written in capital letters, although they may be written in lower case. The basic symbols are:

I (1), V (5), X (10), L (50), C (100), D (500), M (1000)

For example

II = 2	IV = 4
XI = 11	XC = 90
DC = 600	CM = 900
MD = 1500	MMII = 2002

See also **number system**, **numeral**, **symbol**

root

The root of a given number is the number that, when multiplied by itself a specified number of times, equals the given number.

For example

1. The square root of a number is the number "unsquared".

   The number 3 is a square root of 9. This is because $3^2 = 3 \times 3 = 9$.

   We write $3 = \sqrt{9}$

   The negative number –3 is also a square root of 9. This is because $(-3)^2 = (-3) \times (-3) = 9$.

   We write $-3 = -\sqrt{9}$.

2. The cube root of a number is the number "uncubed".

   The number 2 is the cube root of 8. This is because $2^3 = 2 \times 2 \times 2 = 8$.

See also **cube**, **square root**, **square**

rotation

A rotation is the turning of a point or set of points around a fixed point. A rotation is a type of transformation.

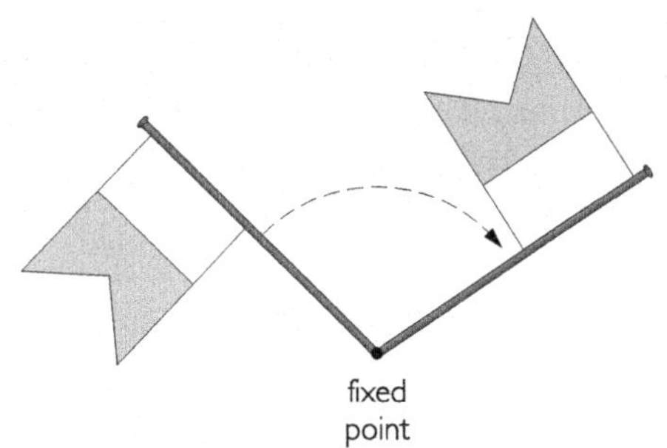

See also **full-turn**, **transformation**

rotational symmetry

A plane or solid shape that still looks the same when it is turned around a fixed point has rotational symmetry. The number of positions in which the shape still looks the same while being turned is the order of symmetry.

For example

1. A square has rotational symmetry of order four.
2. An equilateral triangle has rotational symmetry of order three.

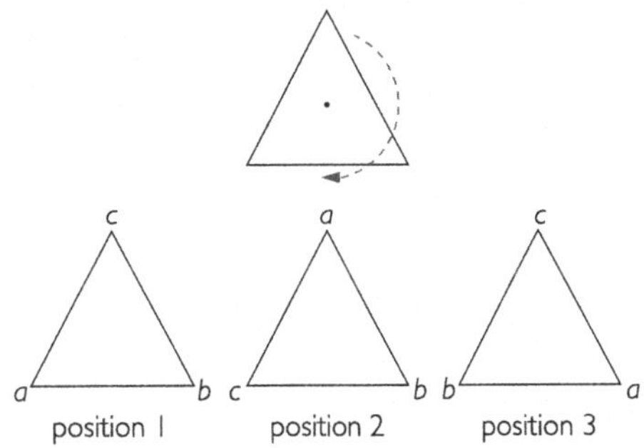

See also **bilateral symmetry**, **line of symmetry**, **rotation**, **symmetry**

rounding

Rounding is an approximation technique. Numbers are rounded to particular place values. Numbers can be rounded to the nearest ten or to the nearest hundred, for example. Money can be rounded to the nearest dollar. Fractions can be rounded to the nearest whole number.

☞

To round a number to a particular place value, look at the digit to the right of that place value. If the digit is 5 or above, then round up. If the digit is below 5, then round down.

For example

1 87 rounded to the nearest ten is 90.

2 2511 rounded to the nearest hundred is 2500.

See also **approximation, decimal place, place value**

row

A row is a horizontal line of objects.

For example

1 There are three counters in each row of the array.

● ● ●
● ● ●
● ● ●
● ● ●

2 In the table below, find the row that contains all the numbers with 2 tens.

0	1	2	3	4	5	6	7	8	9
10	11	12	13	14	15	16	17	18	19
20	21	22	23	24	25	26	27	28	29
30	31	32	33	34	35	36	37	38	39
40	41	42	43	44	45	46	47	48	49
50	51	52	53	54	55	56	57	58	59
60	61	62	63	64	65	66	67	68	69
70	71	72	73	74	75	76	77	78	79
80	81	82	83	84	85	86	87	88	89
90	91	92	93	94	95	96	97	98	99

See also **array, column, horizontal**

rule

A rule is a pattern, a convention or a sequence of steps given as an instruction.

For example

1 The Fibonacci numbers are 1, 1, 2, 3, 5, 8, ... To extend the sequence, the rule is to add the previous two numbers.

2 To multiply a decimal number by 10, the rule is to move the decimal point one place to the right.

ruler

A ruler is a strip of wood, metal or plastic with an edge marked in graduations. Metric graduations are the metre, the centimetre and the millimetre. A ruler is used to measure length and as an aid to drawing straight lines.

See also **graduation**

Ss

sample

A sample is a group of things (or people) taken from a much bigger group. The bigger group is called a population. What is discovered from a study of the sample is applied to the whole population.

For example

By taking a sample of coloured counters from a bag, it is possible to estimate the relative proportions of the different colours of counters in the bag.

See also **estimate**

scale

Meaning 1 A scale is the regular markings along a line on an instrument used for measuring quantities.

For example

The scale on a ruler measures length.

See also **calibration**

Meaning 2 A scale is a standard of measurement.

For example

The Celsius scale measures temperature.

Meaning 3 A scale is the ratio of the unit length on a graph to the real quantity it represents.

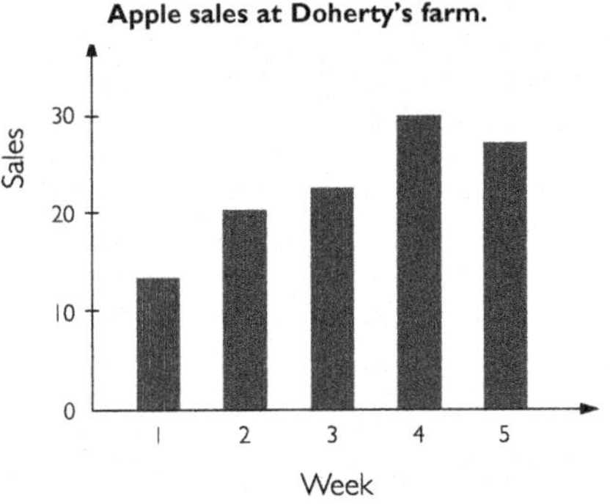

Scale: 1 cm = 10 crates of apples

See also **ratio**

Meaning 4 A scale is the ratio of the length of a line on a map to the real distance on the Earth that it corresponds to. A map cannot be the same size as the region it represents, so it is reduced to a convenient size.

For example

The scale on this map is 1:10 000.

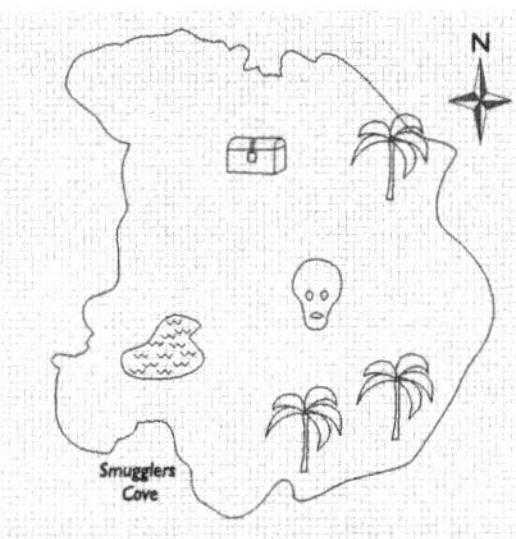

See also **ratio**, **reduce**

Meaning 5 A scale is the ratio of the dimensions of a model or a drawing to the real dimensions of the object that it corresponds to. The dimensions of the real object are reduced.

Scale: 1: 40

See **ratio**, **reduce**

scalene triangle

A scalene triangle is a triangle with each side of a different length.

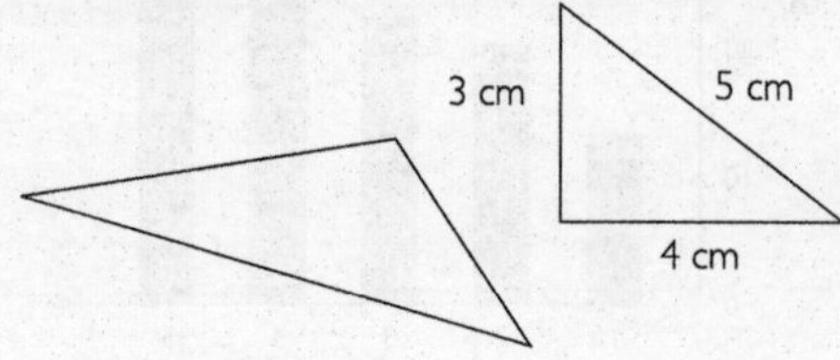

See also **triangle**

scales

Scales are instruments used for comparing masses or measuring weight.

For example

1 A beam balance is a type of scales. A beam balance has a bar poised on a central pivot with pans hanging at the ends. The mass of an object in one of the pans is compared with standard masses placed in the other.

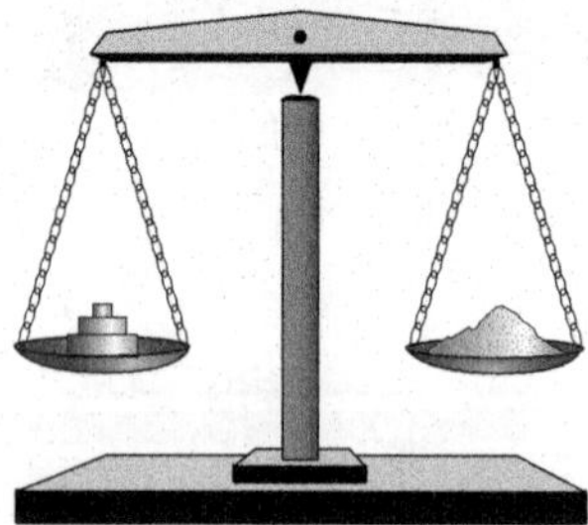

2 A spring balance is a type of scales. A spring balance has a spring that extends with the pull of an object. The measure of that pull on a scale is the weight of the object.

See also **balance, mass, weight**

second

Meaning 1 Second is an ordinal number. Second describes a relative position. It is after the first in order and before the third.

For example

Fiona is the second person in the line.

See also **ordinal number, relative**

Meaning 2 A second is the base unit of time in the International System of Units. A second is one-sixtieth of a minute. The symbol for second is s.

60 seconds = 1 minute

See also **SI system, standard units of measurement**

Meaning 3 A second is a unit for measuring angles. There are 60 seconds in a minute and 60 minutes in a degree. The symbol for second is ″.

See also **degree**

sector

A sector is a region inside a circle that is bounded by two radii and an arc. A pair of radii divides a circle into two sectors. The smaller sector is called the minor sector; the larger sector is called the major sector.

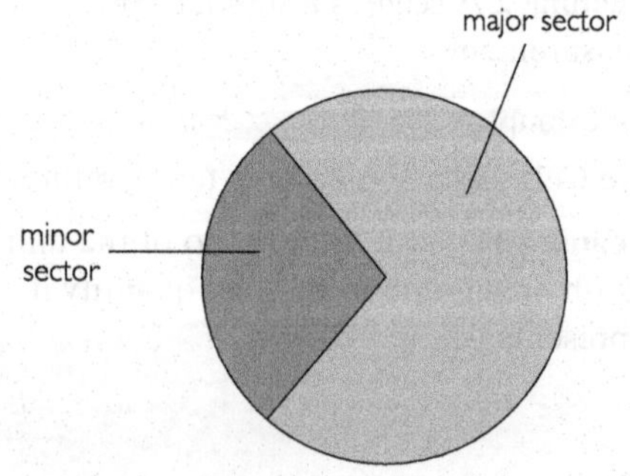

See also **arc, radius**

segment

Meaning 1 A segment is the part of a line between two points. A line segment has finite length, although a straight line can extend infinitely.

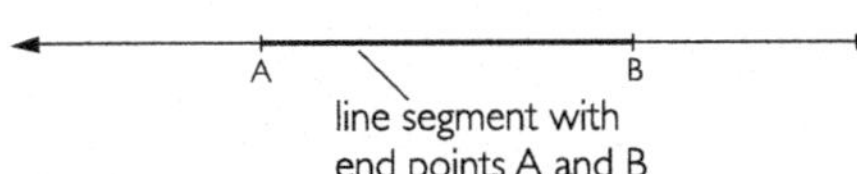

See also **line**

Meaning 2 A segment is the part cut off a plane figure by a line, or the part cut off a solid figure by a plane.

For example

1 A chord divides the region inside a circle into two segments. The smaller segment is called the minor segment; the larger segment is called the major segment.

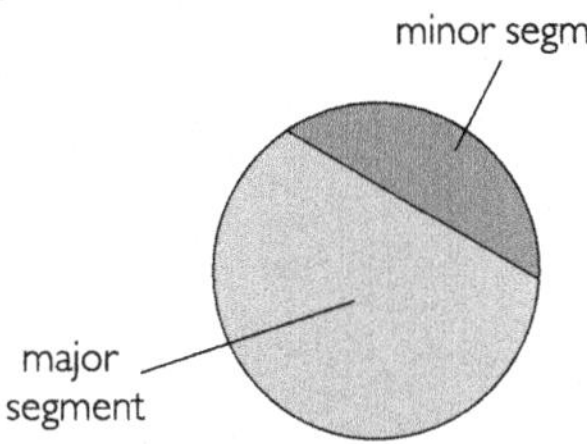

2 A cut through a cube parallel to one of its faces forms two segments. Each segment is a cuboid.

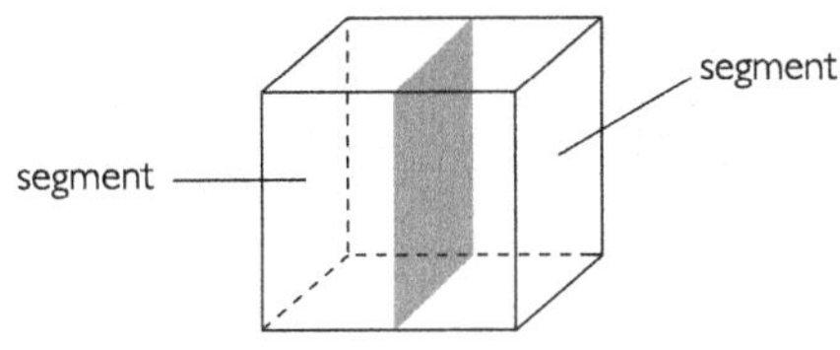

semicircle

A semicircle is half a circle. It is an arc between two opposite points on a circle.

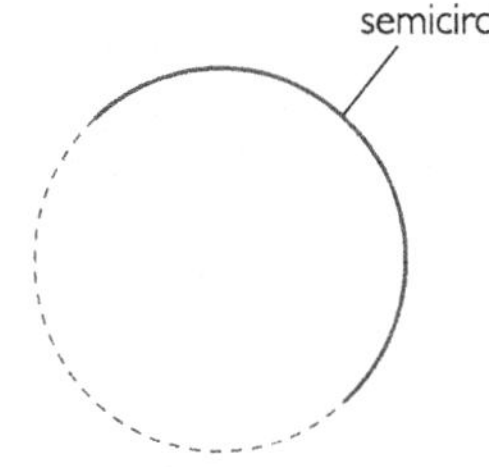

See also **circle**

sequence

A sequence is an ordered list of objects or numbers.

For example

1 The sequence 2, 4, 8, 16, ... follows the pattern "multiply by 2".

2 A special sequence is the Fibonacci sequence: 1, 1, 2, 3, 5, 8, 13, 21, ... In this sequence, each number is the sum of the previous two numbers.

See also **order, series**

series

A series is a list of numbers written like a sum.

For example

1 The finite series 1 + 3 + 5 + 7 has sum 16.

2 The infinite series $\frac{1}{2} + \frac{1}{4} + \frac{1}{8} + \frac{1}{16} + \ldots$ has sum 1.

3 The infinite series 1 – 2 – 3 – 4 – 5 – ... does not have a sum.

set

A set is a collection of things. The things that belong to a set are called elements. Sets can be described by listing their elements inside braces { }.

For example

1 The set of days in the week is {Monday, Tuesday, Wednesday, Thursday, Friday, Saturday, Sunday}. This set has seven elements.

2 The set of natural numbers is {1, 2, 3, 4, 5, 6, ...}. This set is infinite.

See also **element, intersection, Venn diagram**

set square

A set square is an instrument in the form of a right-angled triangle that is used as an aid to drawing angles and parallel lines.

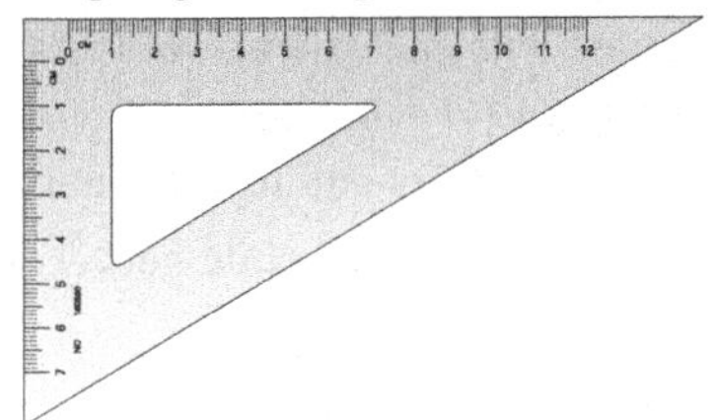

shape

A shape is a form. A shape is determined by its outline or surface. Shapes are studied in geometry.

For example

1 Two-dimensional shapes include triangles and circles.

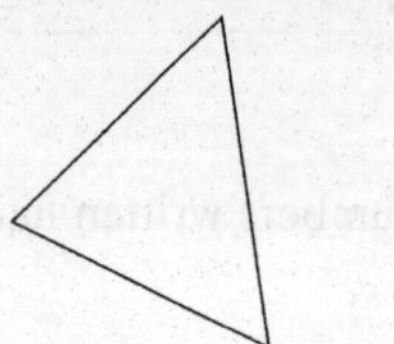

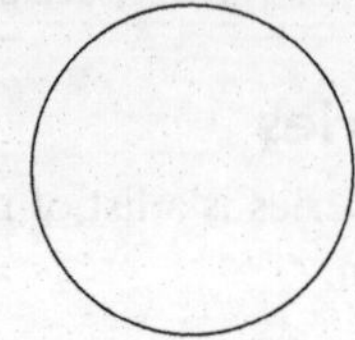

2 Three-dimensional shapes include cubes and spheres.

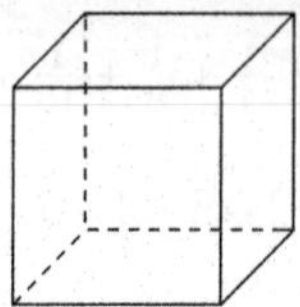

3 Fractal shapes in nature include the shapes of coastlines and clouds.

See also **Euclidean geometry**, **fractal geometry**, **topology**

share

Meaning 1 To share is to deal out in equal-sized parts. Division and fractions use sharing. Another name for sharing is partition.

For example

1 There are 24 cherries in a bowl. Four people will share them. How many cherries for each person?

2 Share the pizza equally between the five children. Each child gets $\frac{1}{5}$ of the pizza.

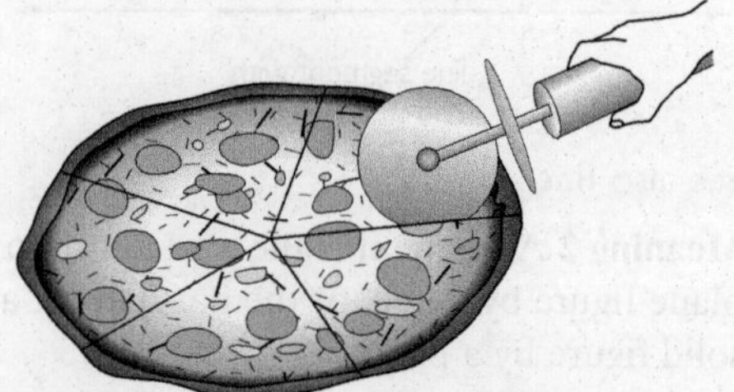

See also **division**, **fraction**, **partition**

Meaning 2 A share is one of the equal fractional parts of a whole, which can be either continuous or discrete.

For example

We can have a fair share of a birthday cake, or a fair share of 12 biscuits.

See also **continuous**, **discrete**, **fraction**

side

A side is one of the straight lines that form a polygon.

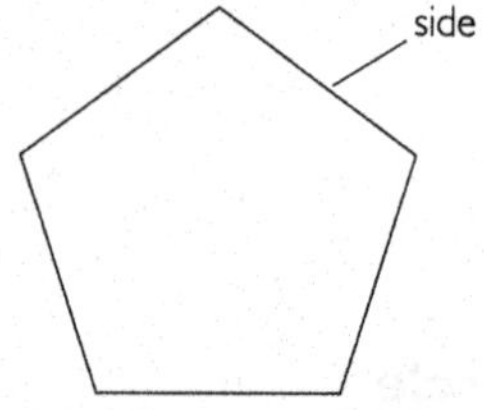

This shape has five sides.

See also **line**

side view

The side view is what you see when looking at an object from beside it.

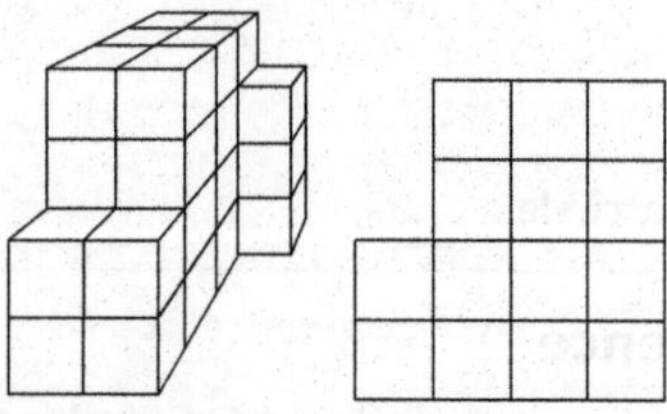

See also **front view**, **top view**

sieve of Eratosthenes

The sieve of Eratosthenes reveals the prime numbers by a process of elimination. The process is to write down all the counting numbers up to, say, 100. Cross out 1. Circle the number 2, then cross out all the other multiples of 2. Circle 3, then cross out all the multiples of 3. Take the next uncircled number, which is 5. Circle 5, and cross out all the multiples of 5. Continue this process until every number is either circled or crossed out. The circled numbers are the prime numbers.

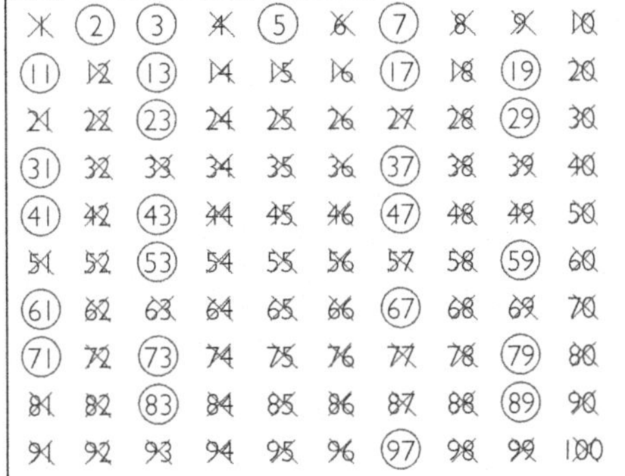

See also **prime number**

sign

A sign is a symbol.

For example

The symbols used to represent the operations are usually called signs:

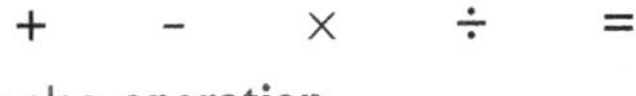

See also **operation**

similar

Similar objects have the same shape, but they do not have to be the same size.

For example

Similar triangles have corresponding side lengths in the same ratio and corresponding angles equal.

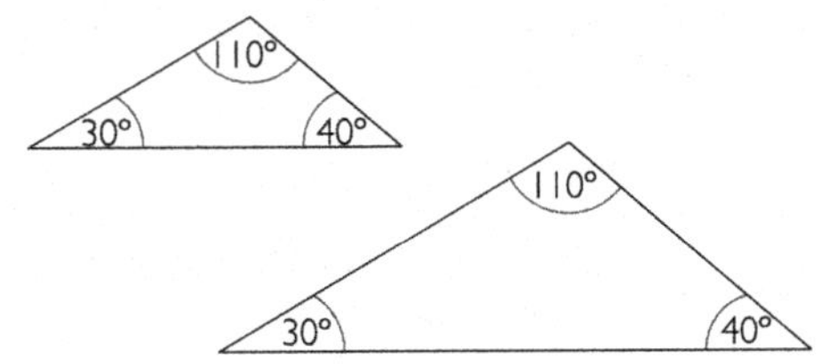

See also **congruent, enlargement, proportional, ratio, reduce**

simple fraction

See **common fraction**

simple shape

A simple shape consists of one part only. It does not cross over itself to create a number of parts.

For example

1 A simple polygon does not have any lines that cross.

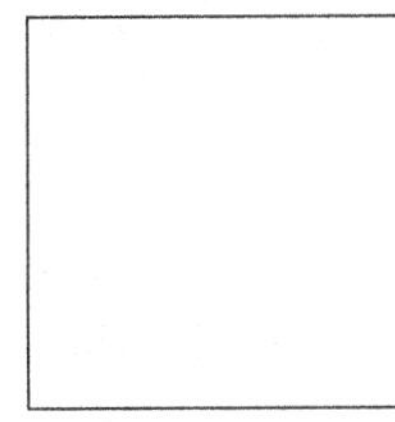

2 A simple polyhedron does not have any tunnels or holes.

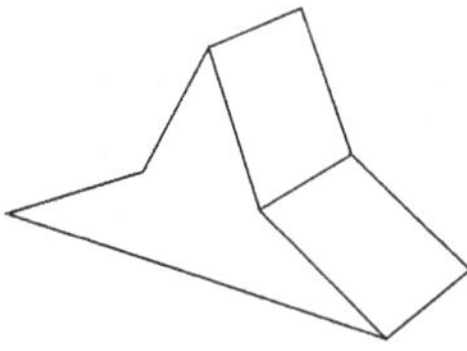

See also **complex shape**

simplify

To simplify is to make less complex. Simplification is applied to algebraic expressions.

For example

1 Expressions may be simplified by combining like terms.

$4a + 3a + 2b = 7a + 2b$

☞

2 Expressions may be simplified by expanding brackets.

$2(4 + 3x) = 8 + 6x$

3 Algebraic fractions may be simplified by cancelling common factors.

$\frac{15m}{10} = \frac{5 \times 3m}{5 \times 2} = \frac{3m}{2}$

simultaneous equations

Simultaneous equations are a group of equations that are meant to be solved together.

For example

$x + y = 5$ and $x - y = 7$ are simultaneous equations.

The solution $x = 6$ and $y = -1$ satisfies both equations.

SI system

The abbreviation SI stands for the French name *Système Internationale d'Unités*, which is the international system of metric units for measuring physical quantities. Base units include the kilogram, the metre and the second.

See **table of commonly used SI units** (page 122)

size

Size means how big or small something is.

For example

1 The size of the angle is 45 degrees.

2 The size of the benchtop is 2 square metres.

3 The size of the population is 20 million people.

See also **dimension, magnitude, quantity**

skeleton model

A skeleton model is a model of the framework of a shape that may be in two or three dimensions. The skeleton is the lines or edges of the shape, joined at the corners or vertices. The model may be made of cardboard strips or rods and fasteners.

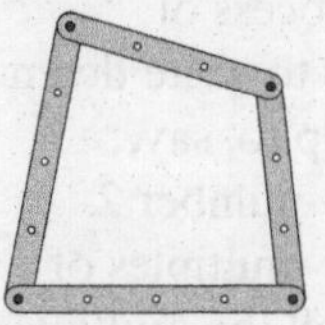

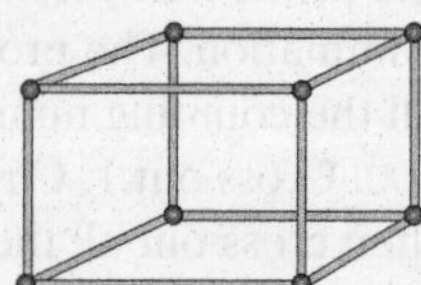

See also **model**

skewed

Meaning 1 Skewed is to appear slanting.

For example

The lines were skewed.

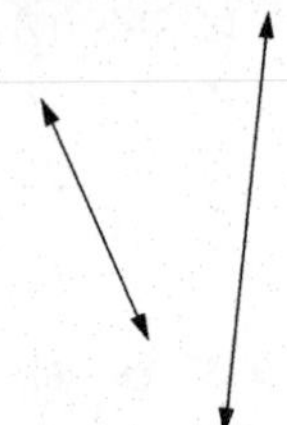

See also **oblique**

Meaning 2 Skewed is to appear asymmetrical.

For example

The curve on the graph is negatively skewed.

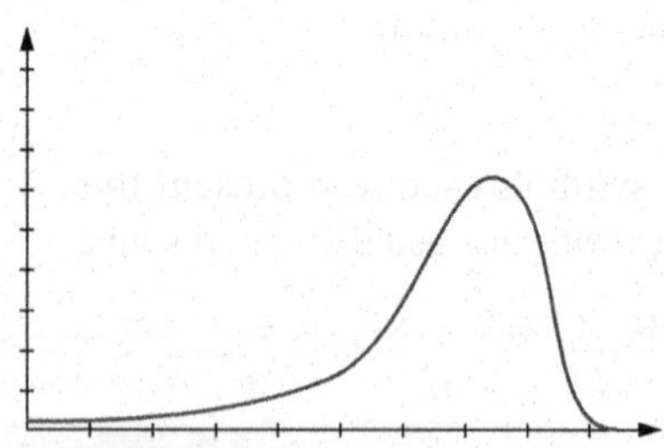

See also **normal distribution**

skip count

To skip count is to count on or count back in equal-sized groups.

For example

1 Counting-on from 10 in groups of ten is 10, 20, 30, 40, 50, 60, 70, ...

2 Counting-back from 20 in groups of two is 20, 18, 16, 14, 12, 10, 8, ...

See also **count**, **counting-back**, **counting-on**

slide

Slide means to move a shape in any direction without any change in orientation. A slide is one kind of transformation.

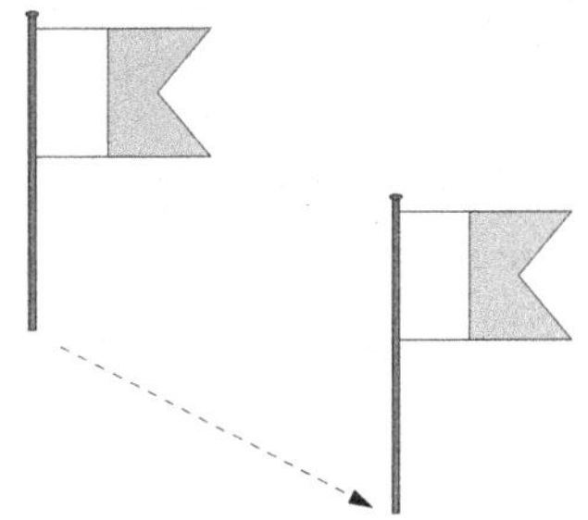

See also **flip**, **transformation**, **translation**, **turn**

slope

To slope is to have a slanting surface. The amount of slope, at any given point, is the gradient.

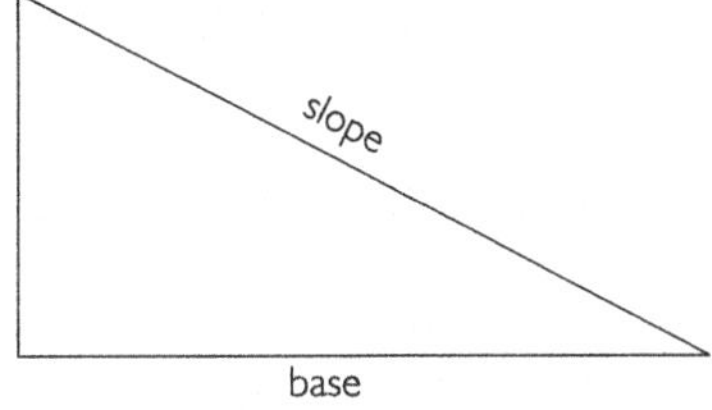

See also **gradient**

snowflake curve

The snowflake curve is a closed curve that looks like a snowflake. It is made by adding equilateral triangles to the sides of existing equilateral triangles. The snowflake curve is an example of a fractal. It was first used by Helge von Koch in 1904, in his study of curves with infinite length that contain a finite area. Another name for the snowflake curve is the Koch curve.

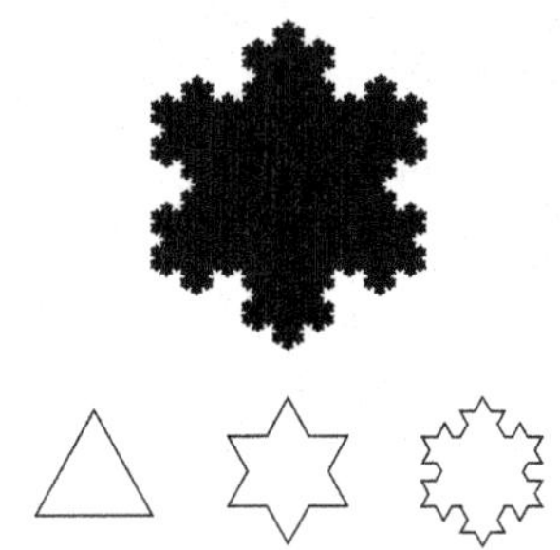

See also **fractal**

solid

A solid is a shape that has three dimensions. A solid shape has length, breadth and height. The solid shape may be hollow or not hollow. Solid shapes can have all their surfaces flat, all their surfaces curved, or some flat and some curved surfaces.

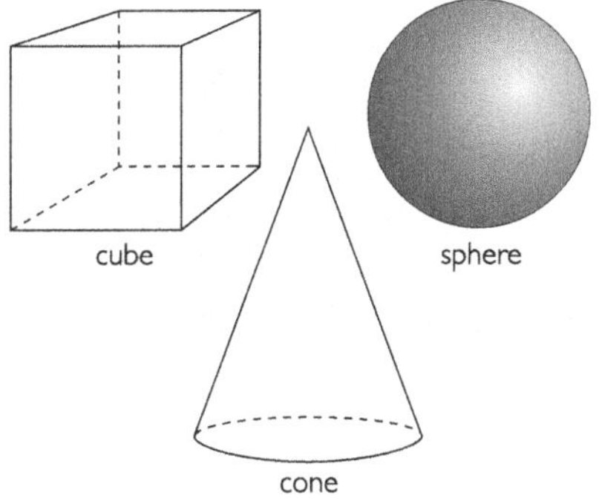

See also **plane**, **three-dimensional**

solution

A solution is an answer to a problem.

For example

1 The solution to the addition problem 15 + 21 is 36.

2 The solution to an algebraic equation makes the left-hand side equal to the right-hand side. For instance, the equation $x + 3 = 7$ has the solution $x = 4$.

solve

To solve is to work out the answer or solution to a problem.

For example

Solve the equation $x - 3 = 5$ by working out the value of x.

sort

To sort is to separate out or group objects according to characteristics.

For example

We can sort objects according to colour, shape, size or number.

See also **attribute, classification**

space

Meaning 1 Space is the area or extent that objects take up.

For example

A three-dimensional cube takes up space.

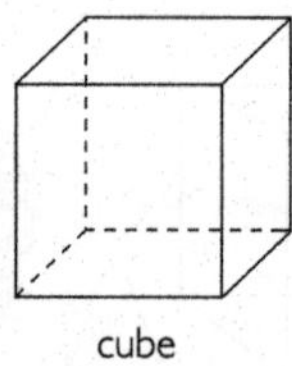

cube

Meaning 2 Space in the school curriculum is an exploration of ideas about space and more formal geometry, and the development of spatial sense and spatial skills.

See also **geometry**

spatial pattern

A spatial pattern uses real objects or drawings of shapes in a repeating sequence, in keeping with a rule. The mathematics-based art of Escher includes spatial patterns.

See also **Escher, number pattern, pattern**

speed

Speed is the quickness of moving. Speed is expressed as the distance travelled in a unit of time. It is the rate of change of position with time.

For example

If the speed limit is 60 kilometres per hour (60 km/h), a car may travel a distance of up to 60 kilometres in one hour.

See also **rate, ratio**

sphere

A sphere is like a perfectly round ball. It is a closed surface in three dimensions. A sphere is made up of the set of points in space that are a fixed distance from a centre point. The fixed distance is called the radius. A sphere is symmetrical.

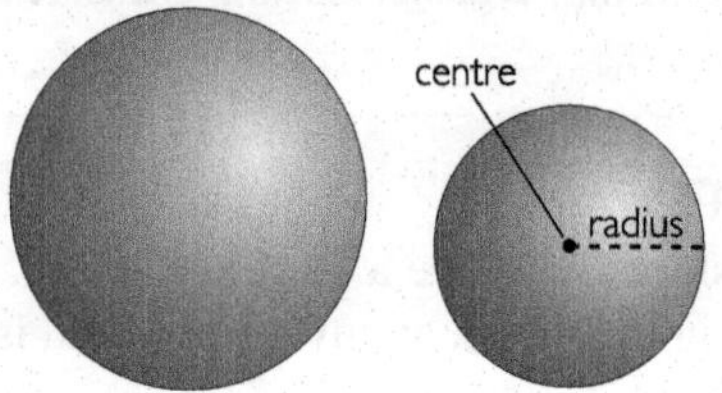

See also **solid**

spinner

A spinner is a turning disc marked with numbers, colours or symbols. When it falls, it shows up a random marking. A spinner is used in chance games.

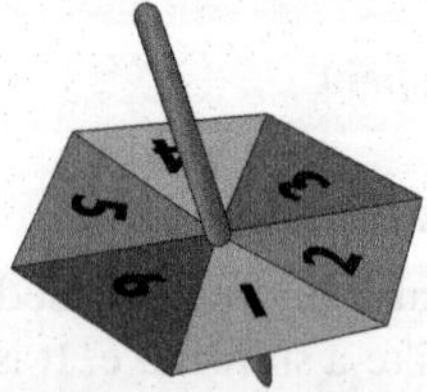

See also **random device**

spiral

A spiral is an open curve. The distance of its points from its starting point is ever-increasing. The curve can wind around endlessly. Spirals can be in two or three dimensions.

For example

Spiral shapes appear in nature. The formation of a fern is a spiral.

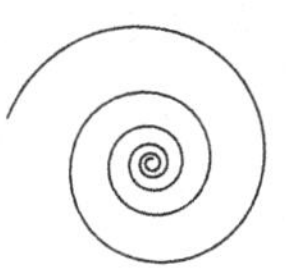
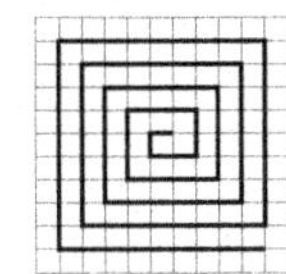

See also **curve, open curve**

spread

Spread is the range or extent, as of data.

See also **range**

square

Meaning 1 A square is a regular polygon, with four straight sides all the same length and four right angles. A square is a type of rectangle.

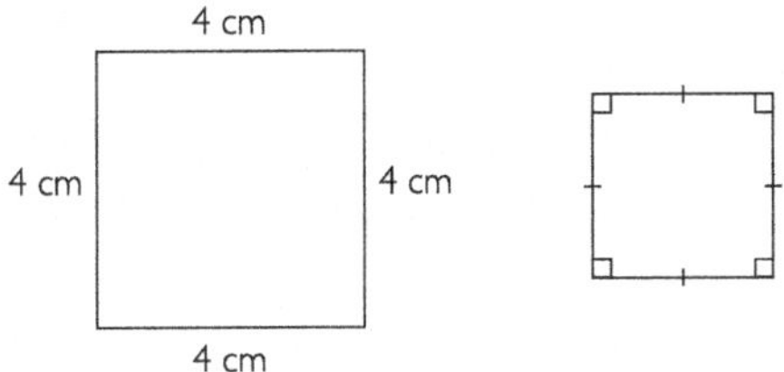

See also **polygon**, **quadrilateral**, **rectangle**, **regular polygon**

Meaning 2 To square a number is to multiply it by itself.

For example

The square of 3 is $3^2 = 3 \times 3 = 9$.

See also **index**, **power**

square number

A square number is a number that can be shown in a square pattern of dots.

For example

9 is a square number.

● ● ●
● ● ●
● ● ●

See also **array**, **rectangular number**, **triangular number**

square root

A square root of a number is a number that, when multiplied by itself, gives the original number.

For example

$3 \times 3 = 9$ and $-3 \times -3 = 9$

So both 3 and –3 are square roots of 9.

We write $3 = \sqrt{9}$ and $-3 = -\sqrt{9}$.

See also **root**, **square**

square unit

A square unit is a unit used to measure area. A square unit is a unit of length squared. The area of a surface is measured by working out how many square units entirely cover the surface. The square centimetre, square metre and square kilometre are examples of standard metric units.

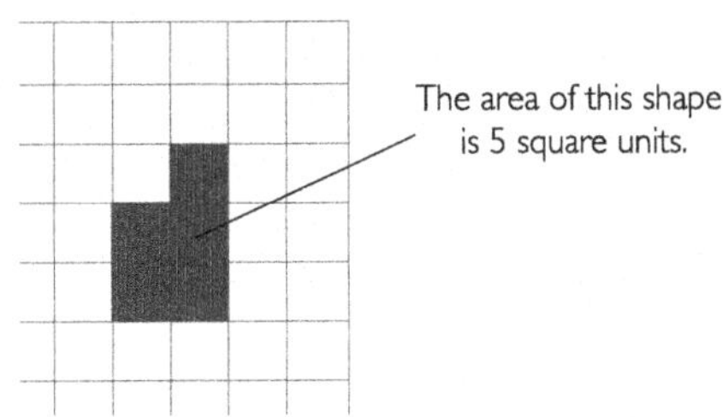

See also **area**, **SI system**, **standard units of measurement**, **unit**

standard units of measurement

Standard units of measurement are the commonly adopted units used to measure physical attributes. Using standard units helps people to understand and compare different measurements. In Australia, most standard units are metric units. Metric units are used to measure length, area, volume and capacity, and mass. Time, temperature and angles are measured with standard units that are not metric.

See also **informal unit**, **measurement**, **metric system**, **SI system**, **unit**, **table of commonly used SI units** (page 122)

statement

A statement is a mathematical sentence. It can be written in words or as an equation.

For example

1 There are more red cubes than blue cubes in the bag.

2 The equation $P = 4 \times l$ is a statement that relates the length of the sides of a square to its perimeter.

See also **equation, formula**

statistics

Meaning 1 Statistics is the study that deals with the collection, the classification and the analysis of data.

Meaning 2 Statistics are numerical data or information. The organisation of the data, as in a table or a graph, helps in examining the data for patterns.

For example

The number of passengers travelling on the metropolitan train system, each hour of each day of the week, gives statistics of train usage. Organisation of the data will highlight patterns of usage throughout the system.

See also **data**

stem-and-leaf plot

A stem-and-leaf plot is a graph that displays data in a table. Each number in the table is broken up into its tens digit in the first column and its ones digit in the second column. The tens digits are called the stems. The ones digits are called the leaves.

For example

The stem-and-leaf plot that follows displays the numbers:

7, 9, 14, 14, 15, 19, 20, 25, 39.

0	7 9
1	4 4 5 9
2	0 5
3	9
Stems (tens)	*Leaves (ones)*

See also **graph**

straight

Something straight has no bends or curves.

For example

A straight line is the shortest line between two points.

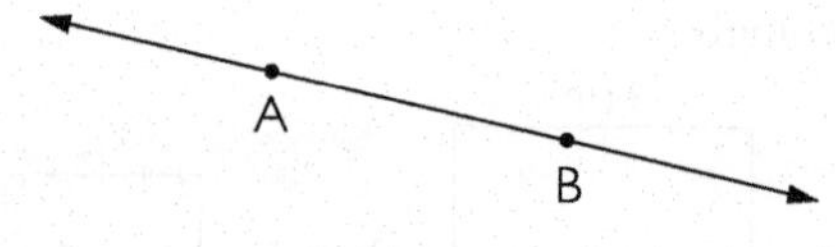

straight angle

A straight angle is an angle of 180 degrees.

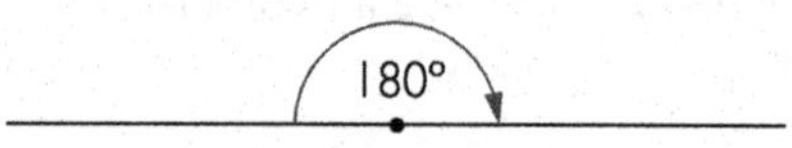

See **angle**

subset

A set is called a subset of another set if all its elements belong to this other set.

For example

1 The larger set of sporting equipment has as subsets: bats, balls and protective clothing.

Sporting equipment

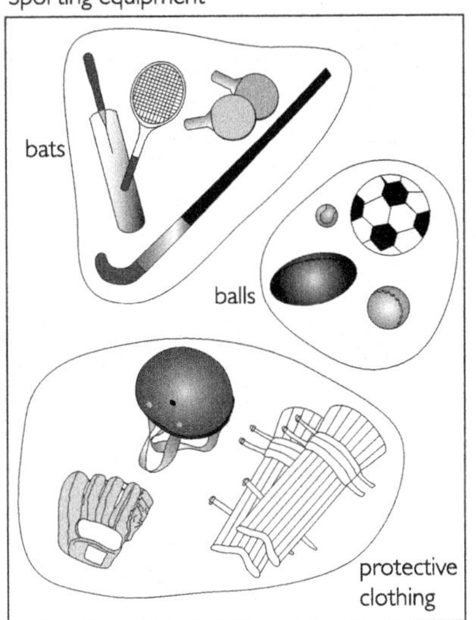

2 The set {B, D} is a subset of {A, B, C, D}, but {A, B, E} is not a subset of {A, B, C, D}.

See also **element**, **set**

substitute

To substitute is to put one thing in place of another. It means to replace.

For example

1 The area A of a rectangle with length L and width W is given by $A = L \times W$. Substitute the pronumerals with numerals to find the area of a rectangle 6 centimetres long and 4 centimetres wide.

2 If $x = 4$ and $y = 9$, substitute to find the value of $2x + y$.

See also **pronumeral**, **variable**

subtend

Subtend means to be opposite to and to mark out a line or an angle.

For example

1 A chord subtends an arc. The two points where the chord touches the circle are the end points of the arc.

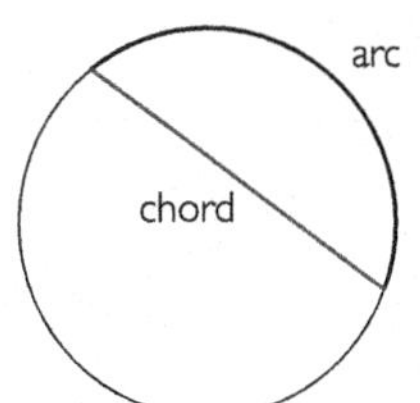

2 An arc of a circle subtends an angle at the centre of the circle. The length of the arc determines the size of the angle.

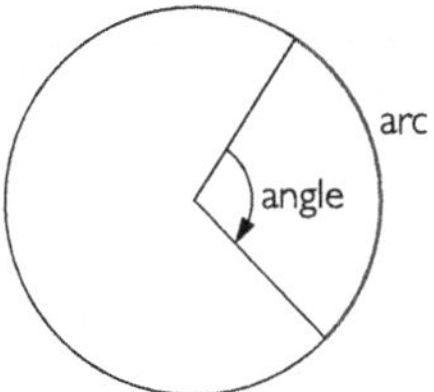

subtract

To subtract is to take one quantity away from another. The whole and one part are known, and the other part needs to be found out.

For example

Subtract 3 from 8 and the answer is 5.

subtraction

Subtraction is the operation of subtracting. Subtraction is the inverse of addition. There are three ways of interpreting the subtraction process:

- as take away
- as missing addend
- as comparison

For example

1 Subtraction as take away: This is to remove one amount from another amount to find what is left, for instance, $10 - 4 = 6$.

2 Subtraction as missing addend: This is to find the missing amount. If you have 3 beads, how many more do you need to make 7 beads?

3 Subtraction as comparison: This involves finding the difference. If there are 15 boys and 18 girls, how many more boys are there than girls?

See also **compare**, **missing addend**, **take away**

subtrahend

The subtrahend is the number that is subtracted from another number.

15	-	3	=	12
minuend		subtrahend		difference

See also **difference, minuend**

sum

Sum is the total of two or more quantities.

For example

12 is the sum of 5 and 7. That is,
5 + 7 = 12.

See also **add, addend, addition**

summary statistics

Summary statistics provide a measure of the "middle" of a group of numbers or quantities. Summary statistics include the mean, median and mode.

See also **average**

supplementary angles

Supplementary angles are two angles that together add up to 180 degrees.

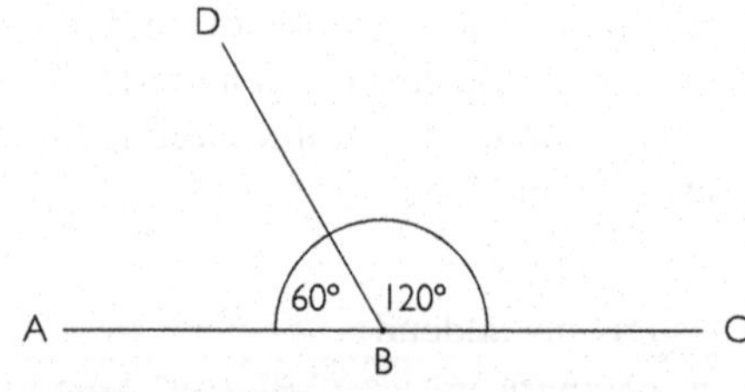

∠ABD and ∠CBD are supplementary angles.

See also **complementary angles**

surface

A surface is the outside boundary of an object. A surface may be flat or curved.

For example

The surface of a sphere is curved.

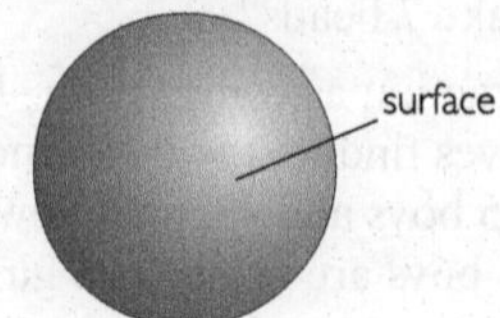

See also **face**

surface area

Surface area is the size of the surface of an object.

For example

To find the surface area of a rectangular prism, add up the area of each of the faces.

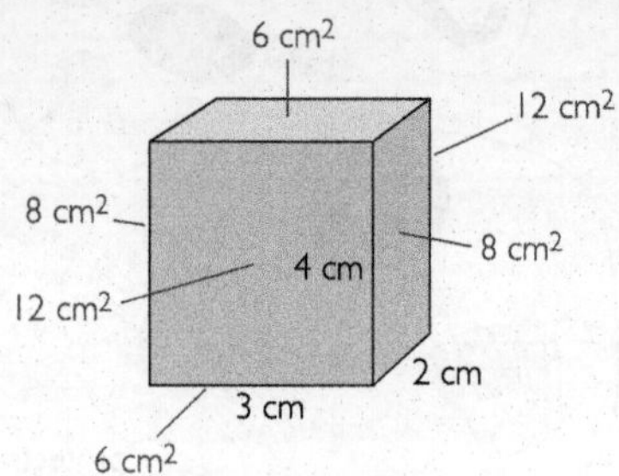

12 + 12 + 8 + 8 + 6 + 6 = 52
The surface area of the prism is 52 cm^2.

See **area, surface**

survey

Meaning 1 To survey is to collect opinions and facts. Data may be collected from a sample or a whole population.

Meaning 2 A survey is a way of collecting data. It may be by interview, questionnaire or observation.

For example

1. Conduct a survey of students, asking them to rate their favourite pop groups on a popularity scale.
2. Carry out a survey of students' eye colour in order to estimate the proportion of the population with brown eyes.

See also **data, statistics**

symbol

A symbol is a figure, sign or letter that represents something.

For example

1. Symbols represent the mathematical processes of addition, subtraction, multiplication and division: +, –, ×, ÷.
2. Symbols represent unknown numbers in algebra. In the equation $2y + 5 = 7$, the symbol y stands for an unknown number.

Common mathematical symbols

Symbol	Meaning	Symbol	Meaning
=	equal	%	per cent
π	not equal	() {} []	brackets
≈	approximately equal	°	degree
<	less than	∠	angle
>	greater than	∟	right angle
≤	less than or equal to	√	square root
≥	greater than or equal to	∛	cube root
•	infinity	π	pi

See also **pronumeral, table of maths symbols** (page 125)

symmetry

An object has symmetry if it matches about a line or a point. There are two different types of symmetry. Bilateral, or line, symmetry is when one half of a shape is a reflection of the other half. Rotational symmetry is when a shape looks the same after it has been turned about a central point. A shape may have one type of symmetry or the other, both types of symmetry or none at all.

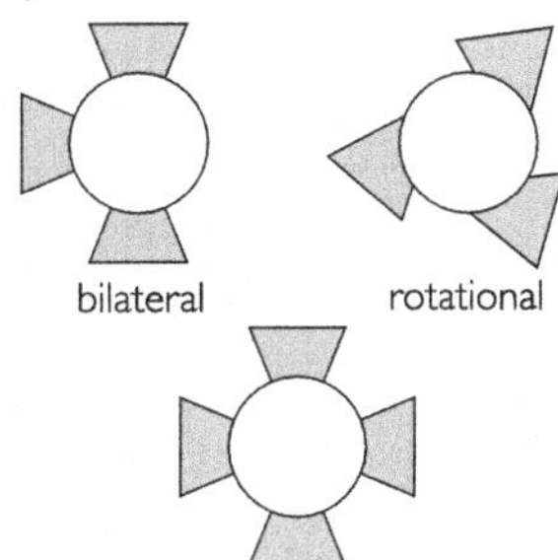

bilateral

rotational

bilateral and rotational

See also **asymmetry, bilateral symmetry, line of symmetry, mirror line, plane of symmetry, rotational symmetry**

system

A system is organised or connected information forming a complex whole.

For example

1. The decimal system is a system of numbers.
2. The metric system is a system of units of measurement.
3. The SI system is a scientific system of units.

Tt

table

A table is the ordered arrangement of numbers or quantities in rows and columns. Putting information into a table can make it easier to interpret the data.

For example

1 multiplication tables

5 ×	6 ×	7 ×
1 × 5 = 5	1 × 6 = 6	1 × 7 = 7
2 × 5 = 10	2 × 6 = 12	2 × 7 = 14
3 × 5 = 15	3 × 6 = 18	3 × 7 = 21
4 × 5 = 20	4 × 6 = 24	4 × 7 = 28
5 × 5 = 25	5 × 6 = 30	5 × 7 = 35
6 × 5 = 30	6 × 6 = 36	6 × 7 = 42
7 × 5 = 35	7 × 6 = 42	7 × 7 = 49
8 × 5 = 40	8 × 6 = 48	8 × 7 = 56
9 × 5 = 45	9 × 6 = 54	9 × 7 = 63
10 × 5 = 50	10 × 6 = 60	10 × 7 = 70
11 × 5 = 55	11 × 6 = 66	11 × 7 = 77
12 × 5 = 60	12 × 6 = 72	12 × 7 = 84

2 a table of ordered pairs of values

x	−2	−1	0	1	2
y	−3	−1	1	3	5

tabular

Tabular means in table form.

For example

A tabular display is a display of data in rows and columns.

See also **table**

take away

Take away means to remove an amount from another amount. Take away is one of the ways of interpreting the subtraction process.

For example

18 − 6 = 12

Taking 6 away from 18 leaves 12.

See also **subtraction**

tally

Meaning 1 A tally is a mark to keep count. Tally marks are usually grouped in fives.

The tally is 12.

Meaning 2 To tally is to count or keep score. To tally is to take account of the number of items.

tangent

A tangent is a straight line or a plane that touches a curve or a surface without cutting through it.

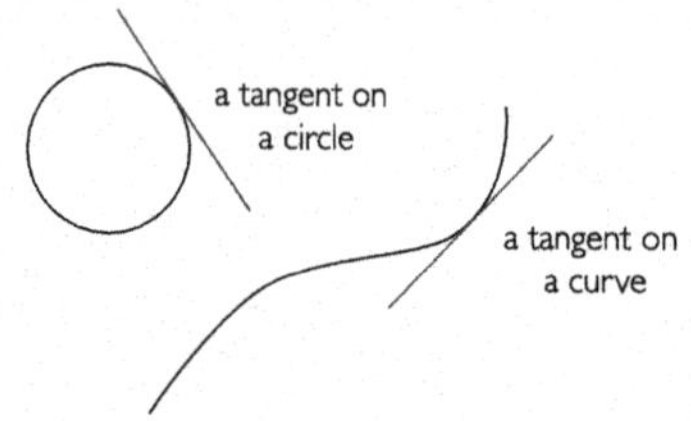

tangram

A tangram is a square cut into five triangles, one square and a parallelogram. The seven shapes can be put together in many different ways to make many different shapes. A tangram is a traditional Chinese puzzle.

temperature

Temperature is the measure of the hotness or coldness of something.

Temperature is usually measured in degrees Celsius.

See also **Celsius scale**

template

A template is a pattern.

For example

1. A template of a shape can be used when drawing the multiple, identical shapes of a tessellation.
2. The template set-up of spreadsheet is a framework for organising data.

See also **pattern**

tens

In a base-ten counting system, tens is the place value that identifies how many groups of ten.

For example

In the number 54, the 5 stands for five tens.

thousands	hundreds	tens	ones
		5	4

See also **decimal system**, **place value**

tenth

Meaning 1 Tenth is an ordinal number. Tenth describes a relative position. It is after ninth in order and before eleventh.

For example

Ann is the tenth person in line.

Meaning 2 One tenth is a fraction. It means one part out of ten equal parts.

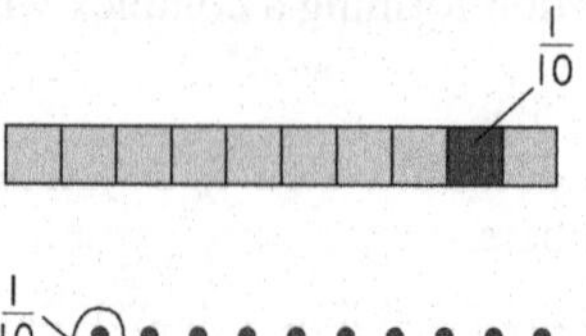

See also **fraction**

tenths

In the base-ten counting system, tenths is a place value position in the decimal-fraction part of a number. It identifies how many tenths.

For example

In the number 0.75, the 7 stands for seven tenths.

thousands	hundreds	tens	ones	tenths	hundredths	thousandths
			0 •	7	5	

See also **decimal place**, **decimal system**

term

Meaning 1 A term is one of the bits that are added together to form an algebraic expression.

For example

1. The expression $5a + bc$ has two terms: $5a$ is a term; bc is a term.
2. The terms of the expression $2(x + y) + x - z$ are $2(x + y)$ and x and $-z$.

See also **expression**

Meaning 2 A term is one of the elements of a sequence or series.

For example

The sequence 2, 4, 6, 8, 10 has five terms.

See also **sequence**, **series**

terminating decimal

A terminating decimal is a decimal fraction with an end to its sequence of digits.

For example

$\frac{3}{8} = 3 \div 8 = 0.375$

See also **recurring decimal**

tessellation

A tessellation is a pattern of identical shapes that fit together exactly. The shapes may be geometric shapes, like equilateral triangles, squares or hexagons. The shapes may be creative shapes, like those of the artist Escher. The shapes can be in two or three dimensions.

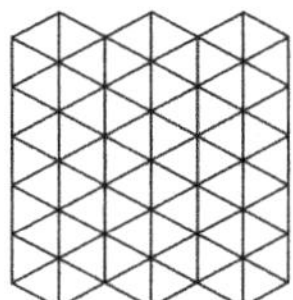
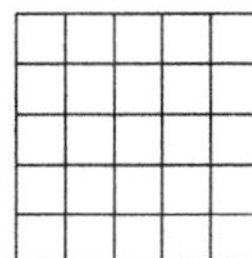
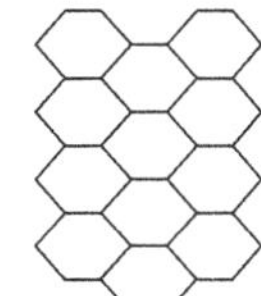

Tessellations in two dimensions.

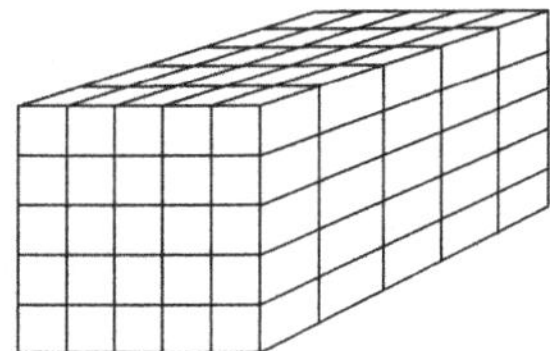

A tessellation in three dimensions.

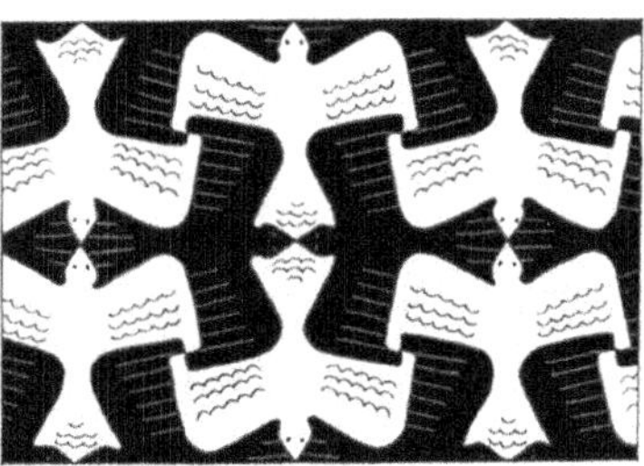

A creative tesselation.

See also **Escher, pattern, spatial pattern**

tetrahedron

A tetrahedron is a polyhedron with four faces. A regular tetrahedron has every face an equilateral triangle and is one of the five Platonic solids. A tetrahedron is also called a triangular pyramid.

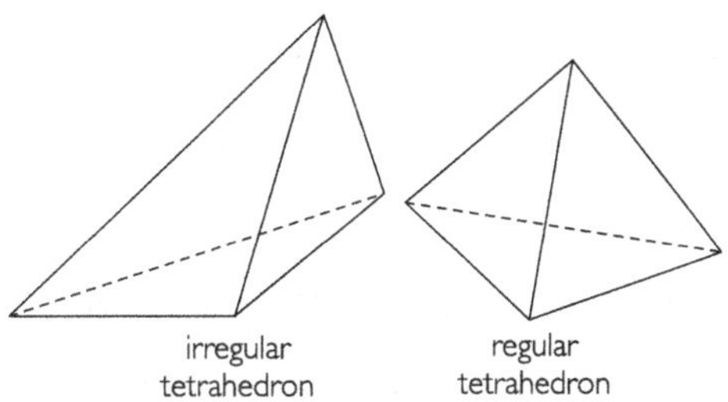

See also **polyhedron**

T

third

Meaning 1 Third is an ordinal number. Third describes a relative position. It is after the second in order and before the fourth.

For example

Mimi is the third person in the line.

See also **ordinal number**

Meaning 2 One third is a fraction. It means one part out of three equal parts.

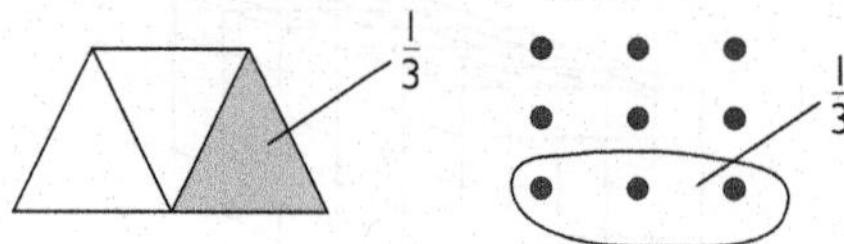

See also **fraction**

thousands

In the base-ten counting system, thousands is the place value that identifies how many groups of one thousand.

For example

In the number 7431, the 7 stands for seven thousands.

millions	hundreds of thousands	tens of thousands	thousands	hundreds	tens	ones
			7	4	3	1

See also **decimal system, place value**

thousandths

In a base-ten counting system, thousandths is a place value in the decimal fraction part of a number. It identifies how many thousandths.

For example

In the number 0.369, the 9 stands for nine thousandths.

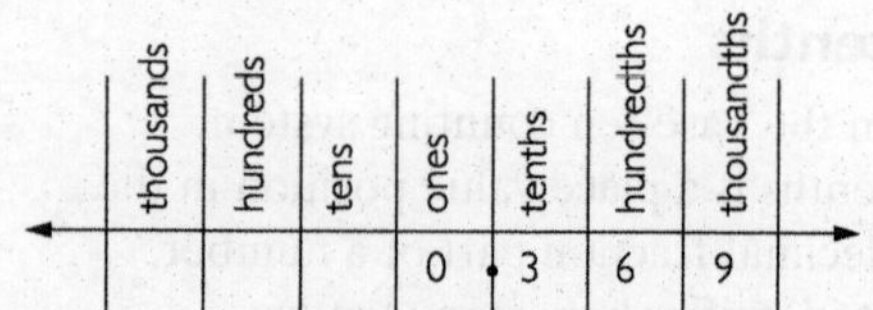

See also **decimal place, decimal system**

three-dimensional

Three-dimensional is having length, width and height. Solid shapes are three-dimensional.

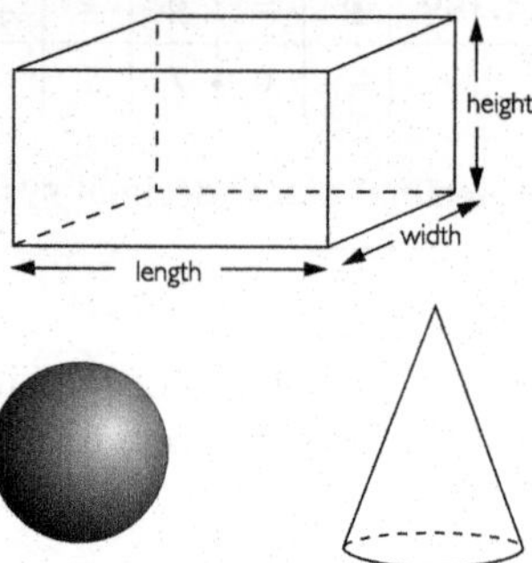

See also **dimension, one-dimensional, two-dimensional, zero-dimensional**

three-quarter turn

A three-quarter turn is three-quarters of a complete rotation. The angle of a three-quarter turn is 270 degrees.

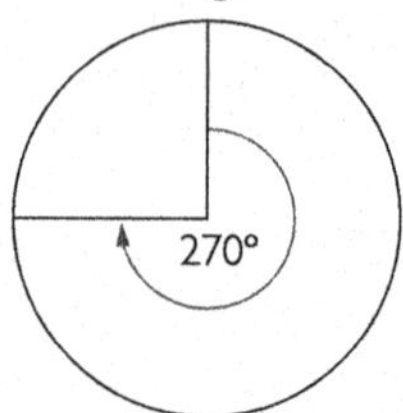

See also **full-turn, rotation**

time

Meaning 1 Time can mean how long it takes to do something. "How long it takes" is called time elapsed or duration.

For example

A stopwatch can be used to measure how long it takes to jump the rope one hundred times.

See also **duration**

Meaning 2 Time can mean the period of time between two events. "The period of time between two events" is called a time interval. A time interval has a start and a finish.

For example

The timetable shows how often trains leave the city. They leave at twenty minute intervals.

See also **interval**

Common units of time are:

60 seconds = 1 minute

60 minutes = 1 hour

24 hours = 1 day

7 days = 1 week

time line

A time line is a scaled line on which events are recorded according to when they happened. A time line helps us to understand the timing of events in relation to each other over a very long period.

For example

Important events in Australia over the last one hundred years were recorded on a time line.

Australian time line 1901–2001.

1901	Australia's population is 3.8 million.
1907	A basic, living wage of 42/- ($4.20) a week is set for unskilled workers.
1923	Vegemite is first manufactured in Melbourne.
1932	40% of Australian workers unemployed.
1948	The first all-Australian car, the Holden, is produced. It costs £733 ($1466).
1951	The first "School of the Air" broadcast is made from the Flying Doctor base at Alice Springs.
1966	Decimal currency is introduced in Australia.
1970	The metric system of measurement is adopted in Australia.
1983	January bushfires destroyed $400 000 000 worth of property.
1990	Virtually 100% of households have a TV.
2000	Australia's population is 19.2 million.
2001	The 100th anniversary of the Federation of Australia is celebrated.

See also **chronological order**, **time**

timetable

A timetable organises the times at which particular things take place.

For example

1. A class timetable lists the days and times of subjects.
2. A bus timetable lists the times that buses arrive and depart.

See also **chart**, **table**

time zone

A time zone is a region of the world that has a common time. The globe is divided into 24 such regions. Each boundary is along a line joining the North Pole to the South Pole. The time at Greenwich, in England, is the time from which times in the other regions are set. Generally, countries east of Greenwich have their midday before Greenwich, and countries west of Greenwich have their midday after Greenwich. The time at Greenwich is called Greenwich Mean Time.

tonne

A tonne is a unit of mass. The symbol for tonne is t.

1 tonne = 1000 kilograms

For example

A small car has a mass of about 1 tonne.

See also **mass**, **metric system**, **standard units of measurement**

topology

Topology is a relatively recent branch of geometry. Topology is the study of the things about objects that stay the same when the objects are changed by stretching or shrinking. It is sometimes called rubber-sheet geometry. Topology does not deal with size or rigid shapes. The properties of "where", "between what", "inside" and "outside" are considered.

☞

For example

1 The Möbius strip is an object studied by topologists.
2 Knot theory is an area of topology.
3 A mug and a doughnut are topologically equivalent.

See also **geometry**

top view

A top view is how an object looks from above it.

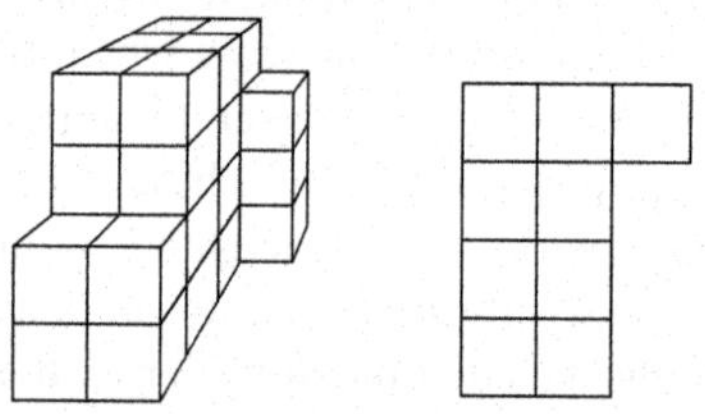

See also **front view, side view**

torus

A torus is a solid that is shaped like a doughnut.

total

Meaning 1 The total is the sum.

For example

The total of $5, $10, $1 and $4 is $20.

See also **sum**

Meaning 2 Total means the whole.

For example

To calculate the amount of paint needed, work out the total area of the four walls.

See also **whole**

transformation

Meaning 1 Transformation is the changing of a figure, with a correspondence between the points of the original and the points of the transformed figure. Slides, flips, turns and enlargements are different kinds of transformation.

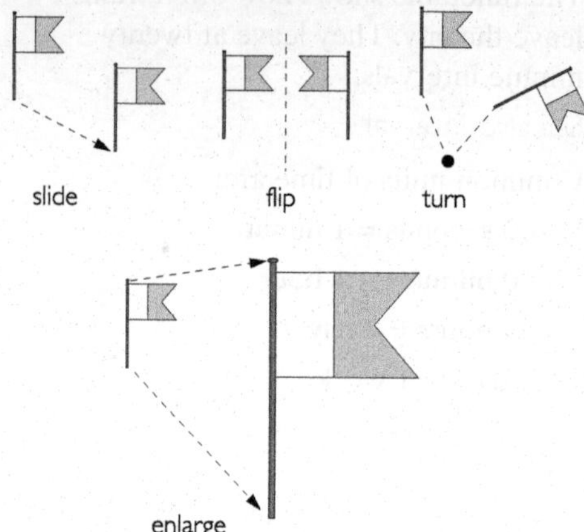

See also **mapping, one-to-one correspondence**

Meaning 2 Transformation is the changing of an expression or statement, without changing the meaning.

For example

The equation $y = x + 5$ can be transformed into the equation $y - 5 = x$ by subtracting 5 from both sides. These two equations have the same meaning.

translation

A translation is a slide. A shape can be translated in any direction, without any change in its size or orientation. Translation is one kind of transformation.

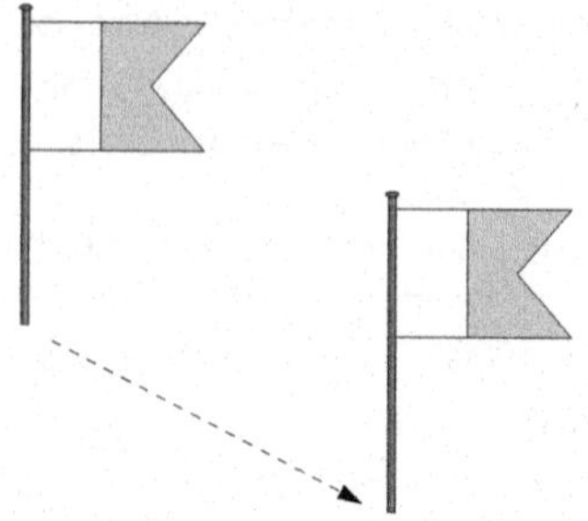

See also **slide, transformation**

transversal

A transversal is a straight line that cuts across one or more lines.

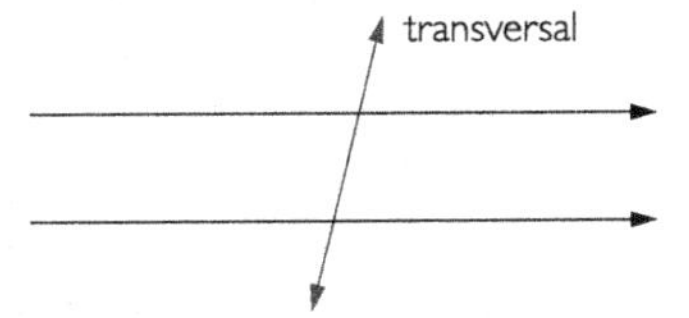

trapezium

A trapezium is a quadrilateral with one pair of opposite sides parallel.

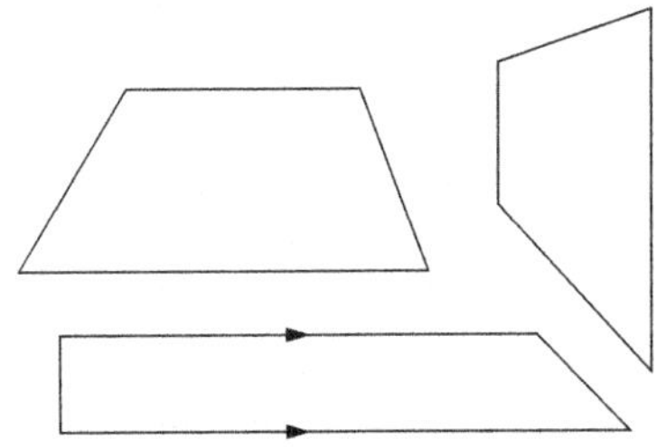

See also **quadrilateral**

treble

Treble means three times as much or three times as many. The quantity is multiplied by 3.

For example

Treble the ingredients to make three times as many pancakes.

tree diagram

A tree diagram is a systematic drawing that records different outcomes by branching.

For example

The possible results of tossing a coin can be shown in a tree diagram.

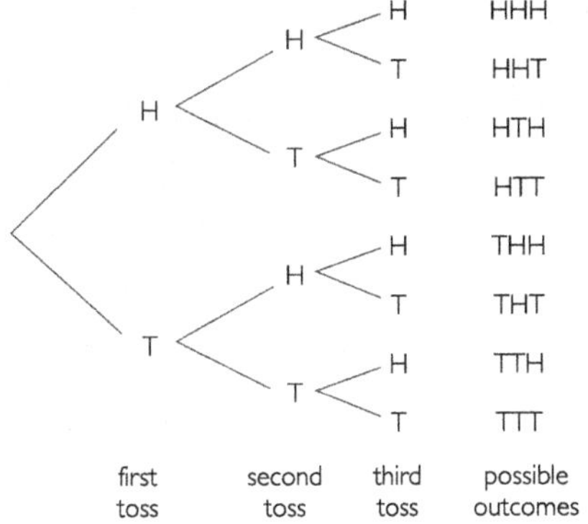

Tossing a coin three times.

See also **factor tree**, **Pascal's triangle**

trend

A trend is a tendency to take a particular direction.

For example

A trend can be identified in the graphing of data.

See also **line graph**

triangle

A triangle is a polygon with three straight sides. The sum of the angles of a triangle is 180 degrees.

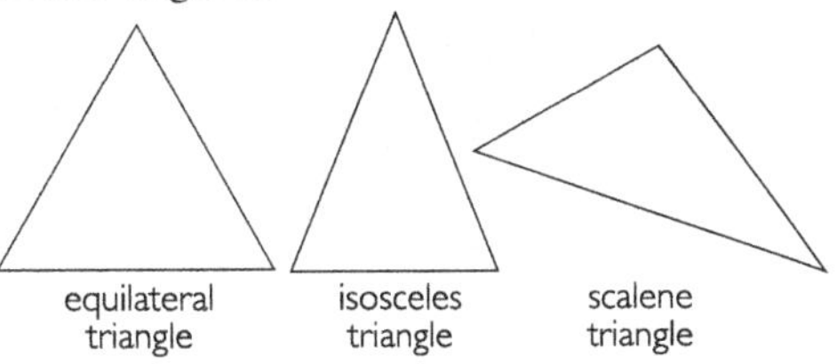

Triangles classified by sides.

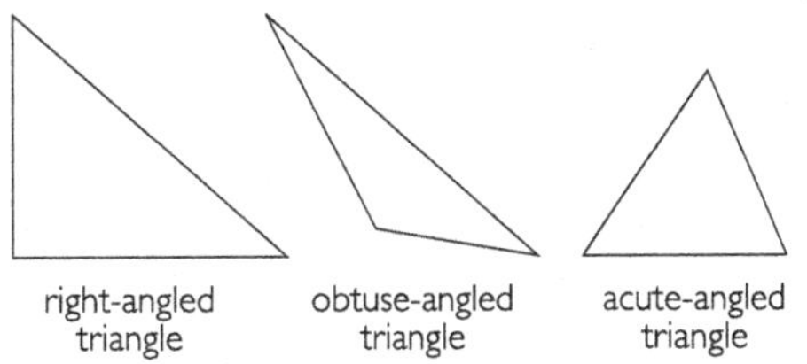

Triangles classified by angles.

See also **acute-angled triangle**, **equilateral triangle**, **isosceles triangle**, **obtuse-angled triangle**, **right-angled triangle**, **scalene triangle**

T

triangular number

A triangular number is a number that can be shown in a triangular dot pattern.

In a triangular dot pattern, each row has one more dot than the row above it.

For example

3, 6, 10 and 15 are triangular numbers.

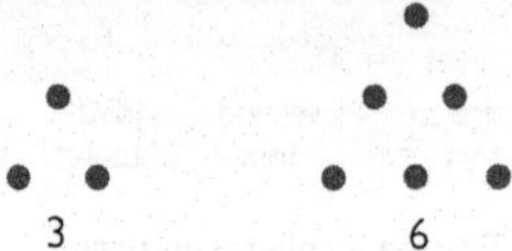

See also **rectangular number**, **square number**

triangular prism

A triangular prism is a prism with two triangles as its bases. The three other faces are parallelograms.

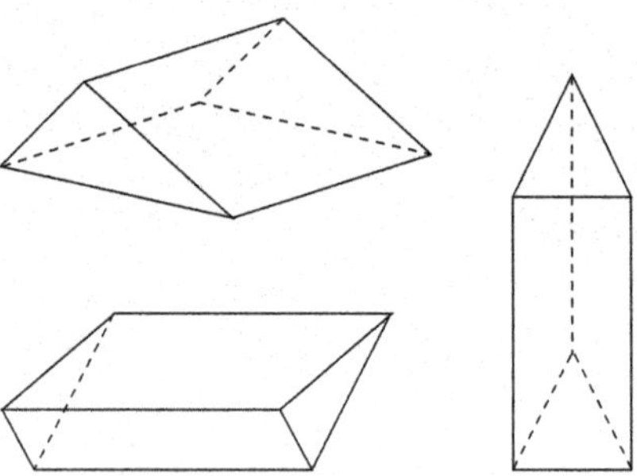

See also **polyhedron**, **prism**

trillion

A trillion is the number 1 000 000 000 000. A trillion can be written as 10^{12}.

See also **place value**

triple

Meaning 1 To triple means to multiply by three.

For example

In the number sequence 3, 9, 27, 81, ..., the rule is to triple the previous number.

Meaning 2 A triple is a group of three things.

For example

Ordered triples can be used to describe points in three-dimensional space.

See also **Cartesian coordinates**

true statement

A true statement is one that is correct. An equation is a true statement if the left-hand side is equal to the right-hand side. An inequation can be a true statement.

For example

1 7 + 5 = 12 is a true statement.

2 9 > 6 is a true statement.

See also **equation**, **inequation**

turn

A turn is when a shape is rotated as a whole, about a point, through an angle. A turn is one kind of transformation.

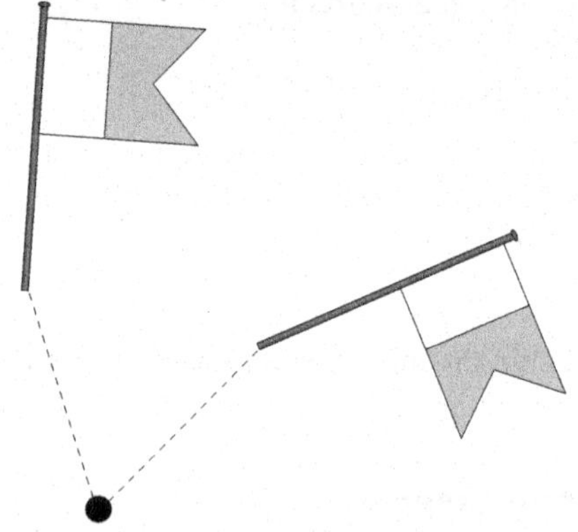

See also **rotation**, **transformation**

twelve-hour time

Twelve-hour time is the method for naming the time that divides a day into two twelve-hour periods: a.m. and p.m. The time shown on a twelve-hour clock needs to be interpreted as either in the period between midnight and midday, or in the period between midday and midnight.

For example

3 a.m. is 3 hours after midnight.

See also **a.m.**, **p.m.**

twenty-four hour time

Twenty-four hour time is the method for naming the time that divides a day into twenty-four one-hour periods, beginning at midnight. Twenty-four hour time is used in airline schedules and on video recorders. In twenty-four hour time, four numbers are used to name a time.

For example

1. 1200 is read as "twelve-hundred hours" and is equivalent to midday.
2. 0900 is read as "o nine-hundred hours" and is equivalent to 9 a.m.
3. 1630 is read as "sixteen thirty hours" and is equivalent to 4.30 p.m.

twice

Twice is two times as much or two times as many.

For example

There are twice as many red counters as blue counters in the bag.

See also **double**

two-dimensional

Two-dimensional means having length and width or length and height. A two-dimensional shape lies within a flat surface.

For example

Plane shapes are two-dimensional.

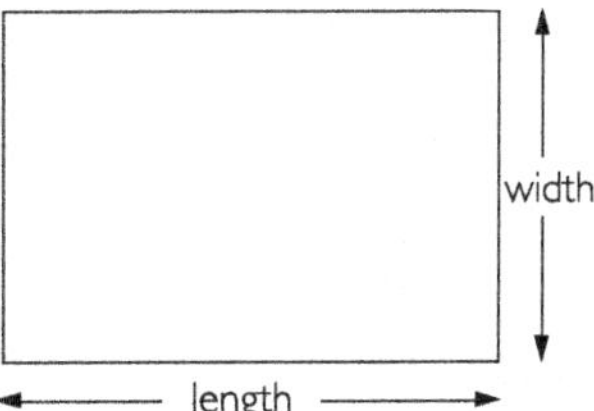

See also **dimension**, **one-dimensional**, **three-dimensional**, **zero-dimensional**

Uu

unequal

Unequal means not equal. Unequal means not of the same quantity or not balanced.

For example

1 $8 < 12$ is an inequation: the sides are unequal.

2 The money was distributed in unequal amounts.

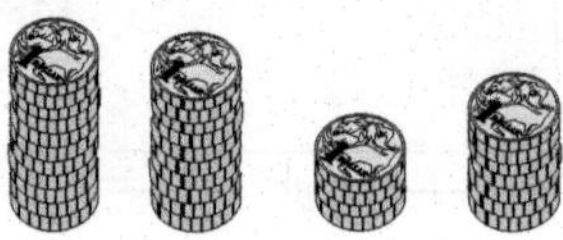

3 The beam balance showed the two masses to be unequal.

See also **equal**

uniform units

Uniform units are the same size units used to measure and compare different objects.

For example

1 Uniform units can be informal units. For instance, use the same size tiles to compare the area of two different book covers.

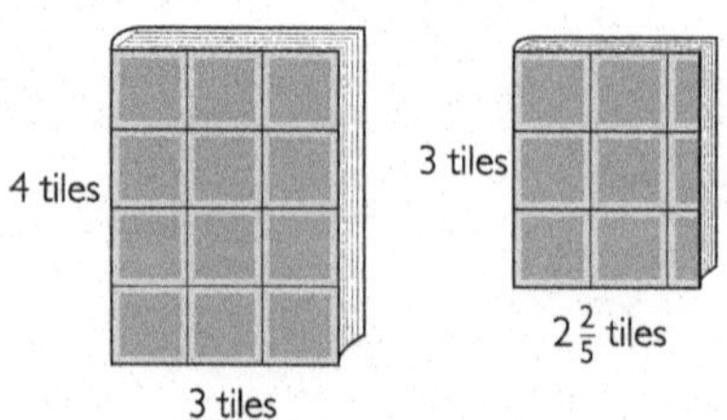

2 Uniform units can be standard units of measurement. For instance, use square centimetres to compare the area of two different book covers.

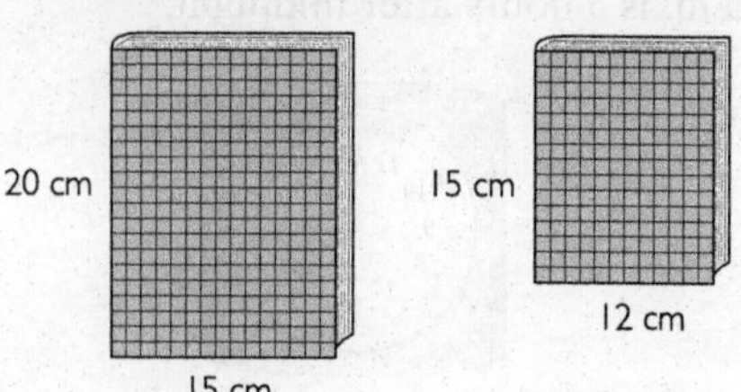

See also **unit**

unit

Meaning 1 A unit means one, being the lowest counting number.

For example

In the number 21, there is one unit in the ones place.

See also **counting number**

Meaning 2 A unit is an amount of a physical attribute that is used to measure and compare the same attribute of other objects. A unit can be an informal unit or a standard unit.

For example

1 Hand spans can be used to measure and compare length. Bricks can be used to measure and compare mass. Hand spans and bricks are examples of informal units.

2 Different standard units are used to measure and compare length, mass and volume. In the metric system, the common units of measurement are cm, m, km, g, kg, mL and L.

See also **informal unit**, **metric system**, **SI system**, **standard units of measurement**

Meaning 3 A unit is a single thing, which in itself can be regarded as a whole.

For example

The maths curriculum is broken down into a number of separate units.

unit fraction

A unit fraction is a common fraction that has 1 as the numerator. A unit fraction is also known as a unitary fraction.

For example

$\frac{1}{2}$, $\frac{1}{4}$ and $\frac{1}{8}$ are unit fractions.

See also **common fraction**, **numerator**

unitary method

The unitary method is a method of working out a problem about quantities by first finding out the value of one quantity.

For example

If five pens cost $2, what will 8 pens cost?

First find the cost of one pen, which is $2 ÷ 5 = 40 cents.

So 8 × 40 cents = $3.20 is the cost of 8 pens.

unknown value

An unknown value is a quantity, represented by a symbol, that is not known. An unknown value is yet to be found.

For example

p is the unknown value in $4 \times 3p - 2p$.

See also **pronumeral**, **symbol**, **variable**

unlike terms

Unlike terms are algebraic terms with pronumeral parts that are not identical. Unlike terms cannot be combined using addition or subtraction.

For example

1 The terms $3b$ and $4c$ are unlike terms: b and c are not identical.

2 A number and a pronumeral are unlike terms: $3a + 1$ cannot be simplified.

See also **algebra**, **expression**, **like terms**, **term**

unrelated attributes

Unrelated attributes are characteristics of an object that are not connected.

For example

There is no necessary connection between the length and the width of a rectangle. Length and width are unrelated attributes.

See also **attribute**

Vv

value

Value is the worth of a thing measured by an amount of money or a number.

For example

1 The value of my car is $2000.

2 Pronumerals can be given a value. If x has the value 5, then $2x + 1 = 11$.

3 Algebraic expressions can have a value. If $x = 5$, then the value of $2x + 1$ is 11.

4 If one tile has a value of 6, what is the value of this shape?

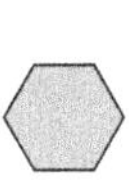

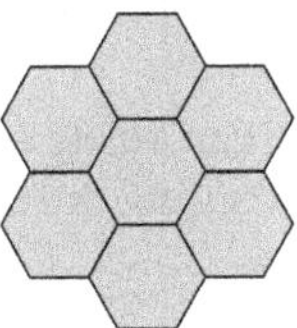

See also **variable**

variable

A variable is a quantity, represented by a symbol, that can take on different values.

For example

In the equation $5a + 4 = 6b$, both a and b are variables. The numbers 4, 5 and 6 are constants.

See also **pronumeral**, **value**

Venn diagram

A Venn diagram is a pictorial representation of sets and the relationship between sets. Sets are represented by circles. Where sets have elements in common, they overlap. The Venn diagram was introduced by John Venn, an English mathematician who lived 1834–1923.

☞

Numbers from 1 to 10

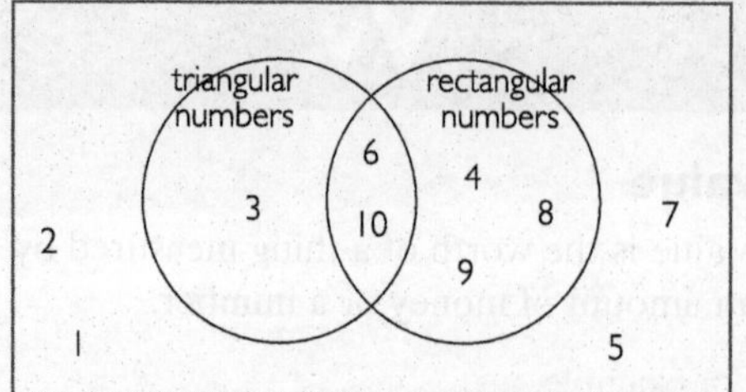

See also **element**, **intersection**, **set**

verify

To verify is to confirm the correctness of something.

For example

1. We can use a calculator to verify an answer to a worked multiplication problem.
2. We can verify an estimate of length by measuring the distance.

vertex

Meaning 1 A vertex is a point in a polygon that is common to two sides.

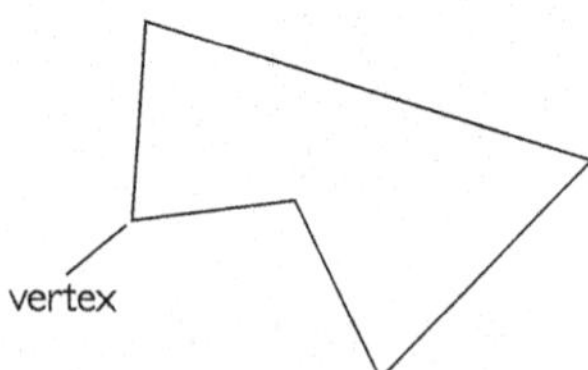

This polygon has five vertices.

See also **plane**, **polygon**

Meaning 2 A vertex is a point in a polyhedron that is common to three or more faces.

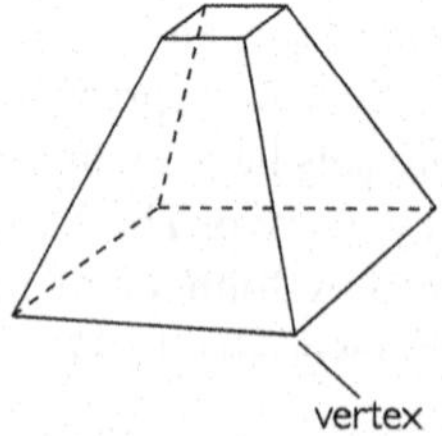

This polyhedron has eight vertices.

See also **polyhedron**, **solid**

Meaning 3 A vertex is the point at which two rays meet to form an angle.

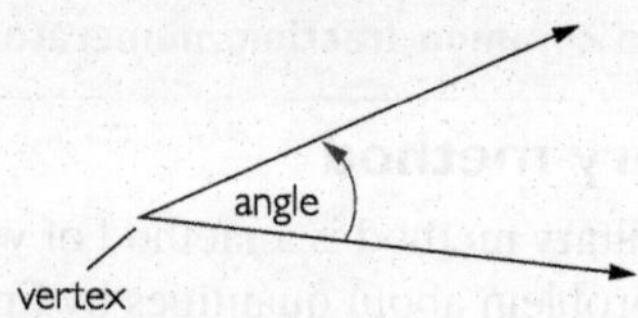

The plural form of the word *vertex* is vertices.

See also **angle**, **ray**

vertical

Vertical means at right angles to the horizon.

For example

1. A line or a plane can be vertical.
2. A Cartesian graph has a vertical axis.

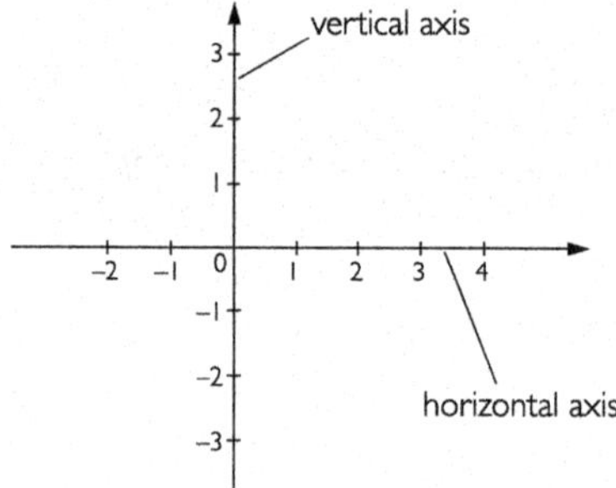

See also **horizontal**, **perpendicular**, **right angle**

vertically opposite angles

Vertically opposite angles are opposite, equal angles that are formed by the intersection of two lines.

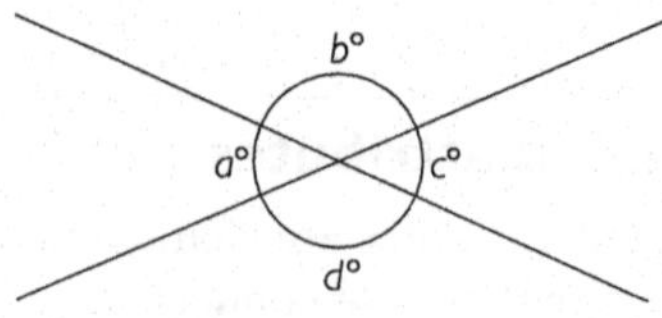

$a°$ and $c°$ are vertically opposite angles.
$b°$ and $d°$ are vertically opposite angles.

See also **angle**, **equal**

volume

Volume is the amount of space occupied by a substance.

For example

1 The volume of a solid brick is the amount of space it takes up, which is measured in cubic units.

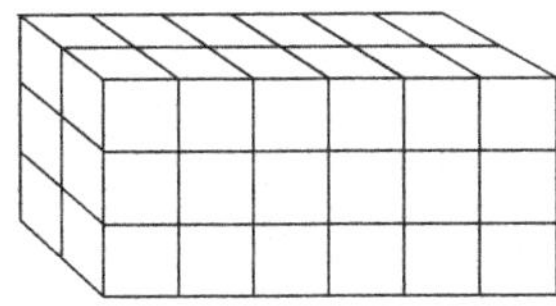

6 x 2 x 3 = 36
The volume of this brick is 36 cubic units.

2 The volume of liquid milk is also the amount of space it takes up, but a liquid is commonly measured in litres.

1000 cubic centimetres = 1 litre

See also **cubic unit, standard units of measurement**, **space**

vulgar fraction

See **common fraction**

week

A week is a unit of time that is equal to seven days. The days of the week are Monday, Tuesday, Wednesday, Thursday, Friday, Saturday and Sunday.

See also **day**, **time**

weight

The weight of an object is how heavy it is. Objects on Earth have weight because the Earth's gravity pulls on them. Weight is different from mass, although the two are commonly confused. An astronaut floating in space has mass but no weight. Kilograms and grams are commonly used as units for measuring both weight and mass. The International System of Units uses newtons to measure weight.

See also **mass, SI system**

whole

Whole is the total amount.

For example

1 The whole class took part in the survey.

2 A fraction represents part of a whole.

3 The whole floor area was measured in square metres.

whole number

A whole number is a number with no fraction in it. A whole number is also known as an integer.

For example

–4, –1, 0, 1, 4, 49 and 706 are all whole numbers.

See also **integer**

width

Width is the distance from side to side.
Width is also called breadth.

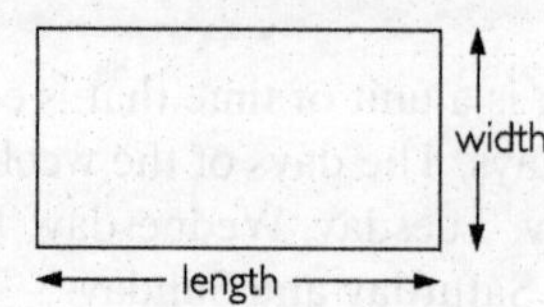

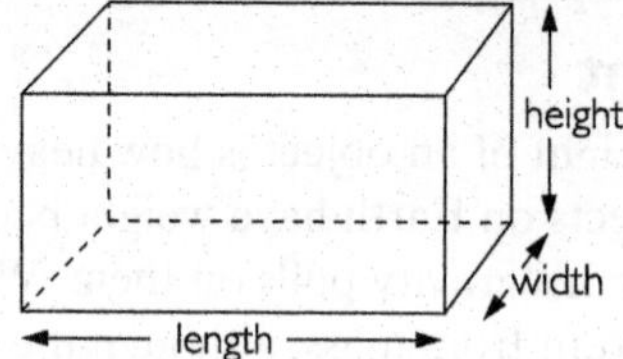

See also **height**, **length**

x-axis

The *x*-axis is a horizontal number line used to state the position of a point on a graph, such as a Cartesian graph. A point on a graph is labelled by a pair of numbers. The first number in the pair, called the *x*-coordinate, gives the position of the point along the *x*-axis. A positive *x*-coordinate means the point is right of the origin. A negative *x*-coordinate means the point is left of the origin.

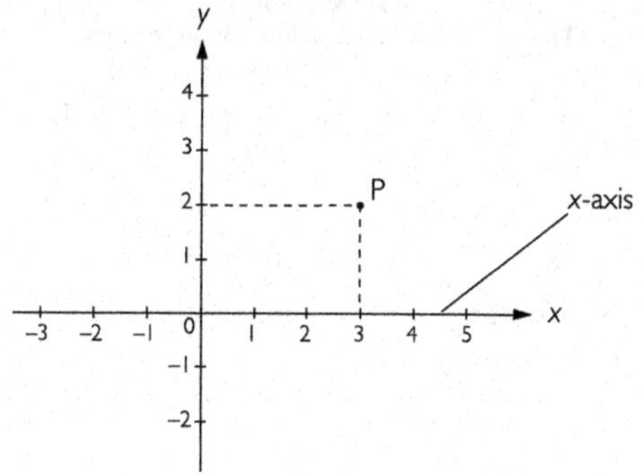

The coordinates of point P are (3, 2).
The *x*-coordinate is 3.

See also **axis**, **Cartesian coordinates**, **Cartesian graph**, **coordinates, origin**, **y-axis**

Yy

y-axis

The y-axis is a vertical number line used to state the position of a point on a graph, such as a Cartesian graph. A point on a graph is labelled by a pair of numbers. The second number in the pair, called the y-coordinate, gives the position of the point along the y-axis. A positive y-coordinate means the point is above the origin. A negative y-coordinate means the point is below the origin.

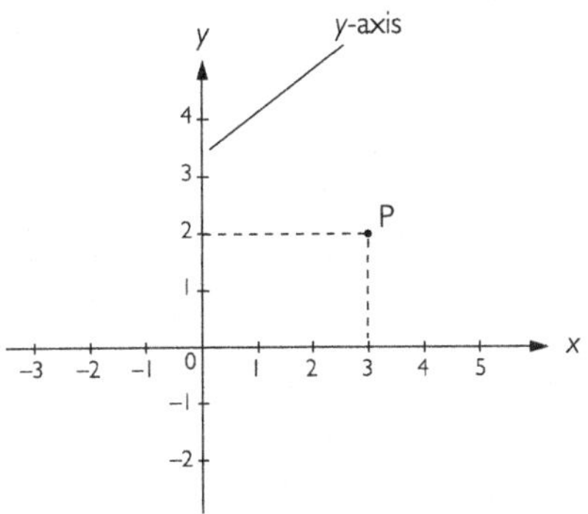

The coordinates of point P are (3, 2). The y-coordinate is 2.

See also **axis, Cartesian coordinates, Cartesian graph, coordinates, origin, x-axis**

year

A year is the time that the Earth takes to make one revolution around the Sun, which is about $365\frac{1}{4}$ days. The calendar year is 365 days, adjusted to 366 days every four years. A year of 366 days is called a leap year. There are 12 months in a year.

See also **calendar, day, month**

Zz

zero

Meaning 1 Zero is the symbol 0. The symbol 0 is used to stand for no quantity in the base-ten counting system.

For example

In the number 605, the 0 stands for no tens.

See also **Arabic numerals, decimal system, nought, place value**

Meaning 2 Zero is a whole number that lies between +1 and –1.

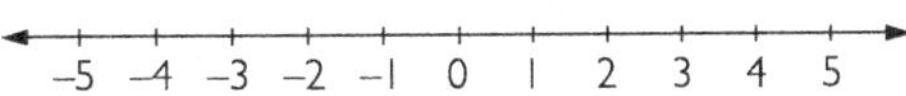

Note how the number zero operates in relation to other numbers:

- $8 + 0 = 8$
- $9 - 0 = 9$
- $5 \times 0 = 0$
- $0 \div 2 = 0$
- $2 \div 0$ has no answer
- $10^0 = 1$

See also **index laws, integer, whole number**

Meaning 3 Zero means not any or nothing. Zero means no quantity or size.

For example

Objects are weightless in zero gravity.

zero-dimensional

Zero-dimensional means of no length, width or height.

For example

A point is zero-dimensional.

See also **one-dimensional, three-dimensional, two-dimensional**

Units of Measurement

Some commonly used units of the SI system

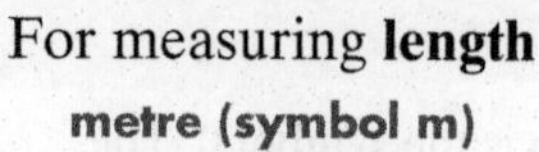

For measuring **length**

metre (symbol m)

For measuring **mass**

kilogram (symbol kg)

For measuring **time**

second (symbol s)

For measuring **force**

newton (symbol N)

For measuring **energy**

joule (symbol J)

For measuring **pressure**

pascal (symbol Pa)

For measuring **power**

watt (symbol W)

For measuring **frequency**

hertz (symbol Hz)

For measuring **electric current**

ampere (symbol A)

For measuring **electric potential**

volt (symbol V)

For measuring **area**

square metre (symbol m^2)

For measuring **volume**

cubic metre (symbol m^3)

For measuring **speed**

metres per second (symbol m/s)

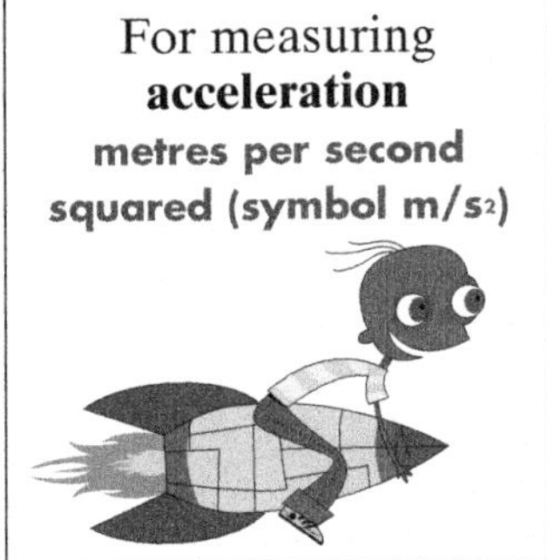

For measuring **acceleration**

metres per second squared (symbol m/s^2)

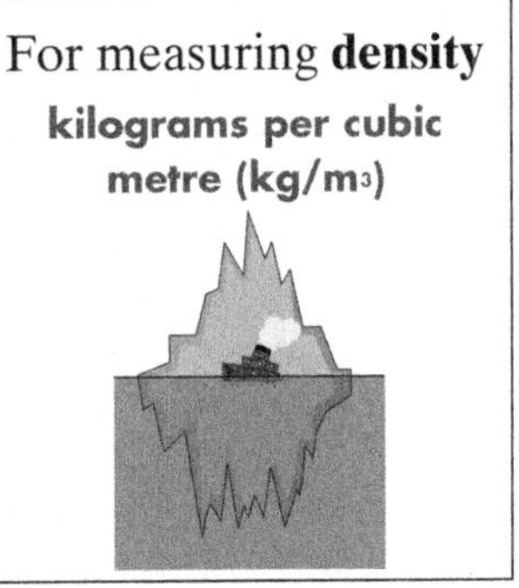

For measuring **density**

kilograms per cubic metre (kg/m^3)

Some non-SI units that can be used in maths and science

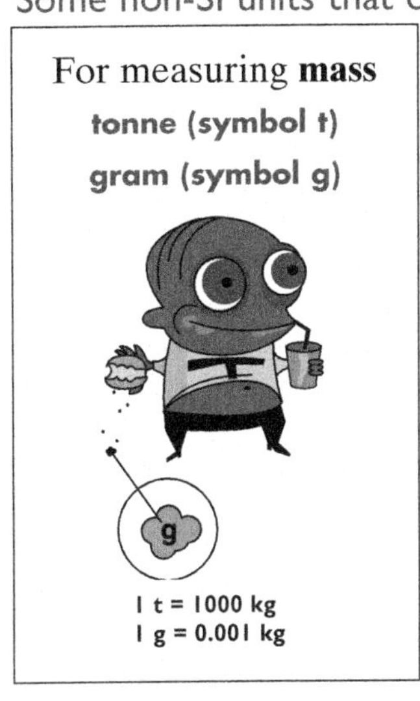

For measuring **mass**

tonne (symbol t)

gram (symbol g)

1 t = 1000 kg
1 g = 0.001 kg

For measuring **time**

day (symbol d)

hour (symbol h)

minute (symbol min)

1 d = 24 h
1 h = 60 min
1 min = 60 s

For measuring **angle**

degree (symbol °)

180° = about 3.142 radians

For measuring **area**

hectare (symbol ha)

1 ha = 10 000 m^2

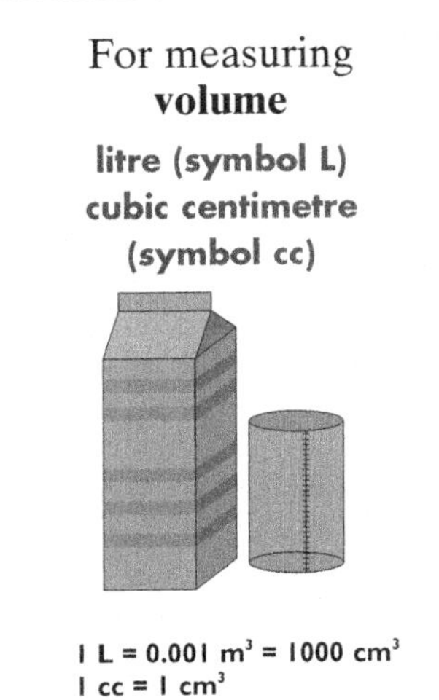

For measuring **volume**

litre (symbol L)
cubic centimetre (symbol cc)

1 L = 0.001 m^3 = 1000 cm^3
1 cc = 1 cm^3

For measuring **temperature**

degree Celsius (symbol °C)

Other units that are sometimes used

For measuring length

mile (symbol **mi**) = about 1609 metres

furlong (symbol **fur**) = about 201 metres

yard (symbol **yd**) = about 0.914 metres

foot (symbol **ft**) = about 30.48 centimetres

inch (symbol **in**) = about 2.54 centimetres

micron (symbol **μ**) = 1 micrometre

Ångstrom (symbol **Å**) = 0.1 nanometres

For measuring mass

hundredweight (symbol **cwt**) = about 50.8 kilograms

pound (symbol **lb**) = about 0.454 kilograms

ounce (symbol **oz**) = about 0.028 kilograms or 28 grams

pound troy (symbol **lb tr**) = about 0.373 kilograms

ounce troy (symbol **oz tr**) = about 0.031 kilograms or 31 grams

pennyweight (symbol **dwt**) = about 0.0155 kilograms or 15.5 grams

For measuring temperature

degree Fahrenheit (symbol **°F**)

To work out the temperature in °F: multiply the temperature in °C by 9, then divide by 5 and finally add 32.

To work out the temperature in °C: take away 32 from the temperature in °F, then multiply by 5 and finally divide by 9.

For measuring area

acre (symbol **ac**) = about 4047 square metres

For measuring pressure

millibar (symbol **mb**) = 100 pascals

millimetre of mercury (symbol **mmHg**) = about 133.32 pascals

atmosphere (symbol **atm**) = about 1013 pascals

For measuring speed

mile per hour (symbol **mph** or **mi/h**)
= about 0.477 metres per second
= about 1.72 kilometres per hour

For measuring energy

calorie (symbol **cal**) = about 4.2 joules

For measuring volume

gallon = about 0.00455 cubic metres or 4.55 litres

US gallon = about 3.79 litres

pint = about 0.57 litres

US pint = about 0.47 litres

fluid ounce = about 0.00284 litres or 28.4 cubic centimetres

Common prefixes for SI and other units

Prefixes are added in front of a unit to change the value. There is no space between the prefix and the unit.

For example:

1 cm = one hundredth of a metre

30 MHz = 30 million hertz

Prefix	Symbol	Meaning
exa	**E**	one million million million
giga	**G**	one thousand million
mega	**M**	one million
kilo	**k**	one thousand
hecto	**h**	one hundred
deka	**da**	ten
deci	**d**	one tenth
centi	**c**	one hundredth
milli	**m**	one thousandth
micro	**μ**	one millionth
nano	**n**	one thousandth millionth
pico	**p**	one millionth millionth
femto	**f**	one thousandth millionth millionth
atto	**a**	one millionth millionth millionth

Maths Symbols

Arithmetic and algebra	
$+$	Add, plus
$-$	Take away, subtract, minus
$\pm$	Plus or minus
$\times$	Times, multiplied
$\div$, $/$	Divided by
$\frac{a}{b}$	Fraction line, divided by
$:$	Ratio, is to
$=$	Is equal to
$\neq$	Is not equal to
$\approx$	Is approximately equal to
$>$	Is greater than
$<$	Is less than
$\geqslant$	Is greater than or equal to
$\leqslant$	Is less than or equal to
$.$	Decimal point
$\sqrt{\ }$	Square root
$\sqrt[3]{\ }$	Cube root
$(\ )$	Parentheses
$[\ ]$	Brackets
$\{\ \}$	Braces
$\dot{}$	Repeating
$\ldots$	Continuing, continues on
∞	Infinity
π	Pi ($\approx$ 3.14)
$\%$	Per cent, percentage

Geometry	
$\longleftrightarrow$	Line
$\bullet\!\!\longrightarrow$	Ray
	Interval
	Lines of equal length
$\parallel$	Parallel
	Parallel lines
$\angle$	Angle
$^\circ$	Degree
$'$	Minute
$''$	Second
	Right angle
$\perp$	Perpendicular
$\triangle$	Triangle
$\square$	Square
$\equiv$	Congruent

Sets	
$\{\ \}$	Braces
ξ	Universal set
$\subseteq$	Subset, is contained in
$\cap$	Intersection
$\in$	Is an element of
$\notin$	Is not an element of

Maths Formulae

Plane shapes	Perimeter (P)	Area (A)
Circle 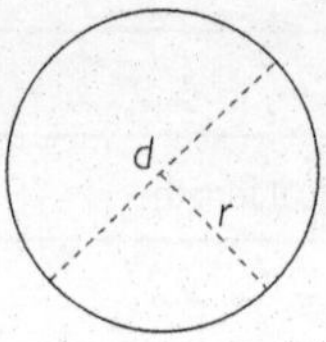	Circumference = pi × diameter Circumference = 2 × pi × radius $C = \pi d$ $C = 2\pi r$	Area = pi × diameter squared ÷ 4 Area = pi × radius squared $A = \frac{\pi d^2}{4}$ $A = \pi r^2$
Square 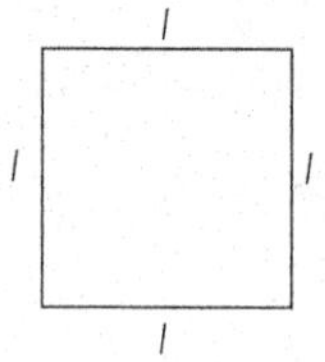	Perimeter = 4 × length $P = 4l$	Area = length squared $A = l^2$
Rectangle 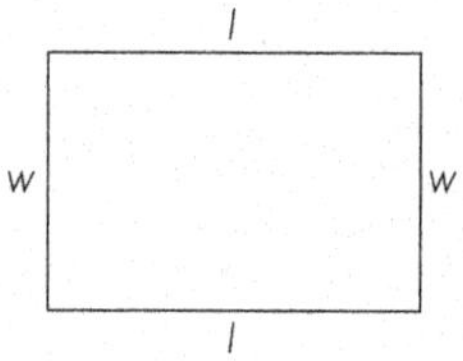	Perimeter = 2 × (length + width) $P = 2(l + w)$	Area = length × width $A = lw$
Parallelogram 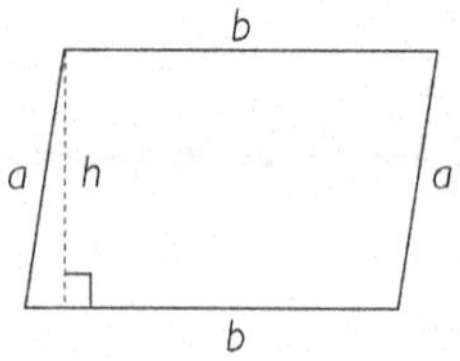	Perimeter = sum of side lengths $P = 2(a + b)$	Area = base length × height $A = bh$
Trapezium 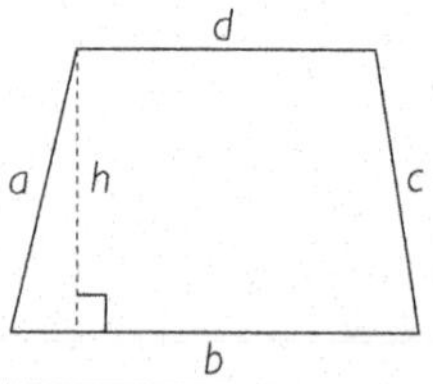	Perimeter = sum of side lengths $P = a + b + c + d$	Area = height × average length of parallel sides $A = \frac{h(b + d)}{2}$
Triangle 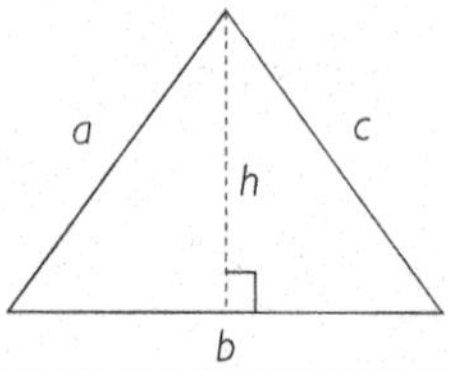	Perimeter = sum of side lengths $P = a + b + c$	Area = $\frac{1}{2}$ base length × height $A = \frac{1}{2}bh$

Solid shapes	Surface area (A)	Volume (V)
Sphere	Surface area = 4 × pi × radius squared $A = 4\pi r^2$	Volume = $\frac{4}{3}$ × pi × radius cubed $V = \frac{4}{3}\pi r^3$
Cube	Surface area = 6 × length squared $A = 6l^2$	Volume = length cubed $V = l^3$
Cuboid	Surface area = sum of areas of faces $A = 2(lw + hl + hw)$	Volume = length × width × height $V = lwh$
Pyramid	Surface area = area of base + sum of areas of triangular faces	Volume = $\frac{1}{3}$ × area of base × height
Right circular cone	Surface area = area of curved surface + area of circular base $A = \pi rh + \pi r^2$	Volume = $\frac{1}{3}$ × area of base × height $V = \frac{1}{3}\pi r^2 h$
Right circular cylinder	Surface area = area of curved surface + sum of areas of circular faces $A = 2\pi rh + 2\pi r^2$ $= 2\pi r(r + h)$	Volume = area of one circular face × height $V = \pi r^2 h$
Right prism	Surface area = 2 × area of base + perimeter of base × height	Volume = area of base × height

A to Z of
Science
&
Technology

Aa

abdomen

The abdomen is the part of the body that contains the organs used for digestion and reproduction.

In humans and other mammals, the abdomen is between the hips and the chest. In many insects, such as ants and grasshoppers, it is at the end of the body.

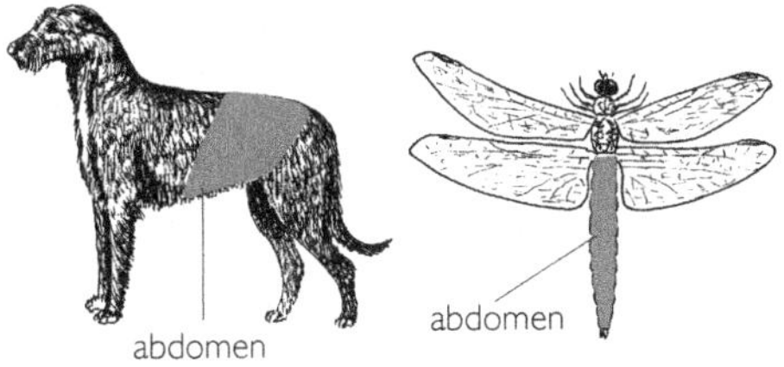

abrasive

An abrasive is a gritty substance used to grind down or smooth out a surface.

Two common abrasives are carborundum and quartz, which are used for grinding and smoothing wood, metal and glass.

absorbent

Absorbent means able to soak up liquids or gases.

Absorbent things include sponges, bath towels, paper towels and the roots of plants.

See also **porous**

AC

AC is short for **alternating current**.

acid

An acid is a substance that turns blue **litmus paper** red.

Acids have a sharp, sometimes burning taste. If you have tasted lemon juice, you will know what an acid is like.

See also **base**, **pH**

acid rain

Some sorts of air pollution can mix with the water in the atmosphere and make it more acid. When this water falls as rain, it is called acid rain.

acoustics

Acoustics means the study of sound. This includes the way that sound behaves in a particular place, such as inside a room or around a loudspeaker.

adaptation

An adaptation is something that an organism has that helps it to survive in its environment.

For example, the thick layer of fat under a seal's skin is an adaptation that helps it to survive in very cold water. An adaptation can also be a sort of behaviour, such as hiding in a burrow during the day.

aerial

An aerial is the part of a radio or television that is used to send or receive radio waves.

aerodynamics

Aerodynamics is the study of how air flows around a moving object, such as a car or an aeroplane.

Smooth, curved surfaces that allow the air to flow easily are called aerodynamic surfaces.

See also **drag**, **lift**, **thrust**

aerosol

An aerosol is tiny particles of a liquid or solid in the air.

When you press the button on a spray can, the fine mist that comes out is an aerosol.

air

Air is the mixture of gases that makes up the Earth's **atmosphere**.

☞

Dry air is about 78% nitrogen and 21% oxygen. There is also about 0.9% argon, 0.03% carbon dioxide and 0.02% neon, plus tiny amounts of other gases.

air pressure

Air pressure is the push of air against its surroundings. It can be caused in three ways.

The first way is by the weight of the air. When you stand outside, there is a huge column of air above you, stretching all the way to the edge of the atmosphere. All this air has a weight, and this weight is pushing down on everything. This sort of air pressure is also called atmospheric pressure.

The second way is by air being trapped inside a container. When you blow up a balloon, the air inside pushes against the inside of the balloon. As you blow air into the balloon, the air pressure inside the balloon becomes greater.

The third way is by air moving. When the wind blows on your face, you feel the air pressure caused by the movement of the air.

albedo

When light shines on an object, some of the light bounces back off the surface. The fraction of light that bounces off the surface is called the albedo.

An albedo of 1.0 means that all the light is reflected; 0.5 means that half the light is reflected; and 0.0 means that no light is reflected.

albumen

Albumen is the clear or white part of an egg.

Albumen surrounds and protects the yellowish **yolk**. The albumen is also called "egg white". It is made of **proteins** that change into a jelly-like substance when they are heated.

algae

Algae are very small to large plant-like organisms that contain **chlorophyll**, but are not true plants. They are classified in the kingdom Protoctista.

Most algae grow in streams, lakes or oceans, or in wet places on land.

See also **blue-green algae**, **lichen**, **seaweed**

alimentary canal

In animals, the alimentary canal is the part of the body where food is digested and absorbed into the body. It is also where wastes are taken way.

In most animals, the alimentary canal is like a long tube with two openings (the mouth and the anus). In humans and other mammals, it is divided into several different parts (see diagram). In insects and other invertebrates, it is much simpler.

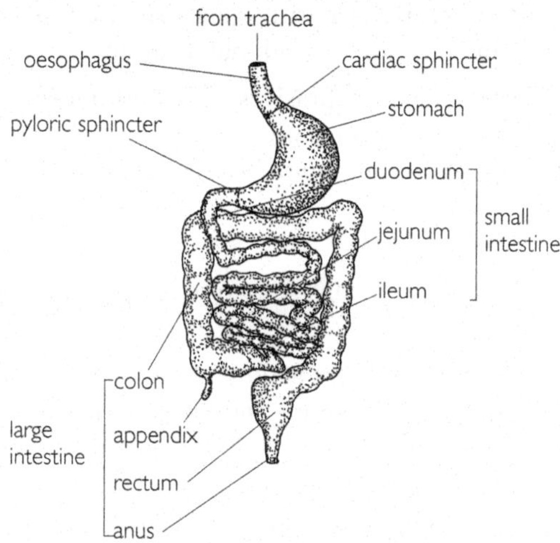

See also **digestion**, **digestive system**

alkali

An alkali is a **base** that can dissolve in water.

Weak alkalis feel soapy, but strong alkalis can cause serious burns. Some common alkalis are toothpaste, soap, baking powder, caustic soda and ammonia cleaners.

When an alkali dissolves in water, **ions** made up of oxygen and hydrogen (called hydroxide ions) are formed.

alloy

An alloy is a substance that is made up of two or more **metals**, or a metal and a non-metal.

Some common alloys are brass (copper and zinc), bronze (copper and tin), pewter (tin and lead) and steel (iron and carbon).

alternating current

Alternating current is an electric **current** that changes direction all the time. It is the sort of current that comes out of the power points in your home. Alternating current is also called AC.

See also **direct current**

altimeter

An altimeter is an instrument that measures height above sea level.

Most altimeters work by measuring the **air pressure**. As you go higher, the air pressure becomes less. Altimeters are used by pilots, skydivers, mountain climbers and other people who have to know exactly how high they are.

altitude

Altitude is sometimes called elevation. It has two meanings.

Meaning 1 Altitude means the height above the ground. It is usually expressed in metres.

Meaning 2 Altitude means the angle between the **horizon** and an object, such as the Sun or the top of a mountain.

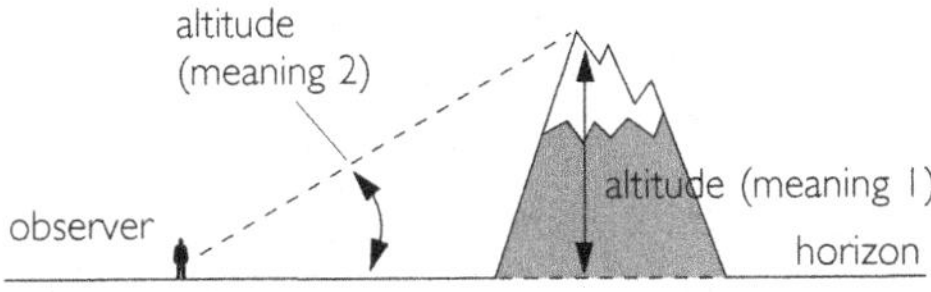

See also **azimuth**

altocumulus

See **cloud**

altostratus

See **cloud**

aluminium

Aluminium is a **metal** that is very light but strong. It is the most common metal on the Earth. It is found mainly in rocks, especially an **ore** called bauxite.

Aluminium can be rolled into thin sheets, and it does not corrode easily. This makes it very useful for making many things, including aircraft, ships, car parts, drink cans, kitchen foil and electrical equipment. The symbol for aluminium is Al.

See also **corrosion**

amphibian

Amphibians are frogs, toads, newts and salamanders.

Amphibians grow from eggs into larvae that have gills for breathing and live in water. The larvae grow into adults that can live in water or on land. The adults have lungs instead of gills, and can also take in air through their skin.

Some examples of amphibians (top to bottom): salamander, toad, newt and frog.

anatomy

Meaning 1 Anatomy is the study of the structure of organisms.

Meaning 2 Anatomy means the actual structure of organisms, especially the internal parts.

anemometer

An anemometer is an instrument for measuring the speed of the wind or flowing water.

The most common sort of wind anemometer has several cups connected to a rotating spindle. Another type has blades connected to a spindle, like a fan. The faster the wind blows, the faster the cups or blades rotate.

animal

An animal is an organism that cannot make its own food, and has a nervous system. Instead, animals feed on other organisms, or the remains of organisms.

All animals can move about at some time in their life, but some stay in one place when they are adults.

See also **kingdom**

annual

Meaning 1 Annual means happening once a year.

Meaning 2 An annual plant is a plant that grows and dies in one year.

See also **perennial**

ant

Ants are small **insects** that live in colonies, usually in the soil. Their bodies are divided into three main parts called the head, thorax and abdomen. The front part of the abdomen is very narrow and forms a petiole or "waist".

Male ants have wings and die soon after mating. Queen ants also have wings for a short time, but the wings eventually fall off. All the other ants are wingless females called workers. Their job is to find food for the colony.

Ants emerging from their underground nest.

antenna

Meaning 1 An antenna is a long, narrow **sense** organ attached to the head of an **arthropod**. It is used for smelling and touching things, and might also be used for sensing vibrations and sound. The plural of antenna is antennae.

Meaning 2 Antenna is another name for an **aerial**.

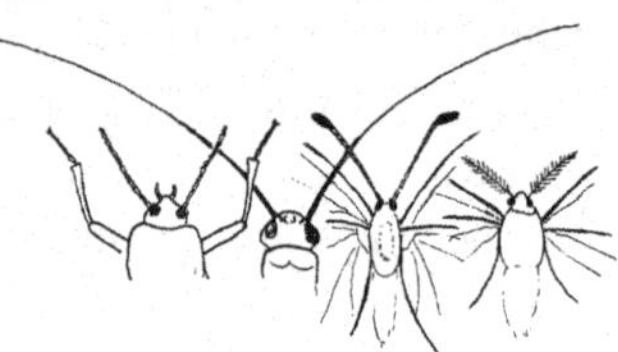

Some insect antennae (left to right): beetle, cricket, butterfly and moth.

antibiotic

Antibiotics are substances that kill **micro-organisms** such as bacteria and fungi.

If you have an infection, the doctor will probably give you some antibiotics. Some common antibiotics are penicillin, streptomycin and tetracyclines.

antiseptic

Antiseptics are substances that kill **micro-organisms** without harming your own body.

Antiseptics are usually liquids or creams that are put on cuts and wounds to stop micro-organisms causing an infection.

See also **disinfectant**

antivenom

An antivenom is a substance that stops a **venom** from killing or injuring an animal.

Antivenoms have been produced for the venoms of many dangerous animals, such as the Diamondback Rattlesnake, Sydney Funnel-web Spider and Box Jellyfish.

aorta

The aorta is the large **artery** that takes blood from the left ventricle of the **heart** and sends it to the rest of the body.

apparatus

Apparatus means all the things you use when you do an experiment or any other scientific or technical project.

appendix

The appendix is a small pocket connected to the **alimentary canal**.

In humans the appendix probably does not have any function. Sometimes the appendix can become infected and swollen. This is called appendicitis.

application

An application is computer **software** that is used for a particular job, such as sending e-mails, connecting to the Internet or scanning photos.

See also **file**

aquatic

Aquatic means living in or happening in water.

See also **arboreal**, **terrestrial**

arachnid

An arachnid is an **arthropod** with a body divided into two parts called the cephalothorax and the abdomen.

Arachnids include spiders, harvestmen, scorpions, ticks and mites. Most arachnids catch and eat insects and other small animals, but some are **parasites**.

arboreal

Arboreal means living in trees. For example, a Koala is an arboreal animal.

See also **terrestrial**

arch

An arch is a curved structure with a support at each end.

Arches are used to support heavy parts of buildings, especially when space is needed underneath for a doorway or entrance. They are also used to support bridges and other long structures.

Bridges often use arches to support a road.

argon

Argon is a gas that makes up about 0.9% of the Earth's atmosphere. It is inert, which means it does not react with chemicals.

Argon is used in light globes (mixed with nitrogen), and in special welding and metal-making methods. The symbol for argon is Ar.

artery

An artery is a blood vessel that carries blood from the **heart** to the different parts of the body.

This blood comes from the lungs, so it is loaded with oxygen. The only artery that does not carry oxygen-loaded blood is the pulmonary artery.

arthropod

An arthropod is an eight-legged **invertebrate** animal with a hard skin called a cuticle.

☞

The body of an arthropod is divided into parts called segments, such as the head, thorax and abdomen. Instead of a **circulatory system**, arthropods have a blood-like liquid called haemocoel that fills the body and surrounds the different organs. Some common arthropods are crabs, spiders, millipedes and insects.

asteroid

Asteroids are large rocks that are orbiting the Sun, between Mars and Jupiter. They are probably the remains of a planet that broke up billions of years ago.

The largest asteroid is Ceres, which is about 1000 kilometres wide. Asteroids are also called minor planets or planetoids.

See also **solar system**

astronaut

An astronaut is a person who travels into space.

Astronaut means star traveller. In Russia and some other countries, an astronaut is called a cosmonaut, which means space traveller. The first astronauts were jet pilots and engineers, but today many different people are astronauts.

astronomical unit

An astronomical unit is the distance from the Sun to the Earth, which is just a little less than 150 million kilometres—149 597 870 kilometres to be exact.

This is a handy unit to use when describing the **solar system** because the distances are so huge. The abbreviation for astronomical unit is AU.

astronomy

Astronomy is the study of everything that is outside the Earth's atmosphere. It includes the study of galaxies, stars, planets, moons and other objects, as well as how objects move in space.

atmosphere

The atmosphere is the layer of air that covers the Earth. It is made up mainly of nitrogen and oxygen, but there are many other gases in it, as well as dust, smoke, water and pollen.

There are different layers in the atmosphere (see page opposite). The layer closest to the Earth's surface is called the troposphere. This is where weather happens, and where most aircraft fly. The top of the troposphere is between 7 and 28 kilometres high.

Above the troposphere is the stratosphere, which goes up to about 50 kilometres high.

The next layer is called the ionosphere. It goes up to about 400 kilometres from the Earth's surface.

Above the ionosphere there is not much atmosphere at all. This region is called the exosphere. It is hard to say where the exosphere ends and space begins.

See also **air**

atmospheric pressure

See **air pressure**

atom

Atoms are the tiny particles that make up all matter. Almost every atom consists of **protons** and **neutrons** surrounded by **electrons**.

The electrons orbit the nucleus very quickly, so they form a sort of cloud around the nucleus. In a normal atom, the number of protons is exactly the same as the number of electrons. **Hydrogen** is the only element without any neutrons in its nucleus.

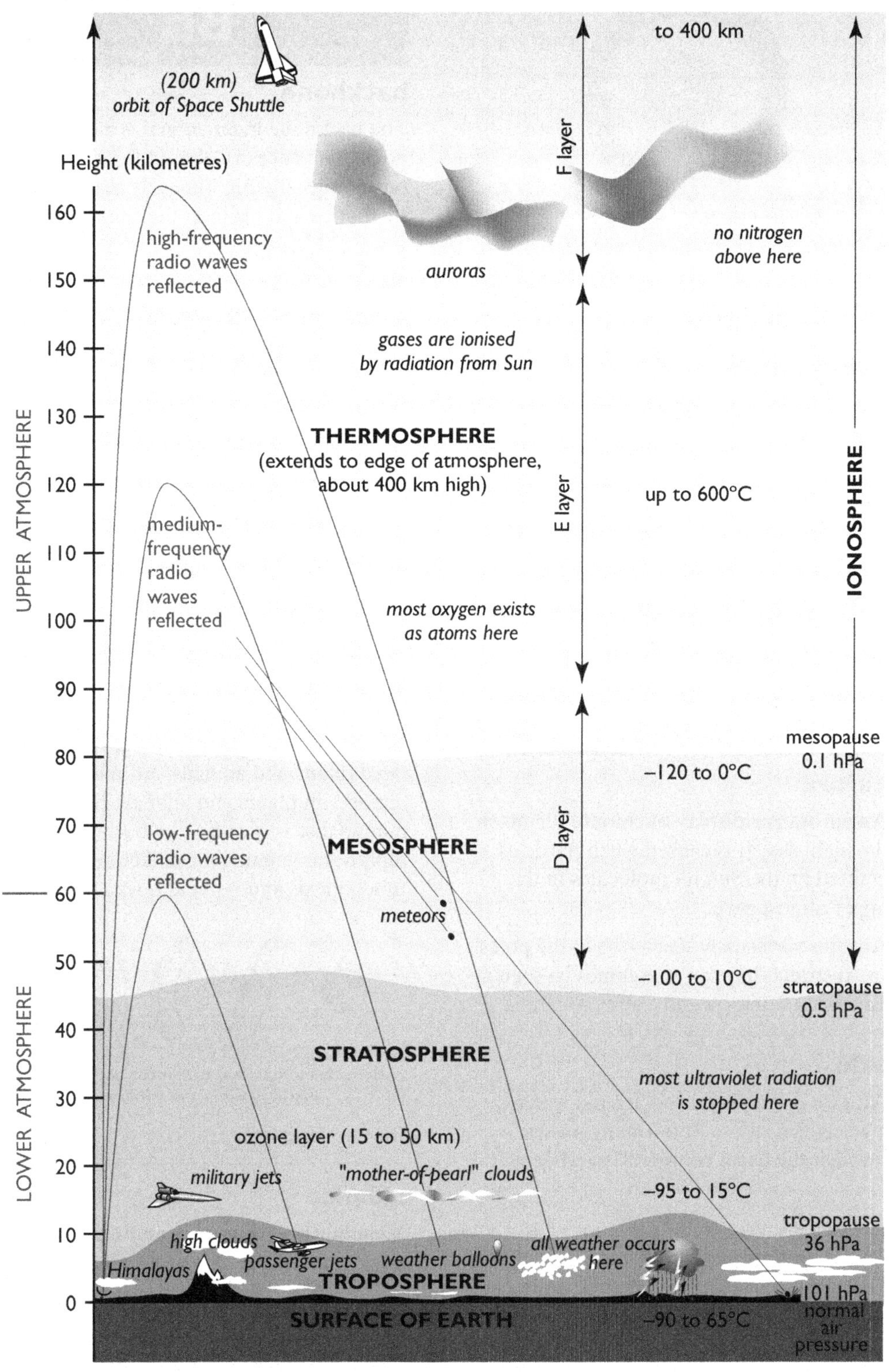

(200 km)
orbit of Space Shuttle
Height (kilometres)
160
150
140
130
120
110
100
90
80
70
60
50
40
30
20
10
0
UPPER ATMOSPHERE
LOWER ATMOSPHERE
high-frequency radio waves reflected
medium-frequency radio waves reflected
low-frequency radio waves reflected
auroras
gases are ionised by radiation from Sun
THERMOSPHERE
(extends to edge of atmosphere, about 400 km high)
most oxygen exists as atoms here
MESOSPHERE
meteors
STRATOSPHERE
ozone layer (15 to 50 km)
military jets
"mother-of-pearl" clouds
high clouds
passenger jets
weather balloons
all weather occurs here
Himalayas
TROPOSPHERE
SURFACE OF EARTH
F layer
E layer
D layer
to 400 km
no nitrogen above here
up to 600°C
−120 to 0°C
−100 to 10°C
most ultraviolet radiation is stopped here
−95 to 15°C
−90 to 65°C
IONOSPHERE
mesopause
0.1 hPa
stratopause
0.5 hPa
tropopause
36 hPa
101 hPa
normal air pressure

☞

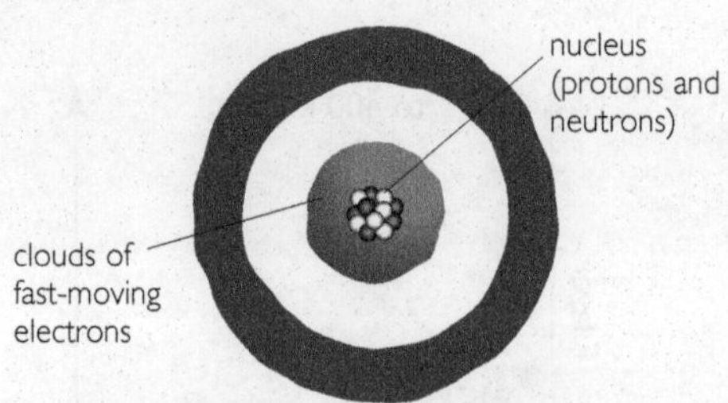

A simple diagram of a carbon atom.

See also **element**, **molecule**, **radioactivity**

atomic energy

Atomic energy is another name for **nuclear energy**.

atrium

An atrium is a chamber inside the **heart**.

attraction

Attraction is when two objects, such as two stars, are pulled towards each other by a force.

Gravity, **magnetism** and **electrical charge** are some of the things that cause attraction between objects.

aurora

An aurora is a display of coloured lights in the night sky. It is caused when particles emitted by the Sun hit molecules in the upper atmosphere.

Auroras are usually seen close to the poles. In Australia they can sometimes be seen in Tasmania, Victoria and South Australia.

axle

An axle is the pin or rod around which a wheel can rotate. It normally passes through the exact centre of the wheel.

azimuth

The azimuth is the angle between north and an object being observed. It is always measured eastward from north.

See also **altitude**

Bb

backbone

The backbone in an animal is a chain of bones that supports the skull, shoulder bones, ribs and hip bones. It also surrounds and protects the spinal cord.

Each bone in the backbone is called a **vertebra**. In some animals, the backbone is made of **cartilage** instead of bone. The backbone is also called the vertebral column or spine.

See also **skeleton**

bacterium

A bacterium is a tiny organism made of only one cell. A bacterium cell does not have a **membrane** around its **nucleus**. The plural of bacterium is bacteria.

Bacteria are usually described according to their shape. Some common shapes are a sphere (coccus), a rod (bacillus), a spiral (spirillum) and a comma (vibrio). Bacteria are very important in recycling dead plants and animals, but many cause diseases in plants and animals. Some human diseases caused by bacteria are diphtheria, meningitis, pneumonia, tuberculosis and typhoid fever.

Some bacterium shapes (left to right): coccus, bacillus, vibrio and spirillum.

See also **kingdom**, **protist**

balance

Meaning 1 A balance is an instrument used to measure mass or weight.

There are many different sorts of balance. Two common ones you might use are the spring balance and the pan balance.

See also **balance** (page 7)

Meaning 2 Balance is a **sense** that tells you the position of your body in relation to your surroundings.

It is controlled mainly by a part of the **ear** called the semicircular canals.

bark

Bark is a layer of dead cells on the outside of the stem of a tree or shrub. It protects the living wood inside.

barometer

A barometer is an instrument used to measure **air pressure**.

There are two common sorts of barometer. The first sort is the mercury barometer, which is a tube filled with **mercury**. The air pressure pushes the mercury up into the tube. The higher the air pressure, the higher the mercury is pushed. The air pressure is read off a scale attached to the tube.

The second sort is called an aneroid barometer. It has a box with all the air removed from inside. The top of the box is connected by a spring to a measuring pointer. When the air pressure changes, the size of the box changes and the pointer changes position. It is not as accurate as a mercury barometer, but it much easier to carry around.

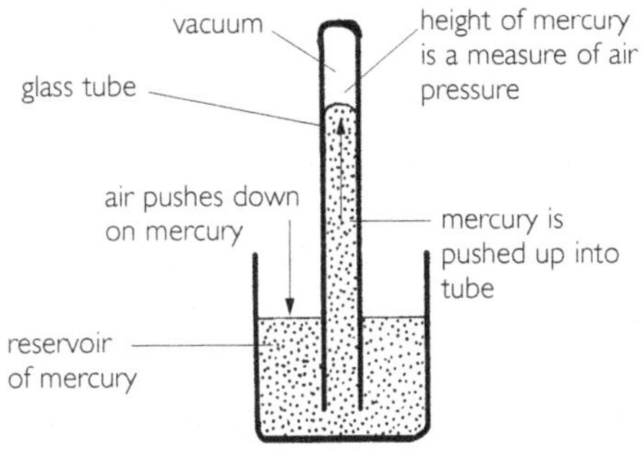

How a mercury barometer works.

basalt

Basalt is the commonest sort of **volcanic rock**. It is formed when lava flows over land and cools slowly. It is dark and very hard, and the crystals of minerals in it are very small.

base

A base is a substance that reacts with an **acid** to produce a salt and water. Bases turn red **litmus paper** blue.

Common bases include **alkalis**.

See also **pH**

battery

A battery is two or more electric **cells** joined together.

An example is a car battery, which has six 2-volt cells, making a total of 12 volts. A battery like an AA battery is really a cell.

Beaufort scale

The Beaufort scale is a scale of numbers used to describe the strength of the wind.

The scale ranges from 0 (dead calm) to 12 (a tropical cyclone). Each number on the scale is equal to a particular range of wind speed. (See the table on the next page.)

beetle

Beetles are **insects** that have a hard covering that protects the abdomen and hind wings. This hard covering is called an elytra and is formed by the front wings.

The Banksia Beetle is one of Australia's largest beetles.

Beaufort number	Description of wind	Wind speed	
		(knots)	(metres per second)
0	Dead calm	Less than 1	0–0.2
1	Light air	1–3	0.3–1.5
2	Light breeze	4–6	1.6–3.3
3	Gentle breeze	7–10	3.4–5.4
4	Moderate breeze	11–16	5.5–7.9
5	Fresh breeze	17–21	8.0–10.7
6	Strong breeze	22–27	10.8–13.8
7	Near gale	28–33	13.9–17.1
8	Gale	34–40	17.2–20.7
9	Strong gale	41–47	20.8–24.4
10	Storm	48–55	24.5–28.4
11	Violent storm	56–63	28.5–32.6
12	Tropical cyclone or hurricane	64 or more	32.7 or more

big bang

Scientists believe that the universe probably began with a huge explosion, which is called the big bang.

The big bang explains why the universe seems to be expanding, like pieces of a firework after it has exploded.

big crunch

The big crunch is one idea about how our universe might end. If it is true, the universe will eventually stop expanding and start becoming smaller again. It will finally collapse into a small space. This final collapse is called the big crunch. This might start a new **big bang**, which would start a new universe.

binary star

A binary star is two stars that are revolving around each other.

Some binary stars are far enough apart to see as two stars, but most are so far away or so close together that they look like one star.

binocular

Binocular means making a separate image for each eye.

Binocular vision makes a three-dimensional image and helps us judge distances better. Binocular instruments include field glasses (also called binoculars) and stereo microscopes.

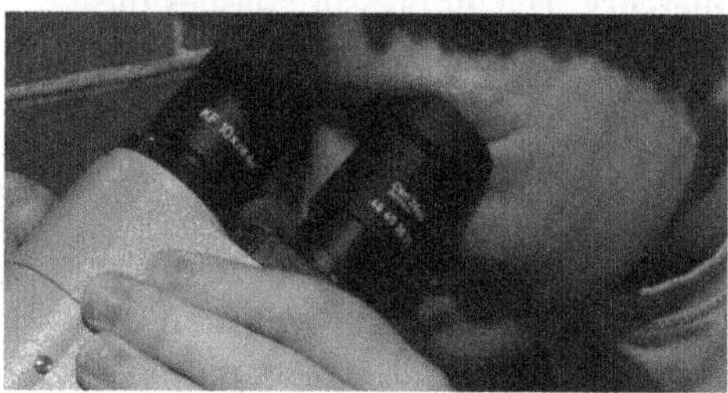

A stereo microscope in use.

biodegradable

Something that is biodegradable can be broken down naturally into harmless substances by bacteria, fungi or other micro-organisms.

biodiversity

Biodiversity means all the different sorts of organisms in an area.

bioengineering

Bioengineering is the use of **technology** in medicine, especially the use of artificial body parts to replace damaged or missing parts.

Some examples are artificial skin, heart valves, pacemakers, **bionic ears**, hip joints and artificial arms and legs.

See also **biotechnology**, **genetic engineering**

biology

Biology is the study of living things. It includes **botany** and **zoology**.

bionic ear

A bionic ear is an electronic device that is put inside the cochlea of the **ear** to help deaf or partly deaf people to hear better. It changes sound into electrical signals and sends the signals to the brain.

biosphere

The biosphere is all the regions of the Earth that are inhabited by living things.

biota

Biota means all the living things in a region.

biotechnology

Biotechnology is the use of organisms to make products or do other things that are useful to humans.

Some examples are the use of bacteria and fungi to make cheeses, the use of fungi to make antibiotics and the use of bacteria to break down **sewage** and oil spills.

See also **bioengineering**, **genetic engineering**

bird

A bird is a **vertebrate** with wings, feathers and a beak.

Birds are **warm-blooded** animals that walk on two legs. Their wings are really modified front limbs.

bit

In computing, a bit is the smallest piece of information that can be stored in a computer system. It is short for **bi**nary dig**it**.

A bit is either the number 0 or the number 1. Larger numbers are made up of more than one bit. Computers process information in groups of bits—usually 16, 32 or 64 bits.

See also **byte**

black hole

A black hole is a huge **star** that has collapsed under its own **gravity**. The gravity is so great that even light cannot escape, so black holes cannot be seen.

blood

Blood is the liquid that carries oxygen and nutrients to the cells of an animal's body and takes away wastes.

Blood carries substances called hormones that control some of the functions of the body. It also contains cells that help defend the body from attack by micro-organisms. In humans, blood consists of three main parts: a clear, watery liquid called **plasma**, red and white blood cells, and platelets. It is pumped through the **circulatory system** by the **heart**.

See also **red blood cell**, **white blood cell**

blue giant

A blue giant is a giant star that is so hot that it shines with a bluish white light.

blue-green algae

Blue-green algae are not algae at all. They are really a sort of bacterium called a cyanobacterium.

Blue-green algae grow in streams and lakes and also on wet soil. If they grow too quickly they can cause problems. This is because when they rot, the water loses oxygen that the plants and animals need. Some blue-green algae also produce poisons.

See also **algae**

boiling point

Boiling point is the temperature at which a liquid boils and turns into a gas. For example, the boiling point of water at sea level is 100 degrees Celsius.

bone

Bone is the hard tissue that makes up the **skeleton** of a vertebrate animal.

There are two sorts of bone: compact bone and spongy bone. Compact bone is very hard and solid, and is found on the outside of a bone. Spongy bone is light and occurs on the inside of a bone.

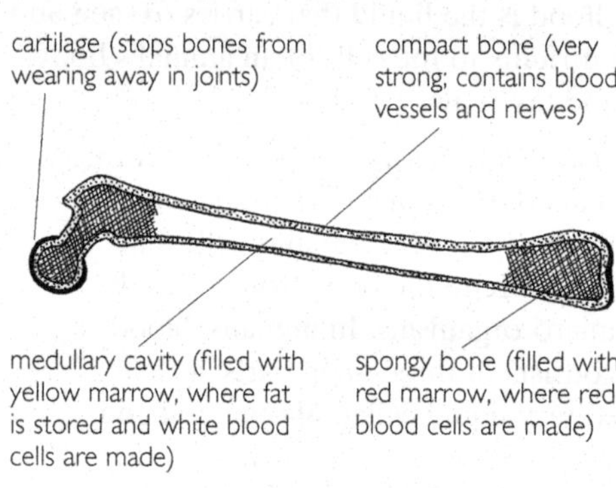

A human femur.

botany

Botany is the study of plants. It includes how they have evolved, how they are classified, their structure, how they function, their role in ecosystems and where they grow.

bowel

Bowel is another name for the **large intestine**.

brain

The brain is the organ that controls most of the functions of the body of an animal. It also controls memory and intelligence.

The brain is connected to the spinal chord. Different parts of the brain control different functions of the body.

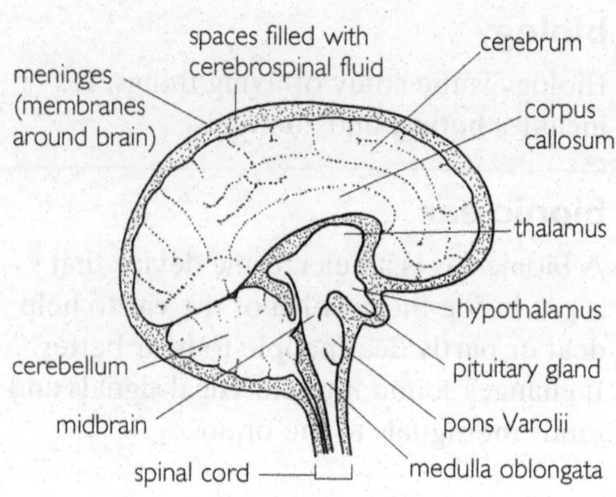

The human brain.

breathing

Breathing is another name for **respiration** (meaning 1).

bronze

Bronze is an **alloy** made from copper and tin. It can also contain small amounts of lead, zinc, phosphorus and other elements.

Bronze is very hard, but it is easily cast into different shapes. It is used to make bearings, valves, shafts and other machine parts that get lots of work. It is also used to make sculptures.

brown dwarf

A brown dwarf is an object in space that is bigger than a planet but smaller than a star. A brown dwarf can give out a small amount of light, which is produced by the huge pressure of its gravity.

browser

A browser is computer **software** that can be used to look at websites on the **Internet**.

bug

Meaning 1 In biology, a bug is an **invertebrate** animal that has a flattened body and a mouthpart that can suck out liquids from plants and animals. There are four wings that fold back over the abdomen when the bug is resting.

Bugs include cicadas, aphids, scale insects, whiteflies, leafhoppers and backswimmers.

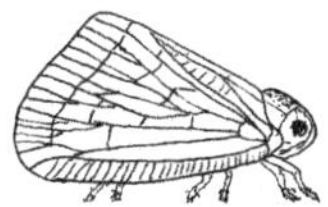

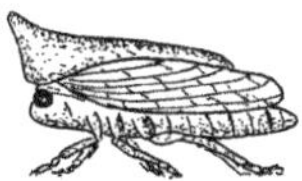

A Passion-vine Hopper (left) and a Green Tree-hopper (right).

Meaning 2 In computing, a bug is a mistake in a **program** that stops the program from working properly.

buoyancy

Buoyancy is an upward force on an object that is in water.

If the buoyancy is greater than the **weight** of the object, the object will float. If the buoyancy is less than the object's weight, the object will sink.

butterfly

Butterflies are **insects** with four large wings covered in tiny scales. They also have a long, thin tube instead of a mouth. This is used to suck up flower nectar or other liquids.

Butterflies are very similar to moths. A butterfly usually rests with its wings folded up, but a moth usually rests with its wings flat or folded over its body. Also, butterflies have tiny "clubs" on the ends of their antennae, but moths do not.

The Cabbage White butterfly.

byte

In computing, a byte is a piece of information made up of eight **bits**. It is the smallest piece of information that can be processed by a computer.

Cc

cactus

A cactus is a fleshy plant that grows in deserts and other dry places. The plural of cactus is cacti.

Many cacti have a waxy surface and are covered in spines, which are really the leaves. These features help them survive without much water.

A typical cactus.

Cainozoic era

See **geological time**

calcium

Calcium is a metal-like **element** that occurs naturally in the Earth's crust.

Calcium is in many sorts of rock, such as **limestone**. Plants get calcium from the soil and use it to build cell walls. Animals get calcium from eating plants, or from eating other animals. They use it mainly to build shells or bones. The symbol for calcium is Ca.

caldera

A caldera is a crater at the top or side of a **volcano**.

A caldera can be formed when the volcano explodes, or when the top collapses after an eruption. Calderas are often filled with water, making a crater lake.

Cambrian period

See **geological time**

camera

A camera is any device that forms an image using a **lens**.

In a film camera, the image falls on a piece of photographic film, which then has to be processed with special chemicals before the image can be seen. In a digital camera, the image falls on a special plate that converts the light into electrical pulses. These are then stored as an electronic **file** that can be viewed on a computer.

camouflage

Camouflage is a feature of an organism that helps it to hide from predators.

Camouflage is often colours that are the same as the surroundings, or a shape that looks like something else.

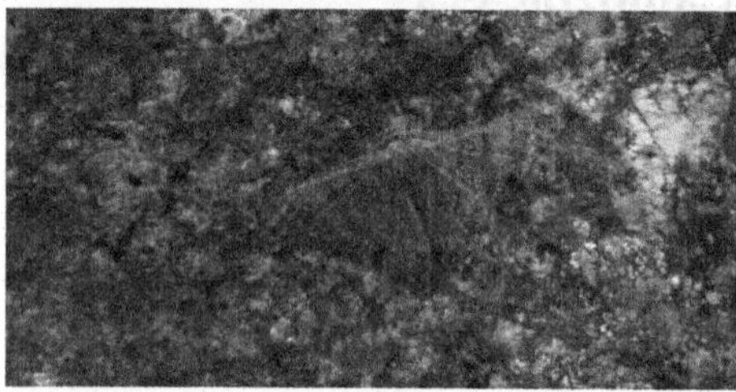

A moth camouflaged against lichens on a rock.

carbohydrate

A carbohydrate is a substance made up of carbon, hydrogen and oxygen atoms.

Some common carbohydrates are sugar, starch and cellulose. Carbohydrates in food are important energy sources for plants and animals.

See also **hydrocarbon**

carbon

Carbon is the **element** on which all life on Earth is based.

Carbon occurs in many different forms, including diamond, graphite and charcoal. The symbol for carbon is C.

carbon cycle

The carbon cycle is the continuous movement of carbon through the environment.

In the main part of the cycle, carbon dioxide is taken from the atmosphere by plants during **photosynthesis**. When animals eat the plants, they absorb the carbon. Some of it is used to build new tissues, and some is breathed out into the environment. When plants and animals die, the carbon in them also returns to the atmosphere or into the soil. Burning also returns carbon to the atmosphere and the soil.

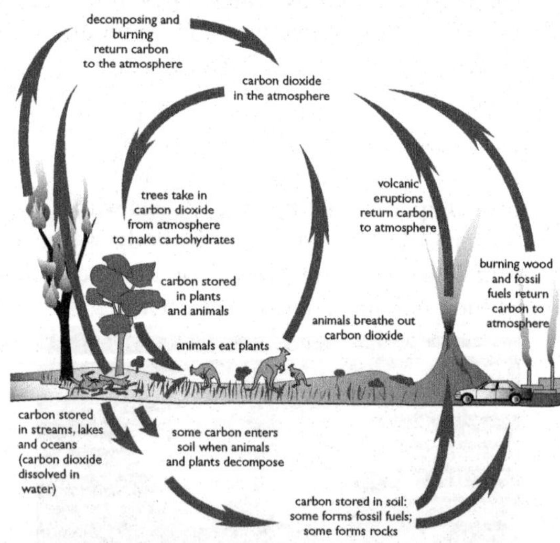

A simple diagram of the carbon cycle.

carbon dioxide

Carbon dioxide is a colourless gas that makes up a small part of the Earth's atmosphere.

Carbon dioxide is essential for life on Earth because plants use it to build **carbohydrates** and other important substances. The formula for carbon dioxide is CO_2.

Carboniferous period

See **geological time**

carnivore

A carnivore is a **vertebrate** animal that feeds mainly on other animals.

Carnivores have sharp teeth for tearing apart flesh, and strong jaws for chewing. Some examples are seals, eagles, bears, wolves, cats and dogs.

carnivorous plant

A carnivorous plant is a plant that catches and feeds on insects or other small animals.

The animal is dissolved by special chemicals made by the plant, and the liquid that this makes is used as food by the plant.

Pitcher plants contain a liquid that dissolves trapped insects.

cartilage

Cartilage is a strong, bendy tissue that forms the skeleton in fishes, sharks and rays.

Cartilage is also found in many other vertebrate animals, especially on the ends of bones in joints. This helps to prevent the bones from wearing, and makes the joints work smoothly.

caterpillar

A caterpillar is the **larva** of a butterfly or a moth.

Caterpillars have soft bodies, with jointed legs on the first three segments and false legs on some of the other segments. Their mouths have a pair of jaws for eating leaves and other plant parts. Moth caterpillars spin a cocoon of silk, in which they turn into a **pupa**. Butterfly caterpillars do not spin a cocoon.

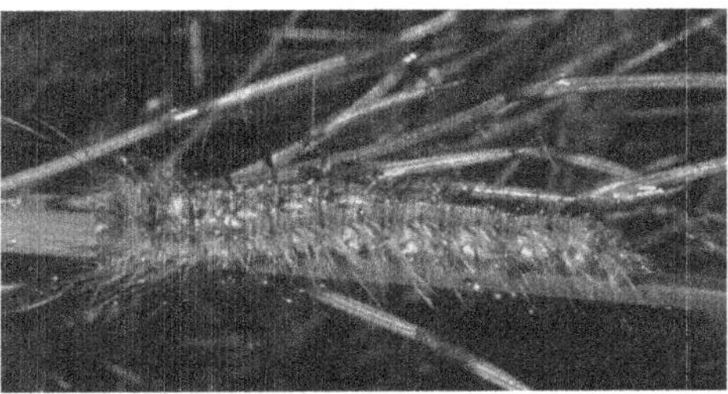

A moth caterpillar.

CD

CD is short for **compact disc**.

A CD-ROM is a CD containing pre-recorded information (ROM stands for read only memory). A CD-R is a CD that is not pre-recorded so that information can be recorded on it (R stands for recordable).

celestial sphere

A celestial sphere is an imaginary sphere on which all the stars, planets, Sun and Moon seem to lie.

The celestial sphere is like the inside of a huge ball, with the Earth at the centre. Before astronomers discovered that the Earth is moving, they thought that this celestial sphere moved instead, turning once every day. The planets, Sun and Moon were thought to move about on this sphere, on a circle called the ecliptic.

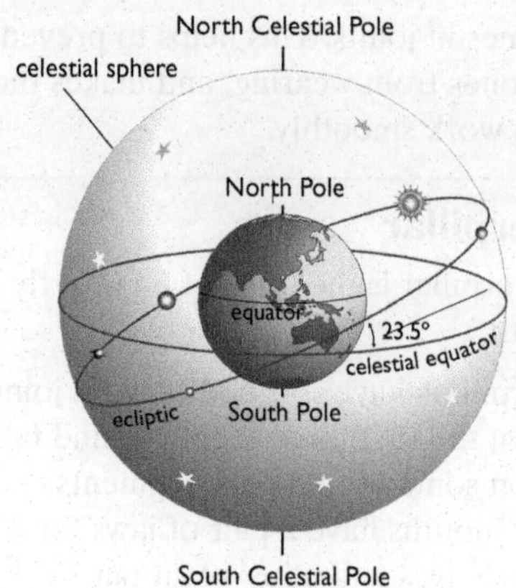

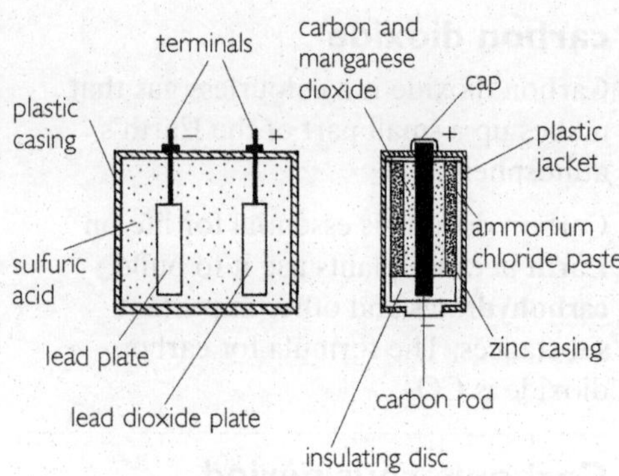

A wet cell (left) and a dry cell (right).

cell

Meaning 1 In biology, cells are the basic units that make up living things.

There are many different sorts of cells. Plant cells have solid walls, and there are green substances called chloroplasts inside the cell. These are involved in **photosynthesis**. Animal cells do not have solid walls, and there are no chloroplasts.

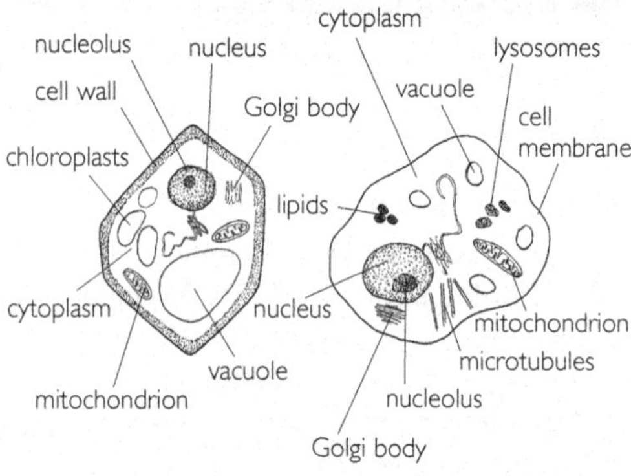

A plant cell (left) and an animal cell (right).

Meaning 2 In chemistry, a cell is a device consisting of two metal or carbon plates or rods (called electrodes) that are set in a liquid or paste (the electrolyte). This arrangement can produce **electricity**.

A cell with a liquid electrolyte is called a "wet cell". An example is a cell in a car battery. A cell with a paste electrolyte is called a "dry cell". An example is an AA battery for a toy.

See also **solar cell**

Celsius scale

The Celsius scale is used in science to measure temperature.

On this scale, ice melts at 0 degrees and water boils at 100 degrees.

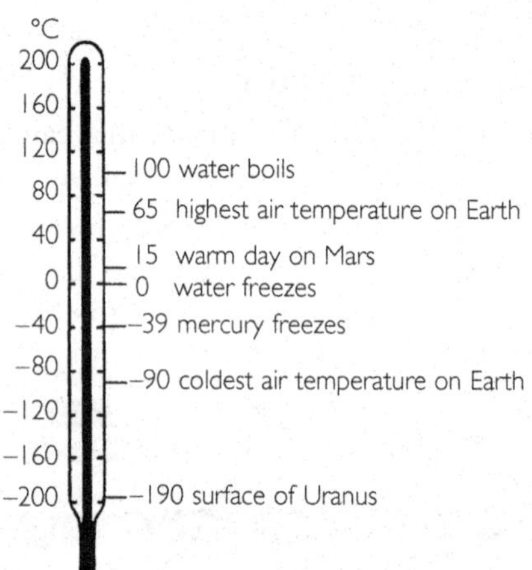

See also **Fahrenheit scale, Kelvin scale**

central processing unit

The central processing unit (also called the CPU) is the **hardware** in a computer that controls how the computer processes information.

The CPU takes data from the memory or from an input device, performs functions on the data and then sends it back to the memory or to an output device.

In modern computers, the CPU is on a single **microchip**.

chemical

A chemical is any substance that is made up of only one sort of molecule. Chemicals can be **elements** or **compounds**.

chemical change

A chemical change is another name for a **chemical reaction**.

chemical reaction

A chemical reaction is the change that happens when chemicals that are mixed together make new chemicals.

See also **mixture**

chemistry

Chemistry is the study of **elements** and all the substances that are formed from them.

chip

A chip is an **electronic circuit** that is made out of a piece of **semiconductor**, such as silicon.

chlorine

Chlorine is an **element** that is present in many natural substances, including common salt.

Pure chlorine is a very poisonous yellow-green gas. The symbol for chlorine is Cl.

chlorophyll

Chlorophyll is the substance that gives the green colour to plants. It is also found in algae.

Chlorophyll is used by plants and algae to trap energy from sunlight to use in **photosynthesis**.

chromosome

Chromosomes are the parts of a cell that contain **genes**. They are thread-like structures inside the cell nucleus. All the members of one species have the same number of chromosomes in their cells. For example, humans have 46 chromosomes.

See also **DNA**, **heredity**

cicada

A cicada is a **bug** that can produce loud drumming or buzzing sounds from a special part of the body.

The cicada **nymph** lives in the soil and sucks sap from plant roots. When it emerges, it climbs into the plant and sucks the sap from the soft parts.

The Greengrocer is a common cicada.

circuit

See **electric circuit**, **electronic circuit**

circuit diagram

A circuit diagram is a drawing that shows the different parts of an **electric circuit**, and how they are joined up.

A circuit diagram is like a map of the circuit. Each device has a special symbol that is used in the diagram.

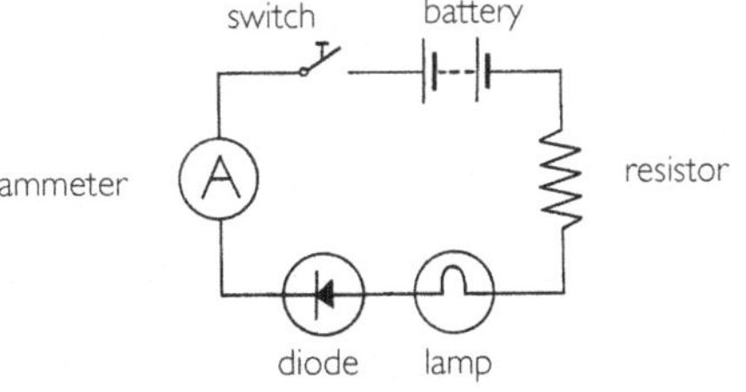

This is a circuit diagram for the circuit on page 159.

circulatory system

The circulatory system is the part of an animal that moves blood around the body.

In humans and other mammals, this is made up of the heart, lungs, arteries, veins and capillaries. The heart pumps oxygen-rich blood into the arteries. The arteries take the blood to the capillaries, where the oxygen is exchanged for carbon dioxide. From the capillaries the blood travels through the veins to the heart, which pumps it to the lungs, where it loses carbon dioxide and picks up oxygen again. Then it flows back to the heart to be pumped around the body again.

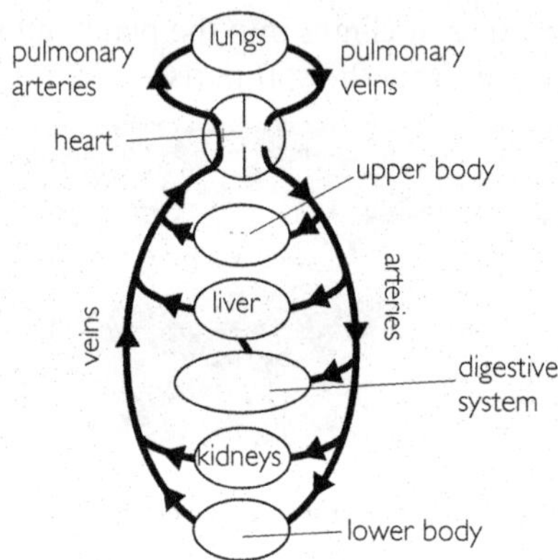

The human circulatory system.

cirrocumulus

See **cloud**

cirrus

See **cloud**

classification of living things

The classification of living things is the way that we group similar organisms so that we can understand how they are related and how they evolved.

All living things are divided into five large groups called kingdoms. Each kingdom is divided into phyla (the singular of phyla is phylum). Then each phylum is divided into smaller and smaller groups called orders, classes, families, genera and species. The modern classification of living things is very different from the one scientists used 20 years ago because we have learnt a lot about the relationships between different organisms since then. For example, we now know that algae are not really plants and that slime moulds are more closely related to algae than to fungi.

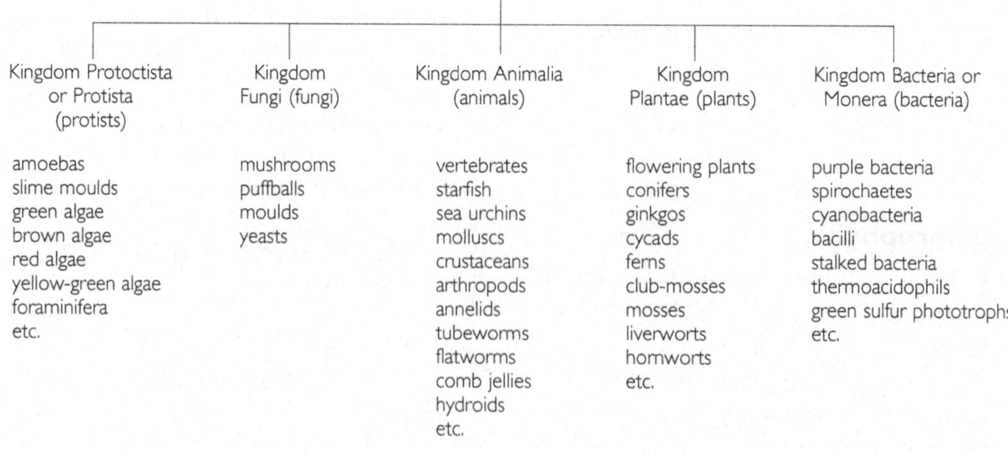

clay

Clay is a sort of soil that is made up of very small particles that swell up when they are wet.

Clay is usually soft and can be easily rolled into a ball when it is wet. When it dries, it shrinks and cracks.

See also **humus**, **loam**

climate

Climate is the sort of **weather** that occurs in an area over a very long time.

The climate of an area is affected especially by the **latitude**, the closeness of oceans and mountains, and the directions of winds at different times of the year.

cloning

Cloning is when a new organism is produced from only one parent. Because there is only one parent, the **genes** of the parent and the offspring are exactly the same.

Cloning can happen naturally in many plants, algae and other organisms. An example is a moss plant that grows from a piece broken off another moss plant. In most animals, cloning does not happen naturally, but it can be made to happen using **genetic engineering**.

cloud

A cloud is a huge number of tiny water droplets or ice crystals that are light enough to drift in the air (see diagram on page 148).

The droplets or crystals form when water vapour in the air condenses onto tiny particles, such as smoke, dust and salt. Clouds are classified according to their height above the ground and what they look like. They can be used to predict the weather, especially rain and winds.

See also **condensation**

coal

Coal is the rock-like remains of plants that have been buried and slowly changed over millions of years.

When the plants died, they formed layers that decayed very slowly. Then they were buried deep under ground by sediments. The great pressure under the ground turned the layers into coal. Coal is used in many countries as a fuel, especially for heating and generating electricity. The best sort of coal is black coal, or anthracite.

cocoon

A cocoon is a protective covering made by the larvae of some insects before **metamorphosis**.

A cocoon is made using threads of silk and often used natural materials, such as leaves or twigs. Some other invertebrates, such as spiders and earthworms, also make cocoons for their eggs.

A moth cocoon. The adult moth has already emerged through the hole in the end.

cold-blooded

Cold-blooded refers to an animal that cannot keep the temperature inside its body steady.

This means that the body temperature tends to go up or down, depending on whether the surroundings are cold or hot. Some examples of cold-blooded animals are snakes, lizards, frogs and fish.

See also **warm-blooded**

colony

Meaning 1 A colony is a group of animals living together so that they depend on each other for survival.

☞

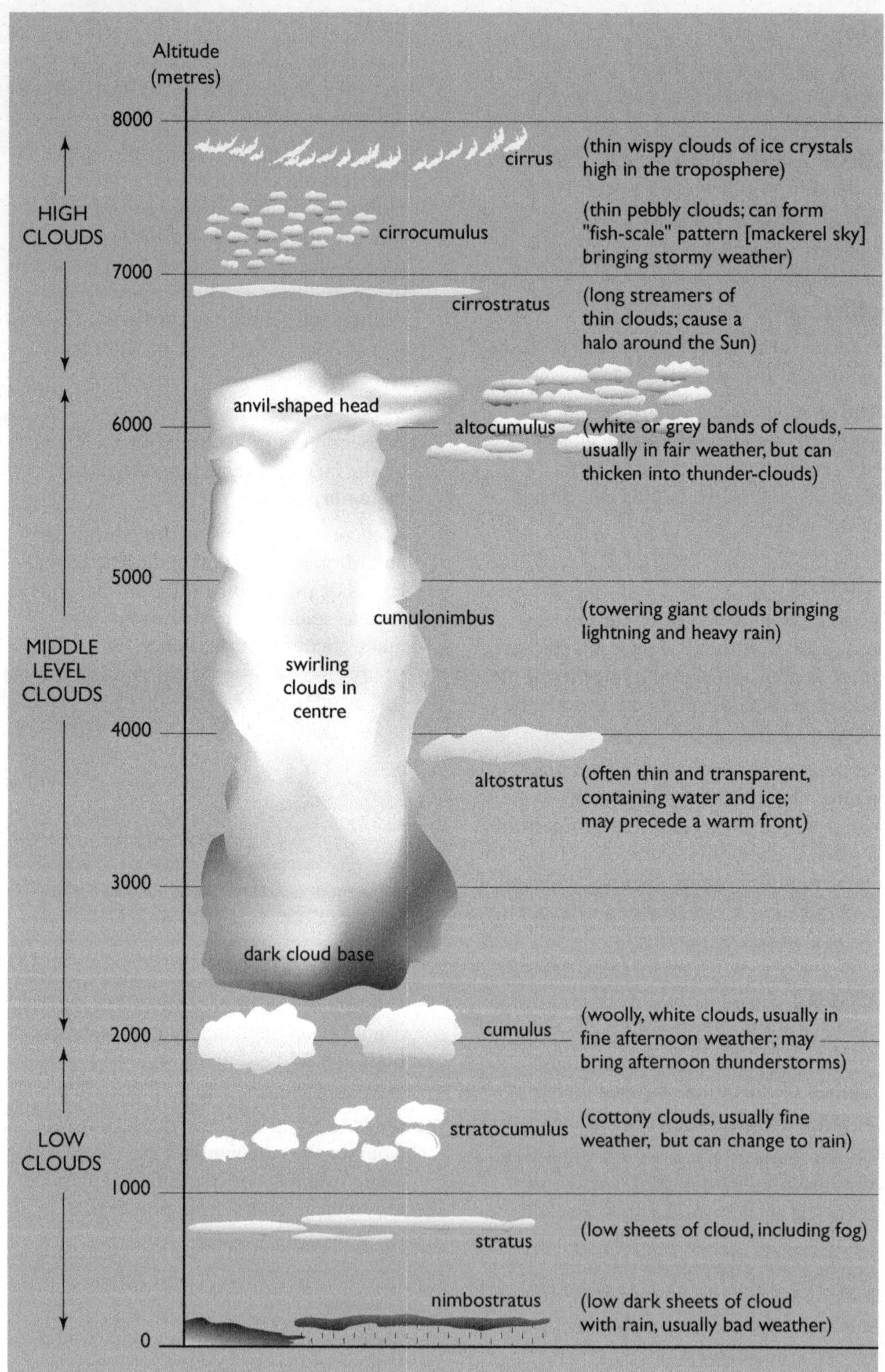
Altitude
(metres)
8000
7000
6000
5000
4000
3000
2000
1000
0
HIGH
CLOUDS
MIDDLE
LEVEL
CLOUDS
LOW
CLOUDS
cirrus
(thin wispy clouds of ice crystals high in the troposphere)
cirrocumulus
(thin pebbly clouds; can form "fish-scale" pattern [mackerel sky] bringing stormy weather)
cirrostratus
(long streamers of thin clouds; cause a halo around the Sun)
anvil-shaped head
altocumulus
(white or grey bands of clouds, usually in fair weather, but can thicken into thunder-clouds)
cumulonimbus
(towering giant clouds bringing lightning and heavy rain)
swirling
clouds in
centre
altostratus
(often thin and transparent, containing water and ice; may precede a warm front)
dark cloud base
cumulus
(woolly, white clouds, usually in fine afternoon weather; may bring afternoon thunderstorms)
stratocumulus
(cottony clouds, usually fine weather, but can change to rain)
stratus
(low sheets of cloud, including fog)
nimbostratus
(low dark sheets of cloud with rain, usually bad weather)

Some examples are ants, hive bees, termites and corals.

Meaning 2 A colony is a large number of micro-organisms, such as bacteria, that have all been produced from a single micro-organism.

colour

Colour is how our eyes see the different parts (wavelengths) of light.

When light hits an object, some of the wavelengths may be absorbed. The colour of the object is the mixture of wavelengths of light that are *not* absorbed. For example, an orange object reflects red and yellow light (red + yellow = orange) but absorbs all the other colours.

See also **spectrum**

combustion

Combustion is a **chemical reaction** in which a substance such as carbon reacts very quickly with oxygen to produce heat energy and light. This is what happens when wood burns.

comet

A comet is a piece of ice, rock and dust that orbits the Sun.

The shape of a comet's orbit is long and narrow, so the comet is sometimes close to the Sun and sometimes far away. As it gets near the Sun, it warms up so that the ice melts. This produces a long trail pointing way from the Sun. This trail is what we see when we see a comet in the sky.

Halley's Comet.

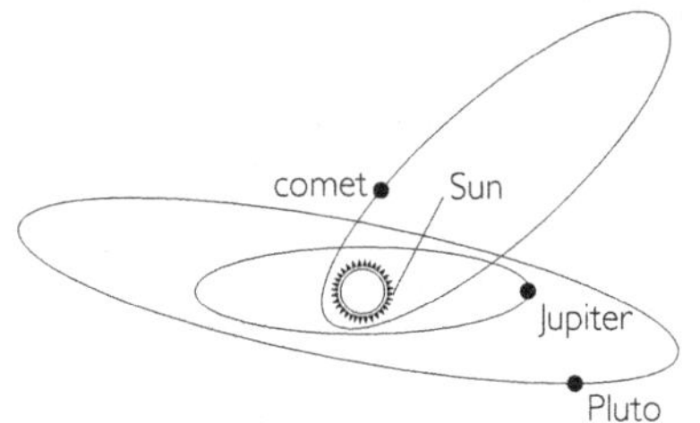

The typical orbit of a comet.

community

A community is a group of plants, animals and other organisms that live in the same place.

A community is named after its main plants (such as a grassland community) or after the main physical feature (such as a sand dune community).

compact disc

A compact disc (also called a CD) is a plastic disc on which information can be stored in **digital** form and read by a **laser** system.

The information on a CD can include music, still pictures, video pictures and text.

compass

A compass is an instrument that shows the direction of the north magnetic **pole**. It is usually a small piece of magnetic metal (called the compass needle) that can spin on a sharp point.

The needle is usually inside a case that is filled with liquid to stop it from spinning too much. Compasses are used to work out the direction you are travelling in, or the direction you need to go to get somewhere.

A compass designed for bushwalkers.

C

compound

A compound is a **chemical** whose molecules are made up of two or more **elements** joined together.

Each molecule of a compound has exactly the same number of atoms of each element. For example, water is a compound in which every molecule has two hydrogen atoms and one oxygen atom. Its formula is H_2O.

compression

Compression is a force that causes something to be squashed or made smaller. It is the opposite of **tension**.

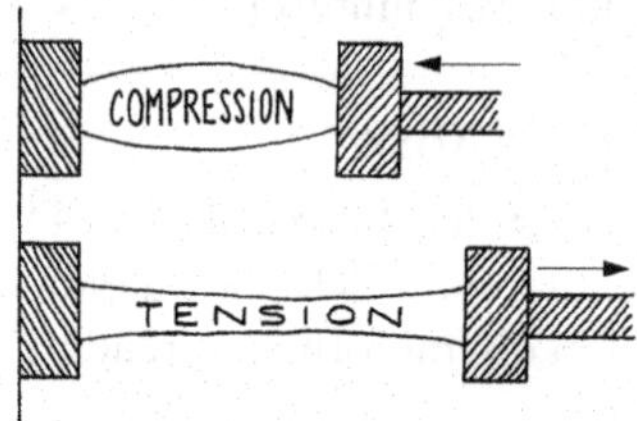

computer

A computer is a machine designed to process information using a set of instructions called a **program**. A computer has three main parts.

The first part is the **central processing unit**, which controls how the computer processes information.

The second part is the memory, which is used to store **software** and data.

The third part consists of the input and output devices, which are used to put information onto the computer and get information out. They include the keyboard, disk drives, modem and monitor. Computers can also have **peripheral devices** added for input and output.

condensation

Condensation is the change of a gas into a liquid, such as **water vapour** into water.

Water vapour in the air condenses when it becomes cold. This happens if the air is lifted up over a mountain range or if it blows over a cold surface, such as cold ground or a cold window.

See also **cloud, dew**

conductor

A conductor is something that allows energy to move easily from one place to another.

For example, metals are good conductors of heat energy and electricity.

See also **insulator**

conifer

A conifer is a sort of **evergreen** plant that has thin, needle-like leaves and seeds in woody cones.

Most conifers are native to the northern hemisphere. They evolved before flowering plants.

Many conifers have a triangular shape.

conservation of energy

The conservation of energy is a rule that says that energy cannot be created or destroyed. It can only be changed from one form to another. Scientists think this rule is true everywhere in the universe.

consumer

In ecology, a consumer is an organism that cannot make its own food and so feeds on other organisms.

Most consumers are animals, fungi and bacteria, but some are plants. Primary consumers are organisms that feed on **producers**. Secondary consumers are organisms that eat primary consumers.

See also **food chain**, **food web**

constellation

A constellation is a group of stars that forms a shape in the night sky.

Different people have different ways of dividing the sky. For example, the constellations recognised by Chinese people are different from the constellations recognised by Native Americans or Aboriginal Australians. Astronomers divide the sky into 88 different constellations.

continental drift

Continental drift is the idea that the Earth's continents were once part of a huge land mass called Pangaea, which began to break up hundreds of millions of years ago.

Pangaea broke up first into two supercontinents called Laurasia and Gondwana. Laurasia became the continents of the northern hemisphere, but Gondwana broke up into Africa, India, Australia, South America and Antarctica.

200 million years ago: the supercontinent of Pangaea

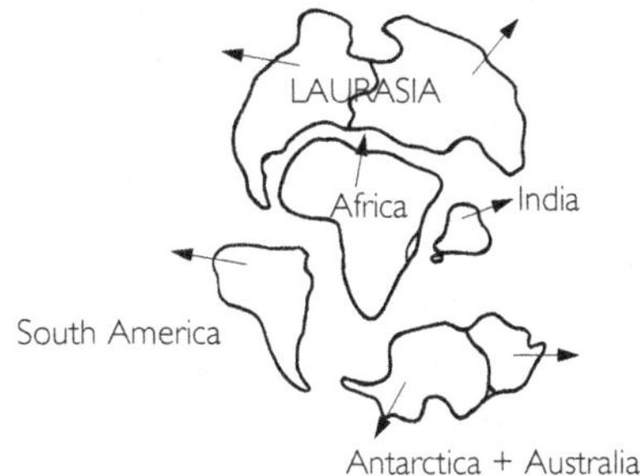

100 million years ago: Laurasia has separated from Gondwana, and Gondwana has begun to break up into the modern-day southern continents

The breakup of Pangaea.

See also **tectonic plates**

convection

Convection is the movement of heat energy in a fluid that is caused by the movement of the fluid itself.

Convection happens because a hot fluid is less dense than a cold fluid, so it tends to rise. The movement of a fluid because of convection is called a convection current.

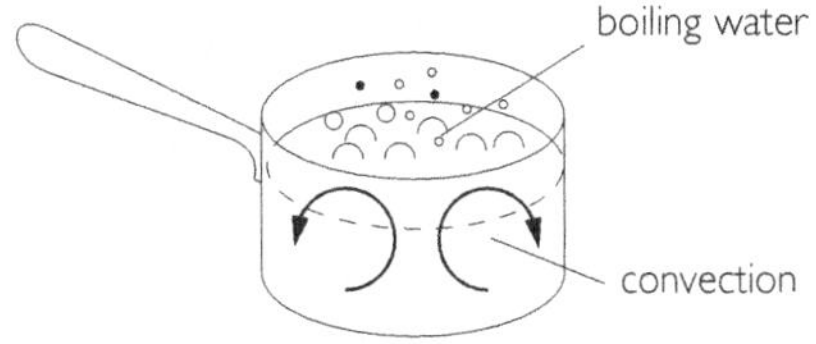

copper

Copper is a reddish-brown metallic **element** that is found as a pure metal in the Earth's crust, as well as in copper **ores**.

Copper is not affected by water, so it is used for water pipes and in ships. It is also a good conductor of electricity and heat energy, so it is used in electrical parts, wiring and cooking pots and pans. The symbol for copper is Cu.

coral

Corals are **invertebrate** animals that live in **colonies** in the warm, shallow parts of the oceans. Each coral colony consists of thousands of coral animals that produce a skeleton in which they live. This skeleton is usually hard and rock-like. Corals feed on **plankton** that drifts around the colony.

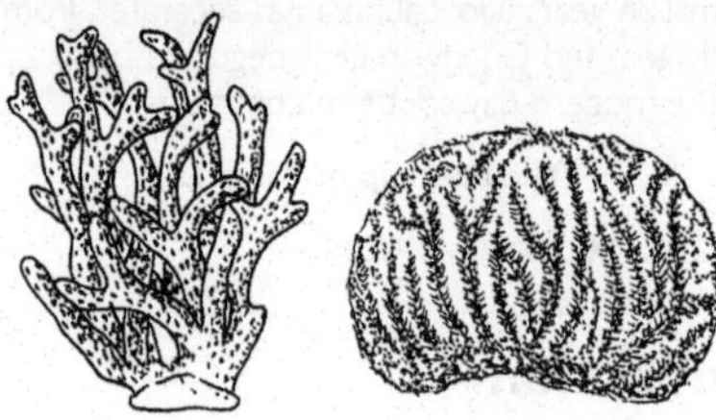

A staghorn coral (left) and brain coral (right).

coral reef

A coral reef is a large group of living and dead **coral** colonies growing in shallow water.

There are three main sorts of coral reefs: fringing reefs, barrier reefs and coral atolls. Fringing reefs are right on the shore. Barrier reefs are further from the land, with a wide lagoon or channel between the reef and the shore. Coral atolls are low, horseshoe-shaped islands with a lagoon in the middle.

cornea

The cornea is a thin, clear layer of tissue covering the front of the **eye** in vertebrate animals. It helps to focus light on the retina.

corona

The corona is the hot outer layer of the Sun and is made up of very hot gas.

There are two layers in the corona. The lower layer is called the K corona. The higher layer, which reaches well out into space and is much cooler, is called the F corona.

corrosion

Corrosion is a **chemical reaction** between a metal and oxygen, or a chemical containing oxygen. This uses up the metal and produces a chemical called a metal oxide, which can form a powdery coating on the surface of the metal.

Metals such as iron can corrode very easily, even in air, so they must be protected with special coatings. Some metals, such as gold and silver, hardly corrode at all.

See also **rust**

CPU

CPU is short for **central processing unit**.

crab

Crabs are **crustaceans** with a shell-like covering (carapace), eyes on stalks and ten legs. The front pair of legs carry claws on the ends for holding prey. The hermit crabs do not have a carapace. Instead, they fit themselves into empty shells for protection.

The Two-spotted Crab.

crater

A crater is any bowl-shaped hole in the Earth's crust caused by the eruption of a **volcano** or by a **meteorite**.

There are also craters on other planets. Volcanic craters are usually on hills or mountains, and are usually not more than a few kilometres wide. A meteorite crater is usually a flat circle surrounded by hills, and can be hundreds of kilometres wide.

Blue Lake, Mount Gambier, is a water-filled volcanic crater.

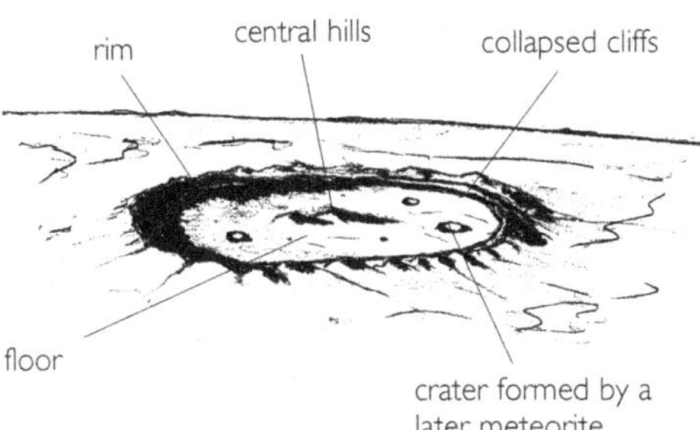

Features of a meteorite crater.

Cretaceous period

See **geological time**

crust

The crust is the outer layer of the Earth. It is made up of solid rock, with a thin layer of rock fragments and soil on top.

See also **mantle**, **regolith**

crustacean

Crustaceans are **invertebrate** animals that include barnacles, crabs, lobsters, prawns and woodlice.

Most crustaceans live in salt water. They have gills for breathing, two pairs of antennae and a pair of eyes. Most crustaceans have a hard outer covering that protects the soft tissues.

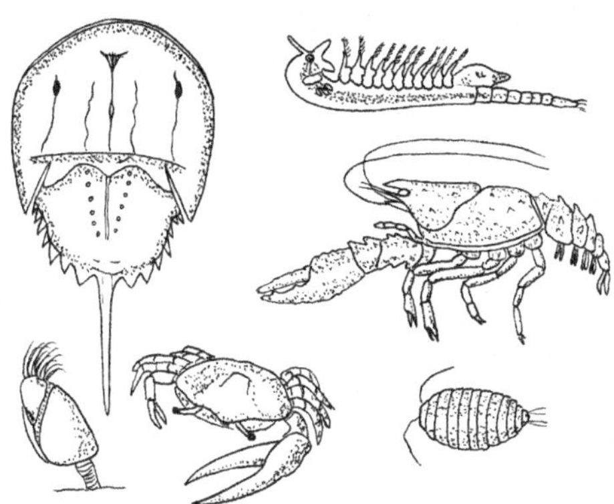

Some examples of crustaceans (clockwise, from top left): horseshoe crab, brine shrimp, lobster, isopod, crab and goose barnacle.

crystal

Meaning 1 In geology, a crystal is a solid piece of rock with a regular, many-sided shape, and made of one substance such as quartz.

The atoms or molecules in a crystal are arranged in a regular pattern.

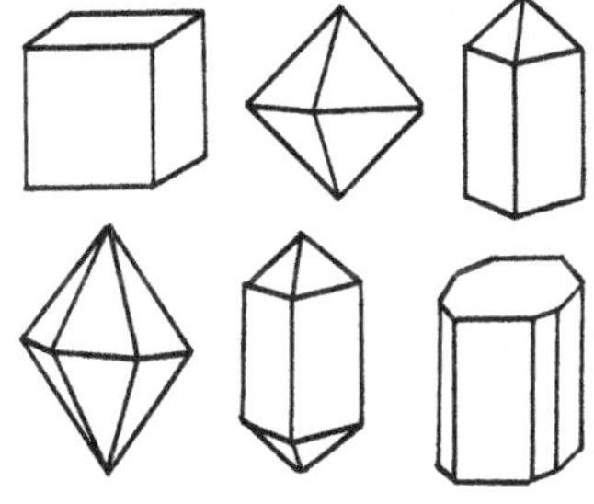

Examples of crystal shapes (clockwise, from top left): cubic, octahedral, pyramidal, dipyramidal, dipyramidal and prismatic.

Meaning 2 In electronics, a crystal is a piece of quartz or a similar substance that is used to control the frequency of a radio transmitter or receiver.

cumulus

See **cloud**

current

Meaning 1 A current is a flow of a liquid or a gas, such as water or air.

Meaning 2 A current is a flow of electrons in an **electric circuit**.

cycad

A cycad is a plant that looks like a palm tree but produces flowers. The leaves are fern-like, and the flowers and fruits are in cones.

Cycads were the first seed-bearing plants to evolve on Earth. There are only a few species of cycads still living.

A cycad.

cycle

A cycle is something that keeps happening over and over again in exactly the same way. The time taken for one cycle is called the period, and the number of cycles that happen in a set time is called the **frequency**.

cyclone

A cyclone is a large area of the atmosphere where the air pressure is low.

Cyclones are also called lows or depressions. The term is sometimes used incorrectly to mean **tropical cyclone**.

Dd

database

Meaning 1 A database is **software** designed to store and use data on a computer.

Meaning 2 A database is data stored on a computer by a database program.

day

A day is the time taken for a planet to turn once on its axis. On the Earth a day is about 24 hours. This is called a mean solar day because it is based on the average position of the Sun.

Astronomers often use two other sorts of days. A sidereal day is based on the position of the stars and is about 23 hours and 56 minutes. A lunar day is based on the position of the Moon and is about 23 hours and 50 minutes.

See also **leap year**, **solar system**, **year**

debug

Debug means to find and fix problems in a computer **program**.

deciduous

A deciduous plant drops its leaves once a year during autumn. Most deciduous plants are native to the northern hemisphere.

Deciduous trees in winter.

See also **evergreen**

decomposer

A decomposer is an organism that breaks down **organic** compounds into simple, non-organic compounds. The most important decomposers are **bacteria** and **fungi**.

See also **consumer**, **producer**

demonstration

A demonstration means showing how an **experiment** is done. Your teacher might give a demonstration before you try an experiment by yourself.

density

Density is the amount of **mass** in a particular **volume** of a substance.

For example, a mass of 1 kilogram that takes up a volume of 1 cubic metre has a density of 1 kilogram per cubic metre (also written as 1 kg/m^3).

denudation

Denudation is the wearing away of the surface of the Earth by natural causes, such as wind, rain, snow, frost, water, waves or heat.

See also **deposition**, **erosion**

deoxyribonucleic acid

See **DNA**

deposition

Meaning 1 In geology, deposition is when **sediment** is laid down by water, wind or ice. This is how **sedimentary rocks** are formed.

See also **denudation**, **erosion**

Meaning 2 In physics, deposition is when a gas turns into a solid without turning into a liquid first.

See also **sublimation**

desalination

Desalination means taking the salt out of salty water so that the water can be used for irrigation or drinking.

desert

A desert is a place where almost no plants grow because there is not enough rain or because it is too cold.

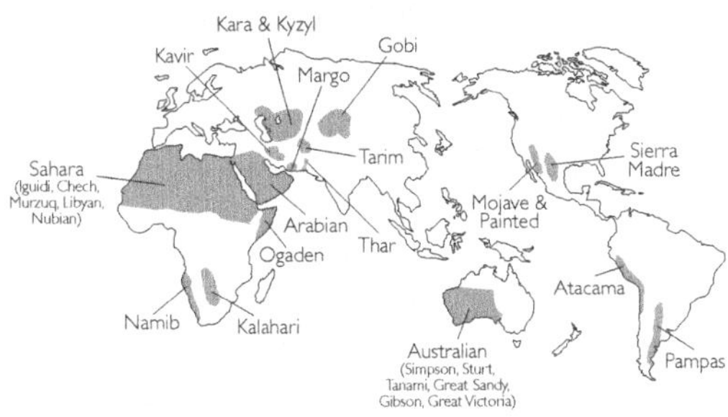

The main deserts of the world.

desktop computer

A desktop computer is a **personal computer** that is meant to be used on a desk.

detergent

A detergent is a substance that causes grease, oil and other substances to **dissolve** in water.

Some examples of detergents are **soap**, washing powder and shampoo.

detritus

Meaning 1 Detritus is rock fragments that have broken off by natural means.

Meaning 2 Detritus is dead plant or animal matter that is floating in water.

Devonian period

See **geological time**

dew

Dew is drops of water that form when **water vapour** in the air cools down enough to become a liquid.

Dew often forms on cold surfaces, such as rocks, plants and windows.

See also **condensation**

diamond

Diamond is a sort of **carbon** that is extremely hard. In fact, it is the hardest substance found on Earth.

Diamond has a **crystal** structure, and is formed by great pressure in the Earth's crust.

See also **Mohs scale**

diaphragm

The diaphragm is a large muscle that makes the **lungs** work. When the diaphragm is stretched, the lungs expand and fill with air. When the diaphragm is relaxed, the lungs are squeezed and old air is pushed out.

digestion

Meaning 1 Digestion is the breaking down of food into substances that can be absorbed by the body.

See also **digestive system**, **enzyme**

Meaning 2 Digestion is the breaking down of dead plants and animals by bacteria and fungi when there is no air. This happens in water and wet soil. It is used in **sewage** treatment plants to remove bad smells from the sewage.

digestive system

The digestive system is all the parts of the body that are involved in breaking down food and absorbing it into the blood. It consists of the **alimentary canal**, the **pancreas** and the **liver**.

digital

Meaning 1 Digital means information or data that is made up of the numbers 0 and 1.

Meaning 2 Digital means information or data that is sent as a series of pulses, instead of a continuously changing signal. This is a sort of binary code, where no pulse equals 0 and a pulse equals 1. Digital television signals are sent in this form.

Meaning 3 Digital means something that shows digits (0 to 9), such as a digital clock.

digital television

See **television**

digital video disc

See **DVD**

digitise

Digitise means to change information, such as text or a picture, into an electronic file that can be stored on a computer.

See also **digital**

dinosaur

Dinosaurs were reptiles that lived on the Earth during the Mesozoic era.

Many dinosaurs were giant animals, but some were quite small. Some were **carnivores**, although most were probably **herbivores**. The dinosaurs died out during the Cretaceous period.

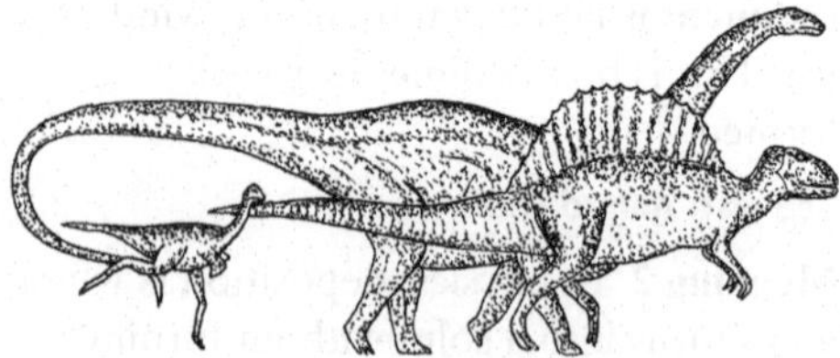

Some examples of dinosaurs (left to right): Avivimus, Apatosaurus and Spinosaurus.

See also **geological time**

direct current

Direct current is an electric **current** that always flows in the same direction. It is the sort of electric current made by batteries. Direct current is also called DC.

See also **alternating current**

displacement

Meaning 1 Displacement is the distance and direction that something has moved.

Meaning 2 Displacement is the volume that is taken up by an object when it is put into a liquid or a gas.

dissolve

Something dissolves if it breaks up completely into **molecules** or **ions** when it is added to a liquid.

The substance that dissolves is called the solute, and the liquid it dissolves in is called the solvent. The mixture is called a solution.

See also **precipitation**

diurnal

Diurnal means happening or being active during the day instead of at night. For example, butterflies and dragonflies are diurnal insects.

See also **nocturnal**

DNA

DNA is short for deoxyribonucleic acid. It is a chemical that is found only in the **chromosomes** of organisms.

The shape of a DNA **molecule** is like a twisted ladder. The ladder is made up of pieces called **genes** that give each organism its different features. The DNA of humans probably contains more than 1000 genes.

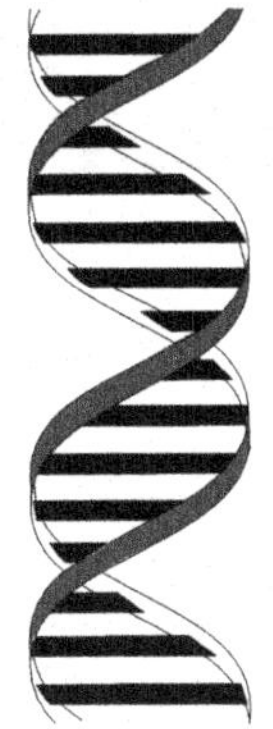

Double-helix structure of DNA.

download

Download means to copy a **file** onto a **computer**.

Files can be downloaded from the **Internet**, another computer or a device such as a digital camera.

drag

Drag is **friction** that slows down an object moving through air or water.

Drag always acts in the opposite direction to the motion. It can be reduced by making the surface smooth so that the air or water flows over the object more easily.

See also **aerodynamics**, **thrust**

DVD

A DVD (short for digital video disc) is a disc like a CD that has high-quality **digital** pictures and sound recorded onto it.

dye

A dye is a coloured liquid that can be used to change the colour of cloth or other material.

In science, special dyes can be used to show how water or other liquids move through plants and animals. They are called dye tracers.

E

Ee

ear

The ear is an organ in vertebrate animals that detects sound and helps with **balance**.

In humans, the ear has three parts called the outer ear, middle ear and inner ear. The outer ear collects the sound. The middle ear changes the sound energy into mechanical energy. Then the inner ear changes the mechanical energy into nerve impulses and sends them to the brain. The inner ear also contains the organs used for balance called the semicircular canals.

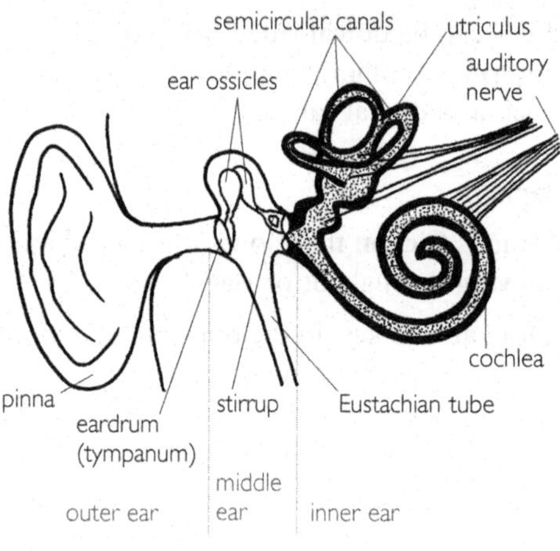

Earth

The Earth, also called Terra, is the third planet from the Sun and is the only place that we know of where life exists. It is a rocky planet made up mainly of nickel-iron and silicates. The Earth is 149.6 million kilometres from the Sun and takes 365 days to make one orbit. The Earth rotates once every 23 hours and 56 minutes, and its diameter is 12 756 kilometres. It has one **satellite** called the Moon, or Luna. The Moon is so big that it and the Earth are sometimes called a "double planet". The Earth's atmosphere is mainly nitrogen and oxygen, and the surface is covered mostly with salt water.

See also **solar system**

earthquake

An earthquake is a sudden movement or break in the Earth's **crust**.

Earthquakes are caused when a part of the crust is squeezed or bent until it suddenly breaks. Most earthquakes are caused by movements of **tectonic plates**, so they tend to happen where the plates meet. But they can also occur in other places, even in the centre of a plate. Small earthquakes are called tremors.

See also **Richter scale, tsunami**

earthworm

An earthworm is a long, soft-bodied **invertebrate** animal with a body made up of segments.

Earthworms live in burrows under ground and feed mainly on **detritus**. They are usually only a few centimetres long, but the Giant Gippsland Earthworm from Australia is almost 2 metres long.

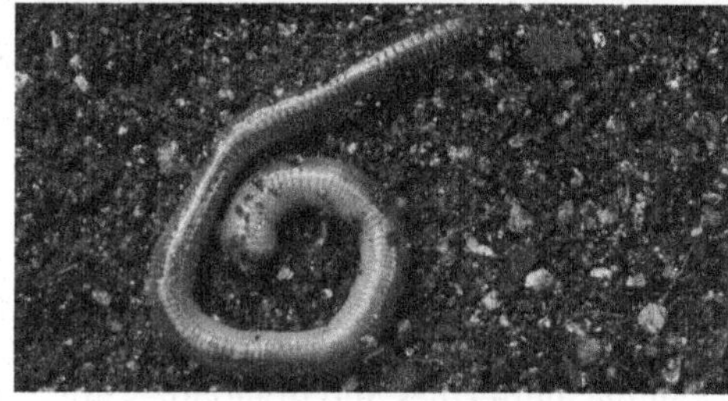

A garden earthworm.

echo

An echo is sound that has bounced off a surface and returned to where it came from.

eclipse

See **lunar eclipse, solar eclipse**

ecology

Ecology is the study of the relationships between organisms and their surroundings. It includes the study of populations, communities and ecosystems.

ecosystem

An ecosystem is a **community** of plants, animals and other organisms, and the environment they live in.

Except for sunlight and rain, an ecosystem produces everything it needs. All the parts of an ecosystem contribute to the health of the ecosystem, and each part depends on the other parts.

effort

The effort is the force needed in a **machine** to move a load.

See also **lever**

egg

An egg is a fertilised **ovum** that can become a new animal.

An egg has a soft or hard layer that protects it while it grows into a new animal. Insects, birds, reptiles and monotremes all lay eggs.

See also **fertilisation**

electrical energy

Electrical energy is the energy carried by electrons flowing around an **electric circuit**.

electric charge

An electric charge occurs when **electrons** are added or removed from **atoms**.

Extra electrons produce a negative electric charge. Fewer electrons produce a positive electric charge.

See also **static electricity**

electric circuit

An electric circuit is a path along which an electric current can flow. The current flows from the source of electrons (such as a battery) and back again.

A circuit can include many different devices. Each device has a different effect on the flow of current.

See also **circuit diagram**

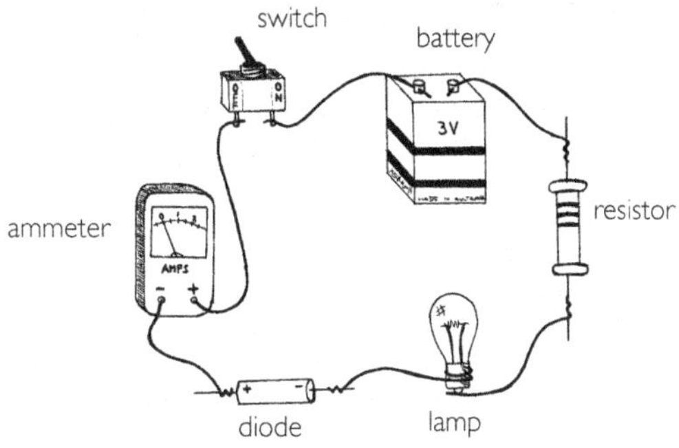

A simple electric circuit.

electric current

An electric current is a flow of electrons through a substance, such as a metal wire.

Electric current is measured in units called amperes.

electricity

Meaning 1 In science, electricity is all the effects caused by the movement of **electrons** through a substance or by the build-up of an **electric charge**.

Meaning 2 Electricity is used as another name for **electrical energy**.

electromagnet

An electromagnet is a piece of iron with a coil of wire around it. When an **electric current** flows through the coil, the iron becomes a powerful magnet.

Electromagnets are useful because they are very powerful and because they can be turned on and off.

Large electromagnets are used for picking up iron and steel in scrap yards.

electromagnetic radiation

Electromagnetic **radiation** is a type of energy made up of both electric and magnetic waves that can travel through a vacuum.

Some examples of electromagnetic radiation are radio waves, infra-red radiation, visible light, ultraviolet radiation, X-rays and gamma radiation. Electromagnetic radiation moves at the **speed of light**.

See also **spectrum**

electron

An electron is the part of an **atom** that carries a negative **electric charge**.

Electrons are usually thought of as circling around the nucleus in an electron cloud.

electronic circuit

An electronic circuit is a **circuit** that includes **semiconductors** or other devices that change the way that electrons flow.

electronics

Electronics is the study and use of devices that change the flow of current in a circuit, especially **semiconductors**.

electrostatic

Electrostatic refers to something that carries **static electricity**.

element

An element is a substance that cannot be broken down into another substance. All its **atoms** have exactly the same number of protons and electrons.

The first 92 elements all occur naturally. The others can only be made in laboratories, and they last only for a very short time.

See also **compound**

El Niño

El Niño is a change in the wind patterns and ocean currents of the Pacific Ocean that happens every four or five years.

El Niño causes droughts in Australia, New Guinea, Indonesia and China, and heavy rain in South America.

The normal weather patterns and ocean currents are called La Niña.

Winds mainly blow east, bringing hot dry desert air to eastern Australia, causing a drought.

Winds blow onto coast, bringing rain to South America.

Wind mainly blows over ocean, evaporating water which falls as rain. Floods and storms are common.

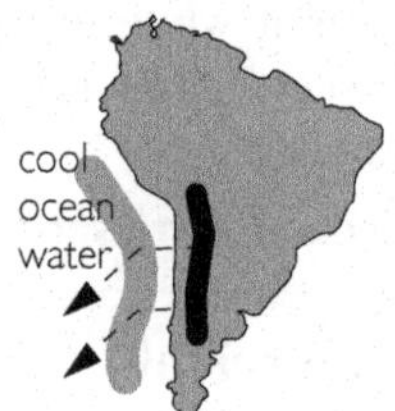

Dry winds blow off land, drought in South America.

El Niño (top) and La Niña (bottom). Black areas experience drought.

e-mail

An e-mail is a message sent from one person to another using a computer.

E-mails can be simple words, or they can include documents, pictures or videos.

embryo

An embryo is a new plant or animal growing inside an egg or a seed.

In animals, the embryo continues to grow until it is born or hatched. In plants, the embryo stops growing until **germination** occurs.

Table of the elements

Atomic number	*Element*	*Symbol*	*Atomic number*	*Element*	*Symbol*	*Atomic number*	*Element*	*Symbol*
1	Hydrogen	H	41	Niobium	Nb	81	Thallium	Tl
2	Helium	He	42	Molybdenum	Mo	82	Lead	Pb
3	Lithium	Li	43	Technetium	Te	83	Bismuth	Bi
4	Beryllium	Be	44	Ruthenium	Ru	84	Polonium	Po
5	Boron	B	45	Rhodium	Rh	85	Astatine	At
6	Carbon	C	46	Palladium	Pd	86	Radon	Rn
7	Nitrogen	N	47	Silver	Ag	87	Francium	Fr
8	Oxygen	O	48	Cadmium	Cd	88	Radium	Ra
9	Fluorine	F	49	Indium	In	89	Actinium	Ac
10	Neon	Ne	50	Tin	Sn	90	Thorium	Th
11	Sodium	Na	51	Antimony	Sb	91	Protactinium	Pa
12	Magnesium	Mg	52	Tellurium	Te	92	Uranium	U
13	Aluminium	Al	53	Iodine	I	93	Neptunium	Np
14	Silicon	Si	54	Xenon	Xe	94	Plutonium	Pu
15	Phosphorus	P	55	Caesium	Cs	95	Americium	Am
16	Sulfur	S	56	Barium	Ba	96	Curium	Cm
17	Chlorine	Cl	57	Lanthanum	La	97	Berkelium	Bk
18	Argon	Ar	58	Cerium	Ce	98	Californium	Cf
19	Potassium	K	59	Praseodymium	Pr	99	Einsteinium	Es
20	Calcium	Ca	60	Neodymium	Nd	100	Fermium	Fm
21	Scandium	Sc	61	Promethium	Pm	101	Mendelevium	Md
22	Titanium	Ti	62	Samarium	Sm	102	Nobelium	No
23	Vanadium	V	63	Europium	Eu	103	Lawrencium	Lr
24	Chromium	Cr	64	Gadolinium	Gd	104	Rutherfordium	Rf
25	Manganese	Mn	65	Terbium	Tb	105	Dubnium	Db
26	Iron	Fe	66	Dysprosium	Dy	106	Seaborgium	Sg
27	Cobalt	Co	67	Holmium	Ho	107	Bohrium	Bh
28	Nickel	Ni	68	Erbium	Er	108	Hassium	Hs
29	Copper	Cu	69	Thulium	Tm	109	Meitnerium	Mt
30	Zinc	Zn	70	Ytterbium	Yb	110	Ununnilium	Uun
31	Gallium	Ga	71	Lutetium	Lu	111	Unununium	Uuu
32	Germanium	Ge	72	Hafnium	Hf	112	Ununbium	Uub
33	Arsenic	As	73	Tantalum	Ta			
34	Selenium	Se	74	Tungsten	W			
35	Bromine	Br	75	Rhenium	Re			
36	Krypton	Kr	76	Osmium	Os			
37	Rubidium	Rb	77	Iridium	Ir			
38	Strontium	Sr	78	Platinum	Pt			
39	Yttrium	Y	79	Gold	Au			
40	Zircon	Zr	80	Mercury	Hg			

endangered

Endangered means in danger of becoming **extinct**.

Species can become endangered for many reasons, but the main cause is the destruction of their habitat.

endocrine system

The endocrine system consists of glands in the body that produce substances for controlling some of the functions of the body.

These glands are the pituitary, thyroid, parathyroid, adrenal and pineal glands, and the pancreas, ovaries and testes. The glands put the substances straight into the blood, which carries them to the places they are needed.

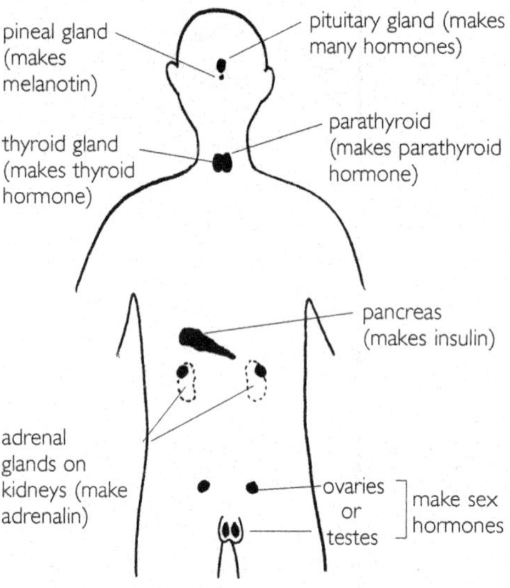

energy

Energy is the ability of something to do **work**.

There are many sorts of energy. Some examples are heat energy, sound energy, electrical energy, kinetic energy, gravitational energy, potential energy, nuclear energy and chemical energy.

See also **conservation of energy**

engine

An engine is a **machine** that produces mechanical energy or kinetic energy from another sort of energy.

Some common engines are petrol or diesel engines in cars, jet engines in aircraft and rocket engines in spacecraft.

engineering

Engineering is the use of science for practical things, such as designing and building machines.

See also **technology**

entomology

Entomology is the study of **insects**.

environment

Environment means the surroundings, including all the living and non-living things there.

enzyme

An enzyme is a substance that speeds up a **chemical reaction** but is not changed by the reaction. Enzymes in the **digestive system** help to break down food before it is absorbed.

epoch

See **geological time**

equator

The equator is an imaginary line around the Earth, midway between the North and South Poles. It divides the Earth into two halves called the northern and southern hemispheres.

See also **lines of latitude and longitude**, **tropics**

equinox

The equinox is the time of the year when the night is 12 hours long and the day is also 12 hours long, everywhere on Earth. This happens twice each year: on about 21 March and 23 September.

See also **season**, **solstice**

erosion

Erosion is the wearing and carrying away of soil and rocks by wind, water or ice.

See also **weathering**

escape velocity

Escape velocity is the speed that a rocket or other object must have to escape from the gravity of the Earth or another body in space.

The escape velocity from the Earth is 11 200 metres per second, or about 40 320 kilometres per hour.

estimate

An estimate is a guess that you make using all the information you have.

evaporation

Evaporation is the change of a liquid into a gas without boiling.

Evaporation happens because there are some molecules of liquid on the surface that are moving fast enough to escape from the surface. Evaporation causes the temperature of the liquid to fall because it is losing energy.

evergreen

Evergreen means having leaves on the branches all year round.

Evergreen trees lose small numbers of leaves all year round, instead of all at once in autumn as **deciduous** trees do. Some examples of evergreen trees are conifers, eucalypts and wattles.

evolution

Evolution is a slow change in the features of a **species** that can result in the creation of new species.

Scientists believe that evolution occurs when the environment changes so much that only some individuals of a species can survive. The ones that survive have features that help them to cope with the change. This is called natural selection. These individuals pass on the features to their offspring.

As more and more changes occur over a long time, the species slowly changes until eventually it may be no longer like the original species. In other words, a new species has evolved.

excretory system

The excretory system is the body system that removes liquid wastes. It includes the **kidneys**, ureter, bladder, urethra, skin and lungs.

exoskeleton

An exoskeleton is a hard, bony covering on the outside of some animals.

Animals with exoskeletons include crabs, insects, molluscs, scorpions and tortoises.

experiment

An experiment is something you do to test a **hypothesis**.

Experiments help us to find out what happens when we do something, or to find out why something happens.

extinct

Extinct means no longer living on the Earth.

See also **endangered**

extrusive rock

See **volcanic rock**

eye

An eye is an organ in animals that detects light.

Most invertebrate animals and all vertebrate animals have eyes, but there are many different sorts of eyes. In humans, the eyes are globular organs filled with a thick liquid called humour. Light enters the eye and is bent by a thin layer called the cornea. It then passes through a thick liquid called the aqueous humour and then through the pupil, which is a hole in the middle of the iris. The size of the pupil gets larger if there is not enough light, or smaller if there is too much light.

After passing through the pupil, the light goes through the lens, which focuses the light. It then goes through another thick liquid called the vitreous humour, and finally reaches the retina at the back of the eye. The retina turns the light energy into nerve impulses, which are passed to the brain by the optic nerve.

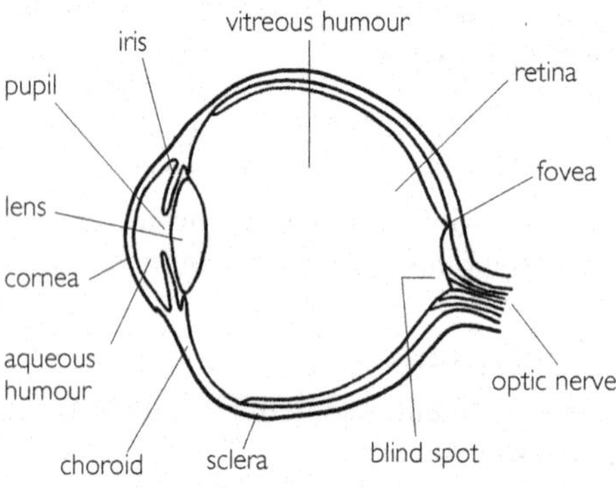

The human eye.

eyepiece

An eyepiece is the lens that you look into when you are using an optical instrument, such as a **telescope**, binoculars or a **microscope**.

See also **objective lens**

extrusive rock

See **volcanic rock**

Ff

Fahrenheit scale

The Fahrenheit scale is a scale used to measure temperature. On this scale, ice melts at 32 degrees and water boils at 212 degrees.

See also **Celsius scale**

fair testing

Fair testing means doing an experiment in a way that makes sure it is the experiment, and not something else, that is causing the results you get.

fats

Fats are solid, oily substances made up of chemicals called lipids.

Many animals and plants contain fats because this is a good way to store energy. Some animals also use fats to protect their organs from cold.

fault

A fault is a crack in the Earth's **crust** where the rock on one side of the crack has moved up, down or sideways.

The amount of upward or downward movement of the rocks is called the throw. The amount of sideways movement is called the heave.

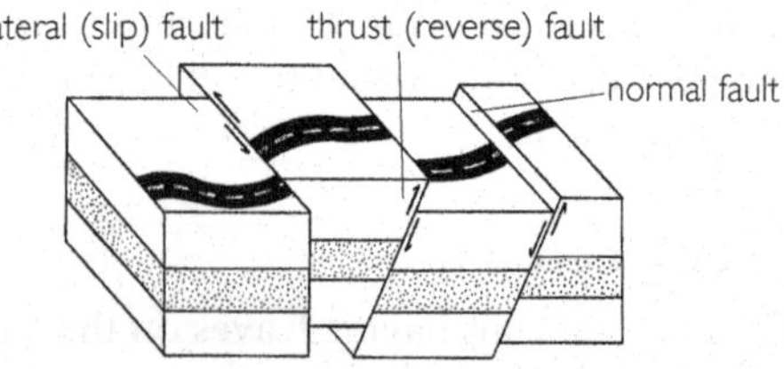

fauna

Fauna is another word for animals. It is usually used when you talk about all the animals that live in one area or that live at one time.

For example, you might talk about the fauna of New Zealand or the fauna of the Jurassic period.

feather

Feathers are the covering on birds, attached into the skin. They are made entirely of a protein called keratin.

Feathers help to protect birds against heat and cold, and the wing feathers play an important part in flight.

femur

The femur is the large leg bone attached to the hip of a vertebrate animal. The other end of the femur is attached to the tibia at the knee. The femur is also called the thigh bone.

See also **skeleton**

fern

Ferns are green plants that reproduce by spores that are made on the underside of the leaves.

Some ferns are very small, but some can grow into tree-sized plants called tree-ferns. Most ferns grow on land, but a few grow in water.

A tree-fern.

fertile

Meaning 1 In animals, fertile means able to breed and produce offspring.

Meaning 2 In plants and other organisms, fertile means having male or female parts, such as flowers.

Meaning 3 In soil science, fertile means containing a lot of **nutrients** that can be used by plants.

fertilisation

Meaning 1 Fertilisation is when a male **nucleus** joins with a female nucleus to produce a new cell. In animals, it happens when a **sperm** joins with an ovum. In flowering plants, it happens when a pollen nucleus joins with an ovum.

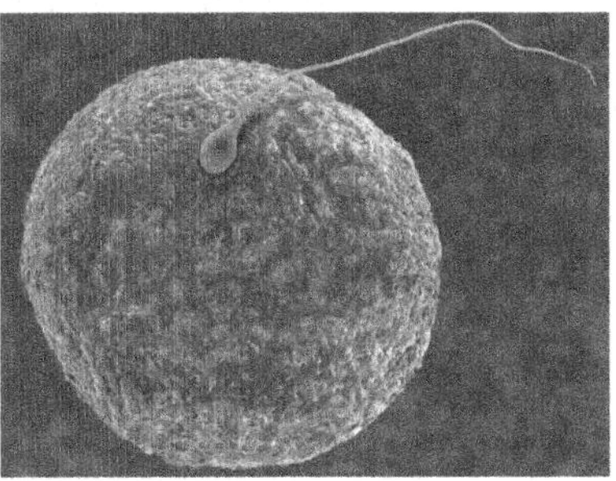

A sperm penetrating the outer membrane of an ovum.

Meaning 2 Fertilisation is the addition of nutrients to the soil or water to increase the growth of plants. Nitrogen, phosphorus and potassium are the three main nutrients added.

fibula

The fibula is one of the leg bones in vertebrate animals. It is the smaller of the two bones in the lower leg.

See also **skeleton**

file

In computing, a file is any electronic document, set of data or other information produced using **software**.

filter

Meaning 1 In chemistry, a filter is anything that is used to remove solid particles from a liquid. The most common sort of filter is filter paper.

☞

F

Meaning 2 In physics, a filter is a piece of coloured glass or plastic that allows only one colour of light to pass through.

fish

A fish is a **cold-blooded** animal that lives in water and has scales covering the body. Fish have fins for movement, and obtain their oxygen from the water using **gills**.

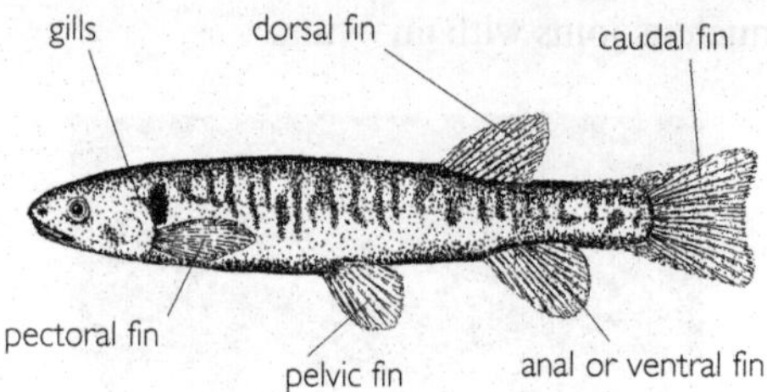

Broad-finned Galaxias, a freshwater fish.

See also **ray**

floppy disk

A floppy disk is a small, magnetic computer disk that can be used to copy files to or from a computer. It fits into a special drive in the computer called a floppy disk drive.

Files on a floppy disk can be erased so that new information can be put onto it. Modern floppy disks can store 1.44 megabytes of information.

flora

Flora is another word for plants. It is usually used when you talk about all the plants that live in one area or that live at one time.

For example, you might talk about the flora of Canada or the flora of the Cretaceous period.

flower

A flower is a part of a flowering plant that has the parts used in **reproduction** and where seeds develop after **fertilisation**.

Some flowers have only male parts, some flowers have only female parts, and other flowers have male and female parts.

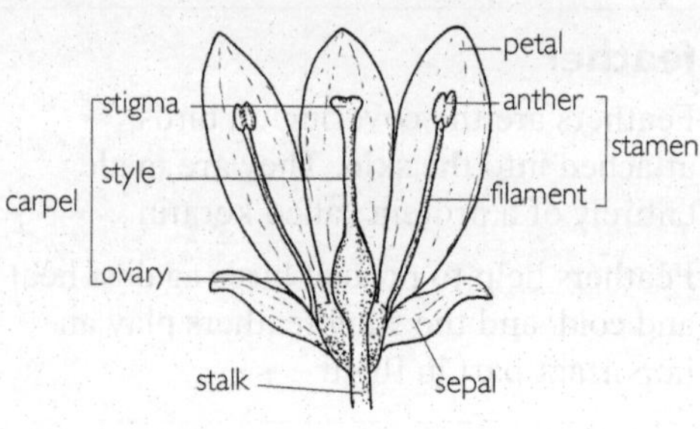

fluid

A fluid is any substance that can flow. Fluids can be liquids or gases.

foam

Foam is a liquid that has bubbles of gas all through it. Some foams, such as foam rubber and polystyrene, can set into a solid.

focus

Meaning 1 In optics, a focus is a point where light rays come together after passing through a lens or reflecting off a mirror.

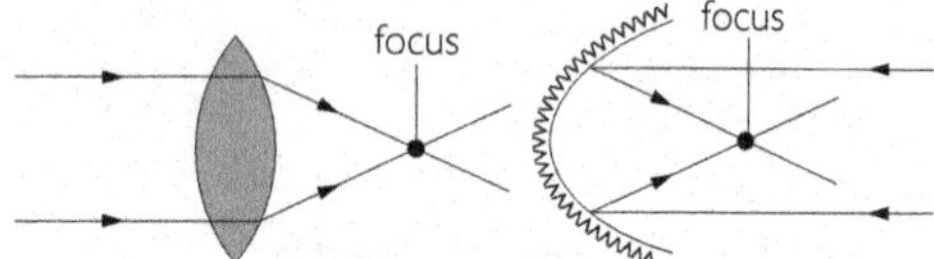

The focus of a lens (left) and a parabolic mirror (right).

Meaning 2 The focus of an **earthquake** is the point in the Earth's crust where the earthquake actually happens.

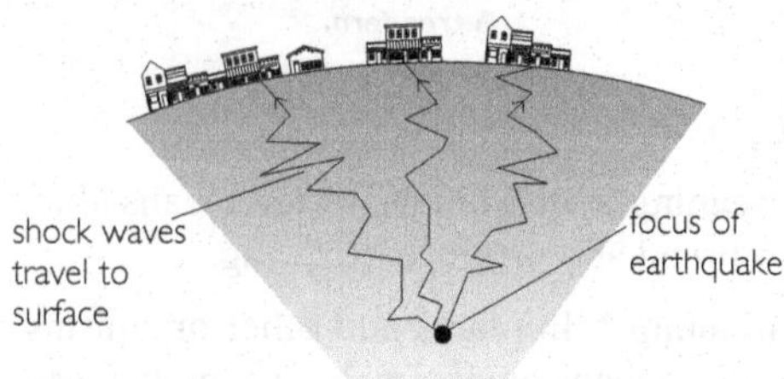

fog

Fog is a cloud that forms at the surface of the Earth.

This happens when warm air that is just above the ground is cooled. This is why fog often forms late in the afternoon or at night.

food chain

A food chain is the path of **food energy** from one organism to another in an ecosystem.

The food chain starts with a **producer**. This is eaten by a **consumer**, which gets some of the energy that the producer contained. The consumer is usually an animal. This consumer is eaten by another consumer and so on. At each step, energy moves up the food chain. The chain ends when a consumer is not eaten.

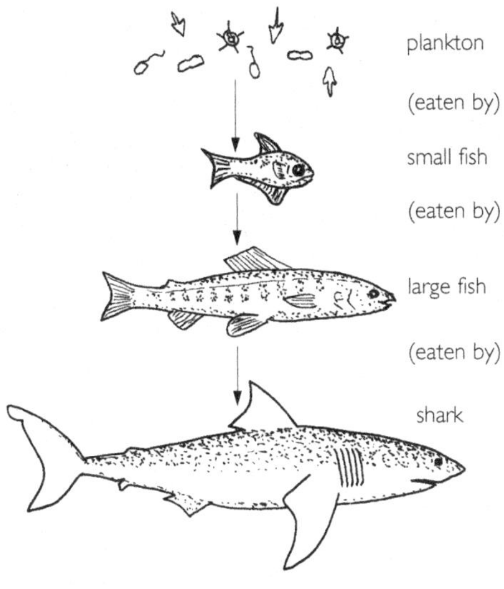

A simple food chain.

See also **food web**

food energy

Food energy is the energy that is contained in the food that an organism eats. It is passed from one organism to another along the **food chain**.

food web

A food web is made up of all the different **food chains** in an ecosystem. Many of these food chains involve the same organisms, so they form a sort of web.

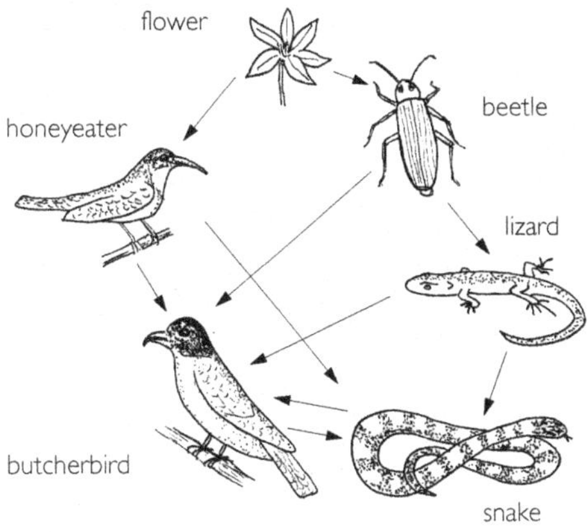

An example of a food web.

force

A force is something that causes a change in the **velocity** of an object. The unit of force is the newton (symbol N).

forest

A forest is an area covered in large trees that are close together.

If the trees are very close together, it is called a closed forest. If the trees are not so close, it is called an open forest.

See also **grassland**, **rainforest**, **woodland**

formula

In chemistry, a formula is a way of showing the **elements** that make up a substance, and how the atoms of the elements are grouped.

For example, the formula for water is H_2O. This means that each molecule of water has two hydrogen atoms and one oxygen atom.

F

fossil

A fossil is the remains or trace of an organism that lived long ago.

The most common sort of fossil is the remains of hard parts of the organism, such as bones, that have turned into rock. Another sort of fossil is just a print made by the organism after it died and slowly decayed.

A polished rock containing a fossil nautiloid, which is about 50 million years old.

fossil fuel

Fossil fuels are **coal**, oil and **natural gas**, or the fuels that are made from them. They are formed from the remains of ancient plants.

Fossil fuels contain a lot of carbon, so they burn very easily.

freezing point

Freezing point is the temperature at which a liquid turns into a solid. For example, the freezing point of water at sea level is 0 degrees Celsius.

frequency

Frequency is how many times something happens in a set time. It is used to describe how quickly a sound wave or an electromagnetic wave is vibrating.

A high frequency means it is vibrating quickly, and a low frequency means it is vibrating slowly. The unit of frequency is called a hertz, and its symbol is Hz.

friction

Friction is a force that makes it harder for an object to move along over another object.

Friction always acts in the opposite direction to the movement. For example, friction makes it hard to push a block of rubber across a table. Friction also happens when an object moves through a liquid or a gas. For example, it is hard to throw a foam ball very far because friction between the air and the ball makes it hard for the ball to move through the air.

frog

A frog is a **cold-blooded** vertebrate animal that breathes air, but can live in water or on land. The skin is smooth or sometimes warty, and can take in oxygen from the air to help the lungs.

Verreaux's Tree Frog.

See also **amphibian, toad**

frost

Frost is small particles of ice that are formed on the ground, plants or other surfaces.

Frost happens when the temperature of the surface falls below the **freezing point** of water.

fruit

A fruit is the part of a plant that contains seeds. It forms from the ovary of a flower after **fertilisation**.

Fruits are not always fleshy, like apples, tomatoes and bananas. They are often dry and cannot be eaten.

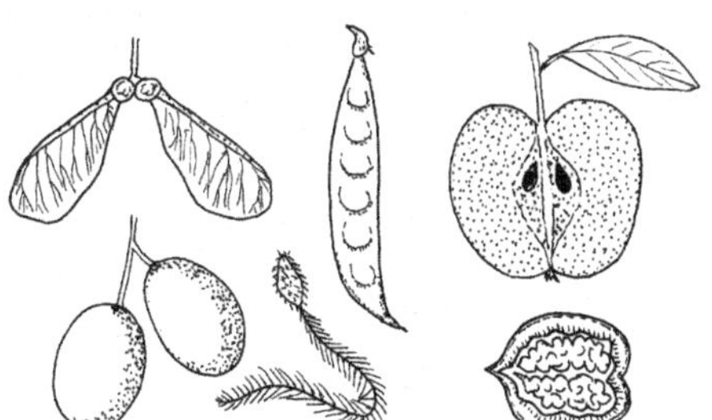

Some types of fruits (clockwise, from top left): samara, pod, pome, nut, achene and berry.

fuel

A fuel is any substance that can be burnt or changed to produce useful energy. Fuels include candle wax, wood, oil, coal, gas, petroleum, kerosene and plutonium.

See also **fossil fuel**

fulcrum

The fulcrum is the point around which a **lever** turns.

fungus

A fungus is an organism that cannot make its own food and reproduces by **spores**.

There are many different sorts of fungi. They include mushrooms, pore fungi, coral fungi, puff-balls, earth stars, stinkhorns, shelf fungi, jelly fungi and cup fungi.

Mushroom fungus.

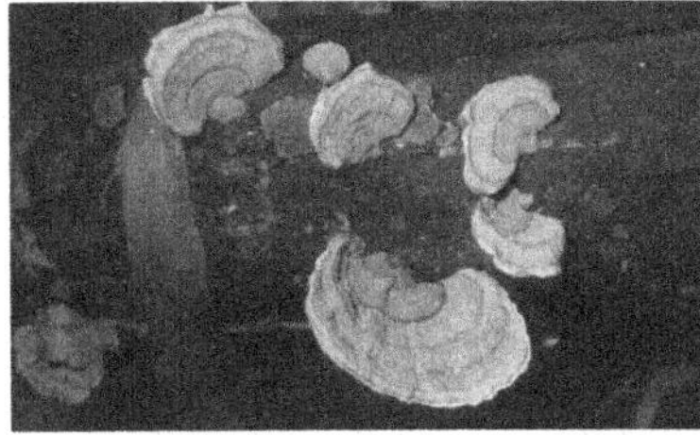

Shelf fungus.

Gg

galaxy

A galaxy is a huge group of **stars**, as well as dust and gas, that is held together by **gravity**.

There are four different sorts of galaxies: spiral, barred spiral, elliptical and irregular. Our own galaxy, called the Galaxy or the Milky Way System, is a spiral galaxy.

The Andromeda Galaxy (M31), a spiral galaxy.

See also **nebula**, **star cluster**

gamma ray

A gamma ray is a sort of energy that can be emitted by the nucleus of a radioactive atom.

gas

A gas is one of the **states of matter**. It has no particular volume or shape. If it is put into a container, it changes shape to fit the container. It also completely fills the container, no matter how big the container is. Most gases are also invisible.

See also **liquid**, **solid**

gaseous

Gaseous refers to something that is a gas.

gear

A gear is a device that transmits a turning motion from one shaft to another.

There are many sorts of gears. The most common is the spur gear, which is a set of toothed wheels that can turn against each other. The wheel that provides the turning motion is called the driving wheel. The other wheel is called the driven wheel.

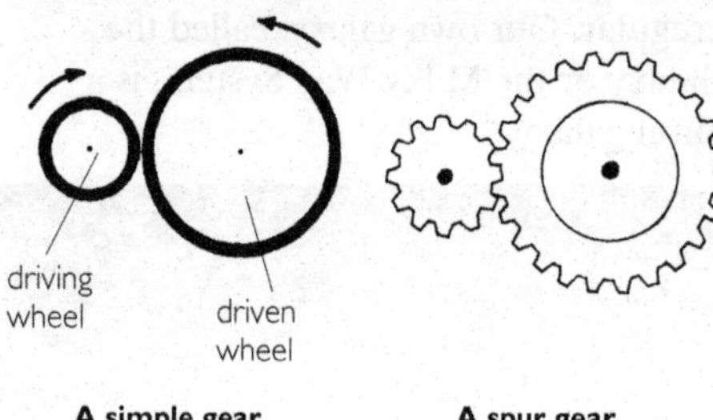

A simple gear. **A spur gear.**

See also **rack and pinion**, **worm gear**

gear ratio

The gear ratio is the number of turns of the driven wheel produced by one turn of the driving wheel in a **gear**.

You can work out the gear ratio of a spur gear by dividing the number of teeth on the driven wheel by the number of teeth on the driving wheel.

gemstone

A gemstone is a **mineral** or rock that is valuable because of its beauty or because it is hard to find.

gene

Genes are the things that determine the features of an organism. They are contained in the **DNA** of organisms.

Different organisms have different genes, and each gene controls one feature, or a group of features, of an organism. For example, there is a gene for your hair colour, another for blood type and another for the colour of your eyes.

See also **chromosome**, **heredity**

genetic engineering

Genetic engineering is the use of technology to put **genes** from an organism into the **DNA** of another organism.

Gene technology can be used to create new sorts of food crops and animals, and to cure diseases caused by faulty genes in humans and other animals. It can also be used for **cloning** animals and plants.

genetics

Genetics is the study of **genes** and their effects on the features of organisms.

genus

A genus is a group of **species** that are closely related to each other. The plural of genus is genera. Genera that are similar are put into groups called families.

See also **classification of living things**

geological time

Geological time means the history of the Earth, from the time it was formed right up until today.

Geological time is divided up into three main parts called aeons or eons. These are the Archaean Aeon, the Proterozoic Aeon and the Phanerozoic Aeon. Each aeon is divided into eras. Eras are divided into periods, and periods are divided into epochs. The geological time scale shown opposite is the modern one used by most geologists today. In this new time scale the old Tertiary period has been replaced by two separate periods called the Neogene and the Palaeogene.

geology

Geology is the study of the structure of the Earth, including how it was formed and what it is made from.

geothermal energy

Geothermal energy is heat energy trapped inside the Earth that can be used

Geological time

Aeon and era	*Period*	*Epoch*	*Began (mya)*	*Events*
Phanerozoic Aeon				
Cainozoic Era	Quaternary	Recent	0.011	Last ice age ended. Sea level rose to present level.
		Pleistocene	1.8	Modern humans evolved and spread across world. Several ice ages occurred. Neanderthals became extinct (30 000 years ago). Last ice age peaked (20 000 years ago).
	Neogene	Pliocene	5.3	Human genus (*Homo*) evolved (2.5 mya).
		Miocene	23.8	First hominids (ancestors of modern humans) evolved (6 mya). Songbirds became common. Grasslands spread and became a major ecosystem.
	Palaeogene	Oligocene	33.7	Carnivores evolved.
		Eocene	54.8	Horses and other ungulates evolved. Tectonic plates reorganised.
		Palaeocene	65	Whales evolved. Australia separated from Antarctica (60 mya).
Mesozoic Era	Cretaceous		141	World-wide increase in temperatures. Flowering plants evolved and quickly spread. Pangaea split into Laurasia and Gondwana. Many new species of dinosaur evolved. Gondwana split into southern continents. Dinosaurs became extinct (65 mya).
	Jurassic		205	Dinosaurs dominated land. Reptiles became common in oceans. Birds appeared (150 mya).
	Triassic		251	Dinosaurs appeared and became dominant (225 mya). First mammals appeared (220 mya). Modern amphibians evolved.
Palaeozoic Era	Permian		298	Cycads evolved. World-wide lowering of temperatures. First great extinction of plants and animals happened. Primitive amphibians evolved.
	Carboniferous		354	Primitive reptiles appeared (300 mya). Winged insects appeared. Spore-bearing plants dominant. Conifers evolved. Coal-forming forests widespread.
	Devonian		410	Trees appeared. Forests spread across land. Oil formed. First four-legged animals appeared (380 mya). Ferns and club mosses evolved. Modern fishes evolved.
	Silurian		434	First plants (liverworts, mosses) appeared on land. First vascular plants evolved.
	Ordovician		490	First coral reefs formed. Primitive fishes evolved.
	Cambrian		545	Great diversity of many-celled organisms produced in oceans. Many invertebrate animals evolved (including molluscs, trilobites, crustaceans, corals, sponges, brachiopods).

G

Aeon and era	*Period*	*Epoch*	*Began (mya)*	*Events*
Proterozoic Aeon				
Late Proterozoic Era			900	Many-celled organisms evolved. Oldest known fossil formed (600 mya).
Middle Proterozoic Era			1600	First living things with cell nucleus (algae) appeared (1500 mya). Many-celled algae appeared (1100 mya).
Early Proterozoic			2500	
Archaean Aeon				
Late Archaean Era			2900	Oxygen began to increase in atmosphere. Oceans became saltier.
Middle Archaean Era			3500	Oceans formed. Oldest fossil evidence of life (stromatolites) from Australia and Zimbabwe.
Early Archaean Era			4600	Earth formed. Atmosphere formed (mainly carbon dioxide). First living things (bacteria) appeared (3900 mya). Oldest known sediments formed (3800 mya).

Note: 'mya' means millions of years ago.

to produce electricity. We can get this energy where the heat energy comes to the surface, such as in **volcanoes**, **geysers**, hot springs and mud pools.

germination

Germination is the time when a **seed** starts to grow into a new plant. The first shoot and root break through the seed coat. After germination, the shoot grows upwards to form the first leaves, and the root grows downwards into the soil.

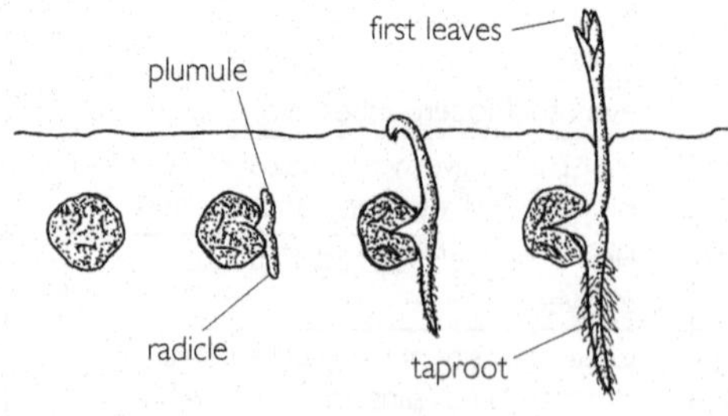

The germination of a pea seed.

geyser

A geyser is a hole in the ground that spurts out steam and hot water. The water is heated by lava far under ground.

gill

Meaning 1 In fish and some other animals that live in water, the gills are used for breathing. They take oxygen from the water and move it into the blood.

Meaning 2 In some **mushrooms**, the gills are the thin tissues underneath the cap, where the spores are produced.

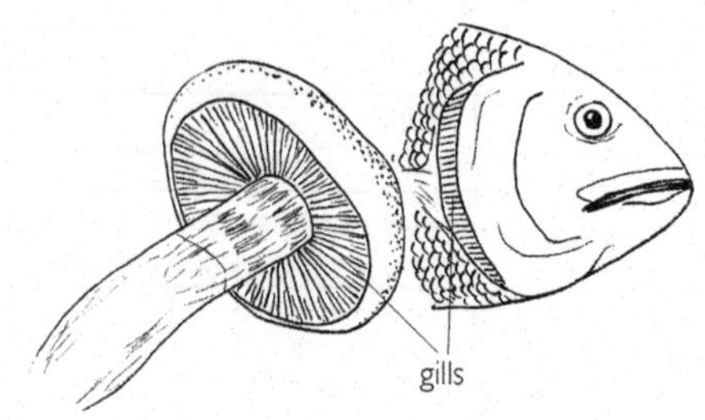

glacier

A glacier is a huge sheet of ice that is moving slowly down a valley.

A glacier in Parque Nacional los Glaciares, Argentina.

gland

In animals and plants, glands are groups of cells that make special substances.

For example, a gland in your **pancreas** makes insulin that controls the amount of sugar in your blood, and the nectary gland of a flower makes nectar.

glass

Glass is a clear solid that is made by melting a mixture of silicon dioxide (sand), sodium carbonate (soda) and calcium oxide (lime).

Glass for special purposes, such as Pyrex and optical glass for telescopes, has other substances added.

global warming

Global warming is the unnatural warming up of the Earth's atmosphere that is caused by human activities.

See also **greenhouse effect**

gold

Gold is a soft, shiny, yellowish metal found in the ground, often in a pure form called nuggets.

Gold is very beautiful and does not react much with other substances. This makes it ideal for use in jewellery, tooth fillings and electronic equipment.

Gondwana

Gondwana was an ancient continent that broke up to form Africa, Antarctica, Australia, India and South America. Many plants and animals in these places are very similar, even today.

See also **continental drift**

granite

Granite is a sort of **plutonic rock** that forms when **magma** cools far underground. It is made up mainly of the minerals quartz, feldspar and mica.

Granite can be coarse or fine, depending on where and how it was formed.

A coarse-grained granite.

graphic

In computing, a graphic is a still image, such as a photograph or drawing, that is stored as an electronic **file**.

An image can be stored in many different graphic formats, including JPEG, TIFF, GIF and PICT.

See also **video**

grassland

Grasslands are areas where grasses are the main sort of plant, and where there are not many trees or shrubs.

A grassland in Kenya.

gravity

Gravity is a **force** that causes two objects to be attracted to each other. It is caused by the masses of the objects. On Earth, gravity is the force that pulls everything downwards.

See also **weight**

greenhouse effect

The greenhouse effect is the way that heat energy is trapped by the Earth's atmosphere. Sunlight passes through the atmosphere and hits the Earth's surface, and then some of it is reflected as heat energy. This heat energy is absorbed by gases in the atmosphere, such as carbon dioxide and methane, so the atmosphere is kept warm.

Because we are putting more carbon dioxide and other gases into the atmosphere, we are adding to the amount of heat energy that is absorbed by the atmosphere. This is called the enhanced greenhouse effect, and causes **global warming**.

growth ring

Growth rings are the darker rings of wood that can be seen when the branch or trunk of a tree is cut through.

Each ring usually counts for one year of growth, so the age of a tree can be worked out by counting the growth rings.

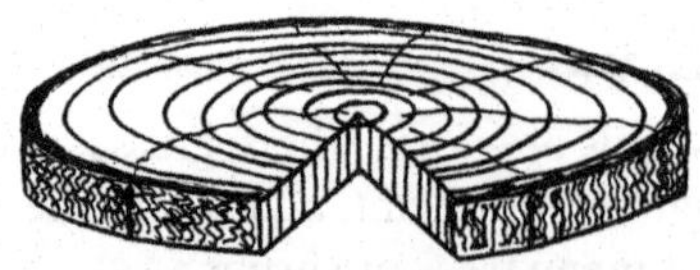

Growth rings in a nine-year-old tree.

grub

A grub is the **larva** of a **beetle** and looks something like a caterpillar.

Hh

habitat

The habitat of an organism means the sort of place it lives in.

For example, the habitat of a whale is the open oceans, and the habitat of snowgrass is high mountains.

hail

Hail is pieces of ice that are formed when rain is blown upwards by strong winds and freezes.

Halley's comet

Halley's comet is a famous **comet** that has been seen since ancient times. It was studied by an English astronomer named Edmund Halley, who discovered that it orbits the Sun once every 76 years. Halley's Comet will not be seen again until 2062.

hardness

Hardness means how difficult it is to scratch something. It is measured using **Mohs scale**.

hardware

Hardware is all the electronic and mechanical parts of a computer, such as the integrated circuits, disk drives, modem, monitor, keyboard, mouse and power supply.

See also **software**

hatching

Hatching is the time when the young of an egg-laying animal breaks out of its egg. Birds, reptiles and many invertebrates hatch from eggs.

heart

A heart is the organ that pumps **blood** through an animal's **circulatory system**.

Different animals have different sorts of hearts. In humans, the heart is a **muscle** with four hollow parts or chambers.

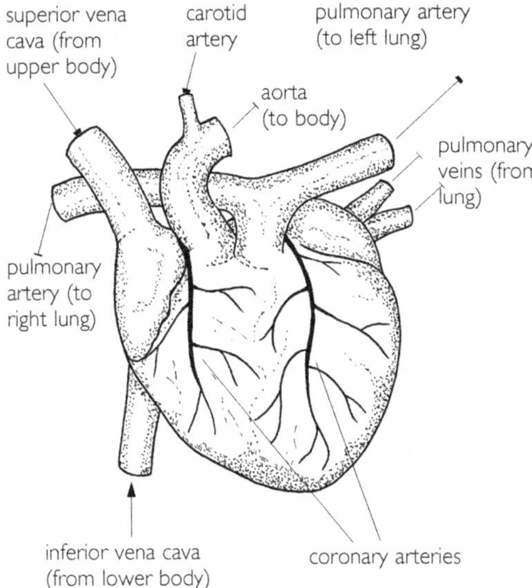

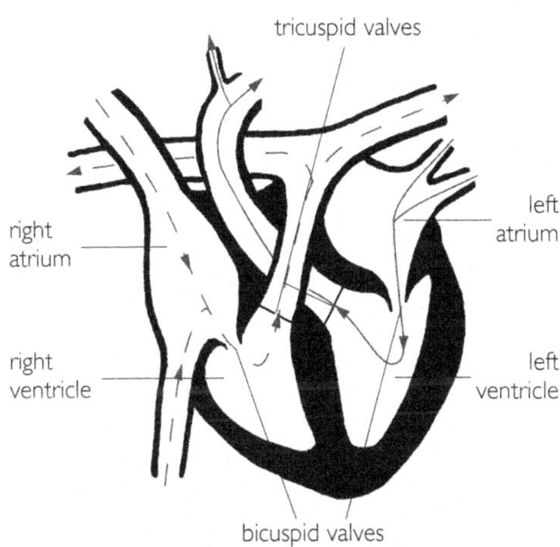

Blood returning from the body travels through the venae cavae into the right atrium. It is then pumped into the right ventricle and out through the pulmonary arteries to the lungs. In the lungs, oxygen is added to the blood. From the lungs the blood enters the pulmonary veins, which take it back to the left atrium of the heart. The blood is pumped into the left ventricle, and then out to the body through the aorta and smaller arteries. The coronary arteries supply blood to the heart itself.

heartbeat

A heartbeat is the beat in your chest that you can feel every time your heart pumps blood into and out of your **circulatory system**.

The heart of a healthy adult human beats about 70 times or less every minute. Another name for the heartbeat is **pulse**.

heat

Meaning 1 Heat is a movement of **energy** from one object to another because they have different **temperatures**.

Meaning 2 Heat can mean the energy that an object has because of its temperature.

helium

Helium is the second most common **element** in the universe, after hydrogen.

It is a gas that occurs in stars and also in gas clouds in space, but on Earth there is not much at all. Helium does not burn or react with other substances. It is used in welding, making electronic parts and in scuba-diving tanks. Because it is lighter than air, it is also used for filling party balloons and also for filling some large balloons that carry people, such as blimps. The symbol for helium is He.

herbivore

A herbivore is an animal that eats only plants.

Some examples of herbivores are kangaroos, possums, cows and horses.

See also **carnivore**, **omnivore**

heredity

Heredity is the passing on of features from parents to their offspring. The information about the features is passed on in the **genes**.

hibernation

Hibernation is a sort of sleep that some animals go into during winter so that they can stay alive without eating or drinking much.

During hibernation, the body temperature of the animals falls, and energy is obtained from fat that has been built up in the body.

hologram

A hologram is a three-dimensional **image** formed by shining light onto a specially made photograph.

Originally, holograms were made using **laser** light, but now ordinary light can be used.

horizon

Meaning 1 The horizon is the line where the sky meets the sea or flat land.

Meaning 2 In geology, a horizon is a layer of soil. The top layer of soil is called the A-horizon, the next one down is the B-horizon and so on.

humerus

The humerus is the long bone in the arm of mammals and many other vertebrates.

In humans, the humerus is joined to the scapula at the top end and to the ulna and radius at the bottom end.

See also **skeleton**

humidity

Humidity is how much **water vapour** there is in the air.

Humidity is usually described as a percentage. A humidity of 0% means that there is no water vapour in the air. A humidity of 50% means that the air contains half the water vapour that it can. A humidity of 100% means that the air contains as much water vapour as it can hold.

humus

Humus is decayed plant and animal matter in the top layer of soil.

Humus has a lot of **nutrients** in it, so it is very important for plant growth.

hurricane

Hurricane is the name used in the USA, Mexico and the Caribbean for a **tropical cyclone**.

See also **tornado**, **typhoon**

hydraulic pressure

Hydraulic pressure is the push of a liquid on its surroundings.

See also **jack**

hydrocarbon

A hydrocarbon is a chemical **compound** that contains only hydrogen and carbon atoms.

See also **carbohydrate**

hydroelectricity

Hydroelectricity is electricity that is produced using flowing water.

The water is used to turn a huge **turbine**, which produces electricity. In most hydroelectric power stations, water from a reservoir is used to turn the turbine. Hydroelectricity is also called hydroelectric power.

Huge pipes bring water from a reservoir to the turbine of a hydroelectric power station.

See also **water energy**

hydrogen

Hydrogen is the most common **element** in the universe.

It is also the lightest element, and pure hydrogen is a gas. It combines with other elements very easily and burns easily too. Most stars are made up of hydrogen, and on Earth it is extremely common. The symbol for hydrogen is H.

hydrosphere

The hydrosphere is the water that lies on the surface of the Earth. It includes the oceans, seas, lakes, streams and underground water.

About three-quarters of the Earth's surface is covered with water, and almost all of it (97%) is in the oceans.

hypothesis

A hypothesis is an idea that explains how or why something happens. In science, a hypothesis should be able to be tested by an **experiment**.

Ii

ice

Ice is frozen water.

At sea level, water freezes at 0 degrees Celcius. When it freezes, it gets bigger, so its **density** is less than the density of water. This is why ice floats.

ice age

An ice age is a time when the Earth was much colder and ice covered huge areas of land and water.

There have been at least three main periods with ice ages since the Earth was formed. There have been at least four ice ages in the last 600 000 years. The last ice age, sometimes called the Ice Age, ended about 10 000 years ago.

Ice age period	*Time*
Pleistocene	10 000 to 600 000 years ago
Carboniferous	250 million years ago
Precambrian	Over 500 million years ago

iceberg

An iceberg is a large piece of ice floating in the sea or a lake.

Icebergs break off from glaciers or from the edges of polar ice caps. Only about one-fifth of an iceberg is above the water.

Icebergs in Lago Argentino, Argentina.

igneous rock

Igneous rocks are rocks that form when **magma** cools. They include **volcanic rocks** and **plutonic rocks**.

ileum

The ileum is a part of the **digestive system** where food is digested and absorbed into the blood. It is at the end of the small intestine, just before the large intestine.

See also **alimentary canal**

image

An image is a picture of an object that is formed using **lenses** or mirrors.

An image can only be seen if our eye or something else (such as a piece of paper or a screen) is placed where the image is made.

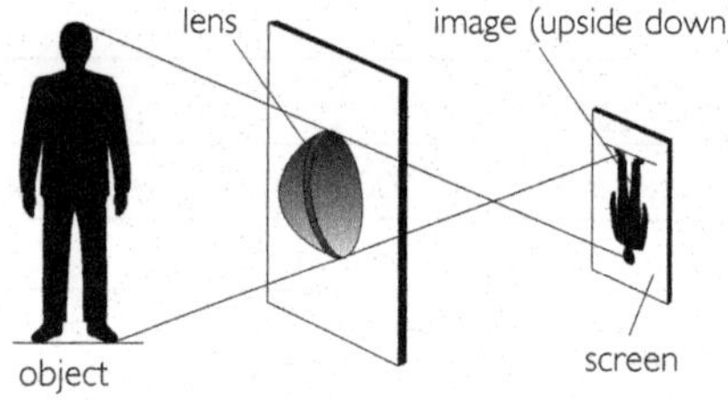

Forming an image.

immature

Immature means not yet grown up to be an adult.

Immature organisms cannot reproduce, and often do not have the same shape or colour as an adult.

immunity

Immunity is the ability of an animal to defend itself against attack from **micro-organisms** and poisons.

In humans, immunity is controlled mainly by **white blood cells** and substances called antibodies, which attack viruses, bacteria and other foreign matter in the body. Immunity can be improved if you are injected with antibodies made in a laboratory. Some examples are injections for tetanus, polio, influenza and snake bite.

inclined plane

An inclined plane is a flat surface that makes an angle with a horizontal surface. A ball that is placed on an inclined plane rolls straight down to the bottom.

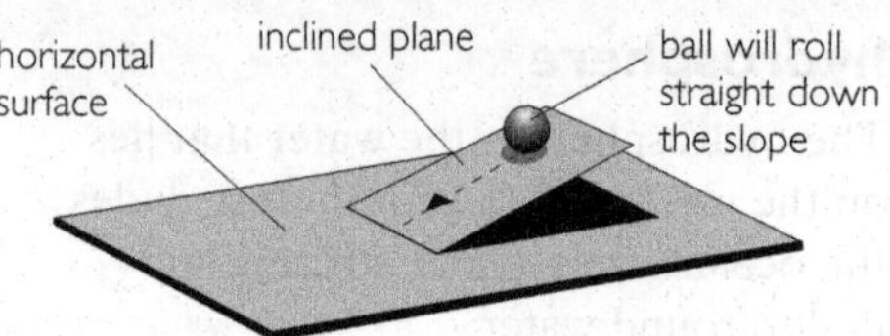

See also **wedge**

inert

In chemistry, inert means not able to combine with another substance to form a **compound**.

For example, helium, argon and neon are inert elements because they do not combine with other elements.

inertia

Inertia means how hard it is to make an object change its **velocity**.

Inertia is measured by **mass**. The more mass the object has, the harder it is to change its velocity.

See also **force**

infection

An infection is an attack on your body by **micro-organisms**.

See also **immunity**

information technology

Information technology is **technology** that is used to store, process and transmit information. Most information technology uses **computers**. Information technology is also called IT.

infra-red radiation

Infra-red radiation is radiation that is past the red end of the light **spectrum**. It is produced by hot objects, and it has a longer wavelength than red light.

inherit

Inherit means to get features that have been passed on by parents.

In humans, inherited features can include skin colour, eye colour, hair colour, nose shape, height, blood type and some diseases.

See also **gene**, **heredity**

inorganic

Inorganic refers to a substance that does not contain carbon.

See also **organic**

insect

Insects are **arthropods** with a head, six legs, wings and an abdomen divided into 11 parts.

In many insects the wings are reduced or lost. Some common insects are flies, bees, ants, wasps, butterflies, moths, cicadas and beetles.

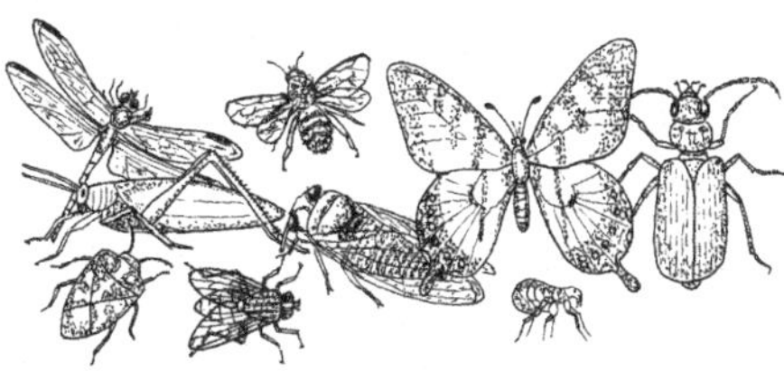

Some insects (clockwise, from top left): dragonfly, bee, butterfly, beetle, flea, cicada, fly, shield bug and grasshopper.

See also **metamorphosis**

insectivore

An insectivore is an animal that eats only insects.

Some examples are echidnas, moles, spiders and dragonflies.

insulator

An insulator is something that stops energy moving from one place to another.

For example, air, plastic and wood are all good heat insulators. Plastic, glass and porcelain are good electricity insulators.

See also **conductor**

interactive

An interactive device is a device that lets the user choose how they want it to work.

For example, a computer, a car and a video game are all interactive.

Internet

The Internet is a world-wide system that can be used to connect a computer with another computer in a different part of the world.

See also **World Wide Web**

intrusive rock

See **plutonic rock**

invent

Invent means to work out a new way to do something.

Inventing can be about real things such as machines or about imaginary things such as mathematics or time. For example, Georges de Mestral invented Velcro and Stephen Hawking invented new ideas about space and time.

invertebrate

An invertebrate is an animal without a **backbone**. Invertebrates also do not have a well-developed brain inside a skull.

Most animals in the world are invertebrates. There are millions of species. They have many different forms, ranging from very simple to very complicated. Some examples are ants, butterflies, snails, jellyfish, worms,

☞

grasshoppers, shrimps, crabs, sponges and centipedes.

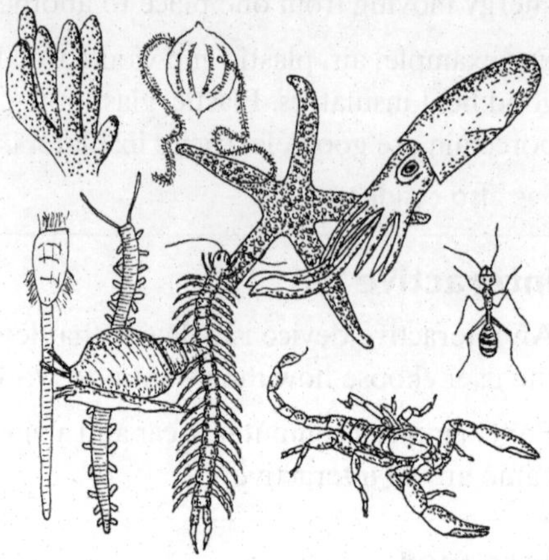

Some invertebrates (clockwise, from top left): sponge, comb jelly, starfish, squid, ant, scorpion, centipede, mollusc, peripatus and brachiopod.

See also **vertebrate**

investigate

Investigate means to study something to find out about it.

You can investigate something that has already happened, such as an earthquake. You can also investigate ideas, such as what will happen if you mix oil with water.

See also **experiment**

iodine

Iodine is an **element** that is needed by the body to control **metabolism**.

Some important sources of iodine are fish, fruit and vegetables. The symbol for iodine is I.

ion

An ion is an **atom** with an **electrical charge**.

If an atom has more electrons than normal, it is a negative ion (or anion). If it has fewer electrons, it is a positive ion (or cation).

ionosphere

See **atmosphere**

iris

The iris is a coloured circle of tissue in the **eye** that controls how much light reaches the retina. In humans the iris is usually blue or brown.

iron

Iron is a metal that is very common on the Earth.

It is obtained from ores such as haematite. Iron is used to make many things, but because it rusts it is usually made into **steel**. We also have iron in our blood.

isobar

An isobar is a line on a map that joins up places where the **air pressure** is the same. Isobars help us to make **weather forecasts**.

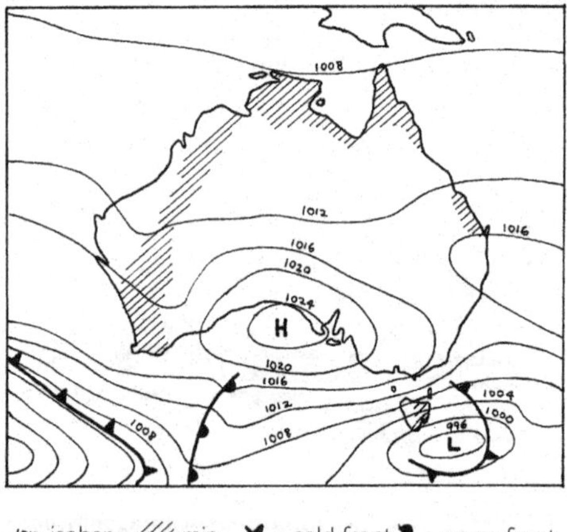

1012 isobar   rain   cold front   warm front

A typical weather map for winter in Australia.

IT

IT is short for **information technology**.

Jj

jack

A jack is a **machine** for moving a large mass a short distance.

Most jacks are designed for lifting. A common example is a car jack. Jacks can be simple machines, such as a **lever** or a **screw**, or complex machines operated by **hydraulic pressure**.

A car jack in use.

jejunum

The jejunum is the middle part of the **small intestine** between the duodenum and ileum.

See also **alimentary canal**

jellyfish

A jellyfish is an **invertebrate** sea creature with a simple bell-shaped body that looks and feels like a jelly.

The mouth is underneath in the centre of the body, and opens into a simple stomach where prey are digested. Many jellyfish have stinging tentacles or threads hanging down from the edge of the bell. Jellyfish are closely related to **corals**, sea anemones and hydras.

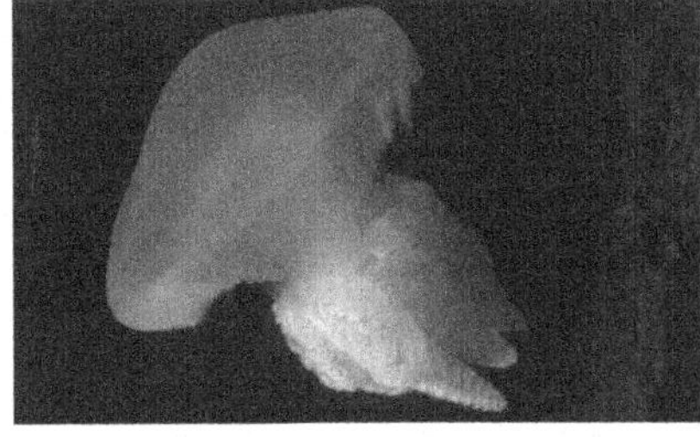

Catostylus mosaicus, the Jelly Blubber.

jet propulsion

Jet propulsion is motion that happens when a stream of gas or liquid is ejected from an object.

The stream of gas or liquid produces a force that pushes on the object, making it move. Jet propulsion engines are used in modern speedboats, airliners, rockets and spacecraft.

jet stream

Jet streams are strong, narrow winds that blow mostly eastwards for great distances in the lower part of the atmosphere.

Jet streams are mainly between about 5 and 14 kilometres above the Earth's surface. Their speeds range from about 60 to 380 kilometres per hour, depending on the latitude and time of the year.

joint

Meaning 1 In anatomy, a joint is a connection between two bones.

Some joints cannot move (such as skull bone joints), some can move a little (such as vertebrae joints) and other can move a lot (such as elbow joints). Moving joints are usually held together by muscles and ligaments, and the ends of the bones are covered in **cartilage** to make them move smoothly and not wear out.

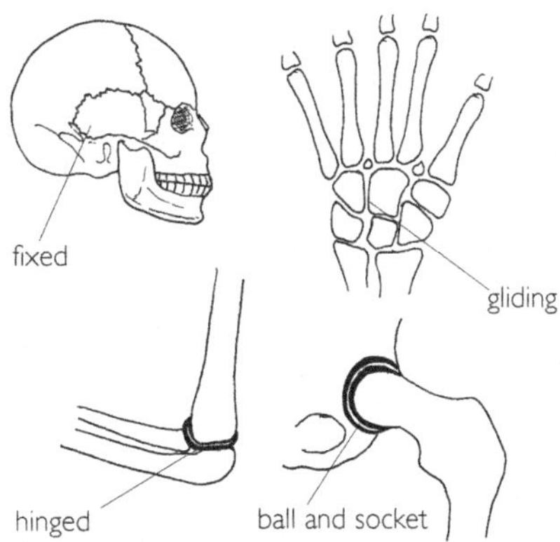

The four types of bone joints.

Meaning 2 In technology, a joint is any sort of connection between two things.

There are many different sorts of joints, depending on how strong or movable they have to be.

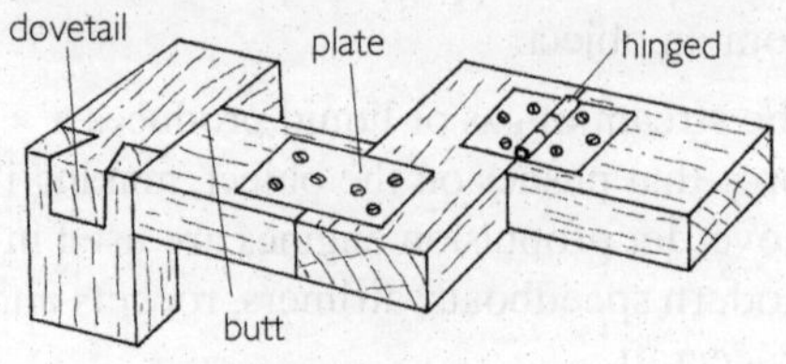

Some common joints.

Meaning 3 In geology, a joint is a crack in rock by cooling.

joule

A joule is the unit used to measure **work** and **energy**. One joule of work is done if a force of 1 newton moves an object 1 metre. The symbol for joule is J.

Jupiter

Jupiter is the fifth planet from the Sun, and it is the largest planet in the solar system. It is made up mainly of hydrogen and frozen gases, probably with a rocky core. It is 778 million kilometres from the Sun and takes 11.9 years to make one orbit. Jupiter rotates once every 9 hours and 55 minutes. Its diameter is 143 200 kilometres, and the gravity at the surface is about $2\frac{1}{2}$ times the gravity on Earth. Jupiter has 16 moons, as well as a thin ring system made up mostly of dust. The ring system was discovered by the *Voyager* spacecraft in 1979. The surface of Jupiter is liquid hydrogen, like a huge sea. Its atmosphere is very thick and is made up of 88% hydrogen and 11% helium, plus some other gases.

See also **moon, solar system**

Jurassic

See **geological time**

juvenile

Juvenile means not fully grown up. It usually refers to young animals but can also be used for other young organisms, such as plants.

Kk

keeper

A keeper is a piece of iron that is placed across the poles of a U-shaped **magnet** or two bar magnets when they are not being used. It stops the magnet from losing its magnetism.

Kelvin scale

The Kelvin scale is a scale of temperatures just like the **Celsius scale.** A temperature of 0 kelvin (0 K) is the coldest temperature possible anywhere in the universe, which is about –273 degrees Celsius. Each degree kelvin is equal to 1 degree Celsius.

kidney

In **vertebrate** animals, the kidneys are the main organs of the **excretory system**.

The kidneys remove substances containing nitrogen and other wastes from the blood and make urine, which is passed to the bladder to be excreted. Humans and other mammals have two kidneys.

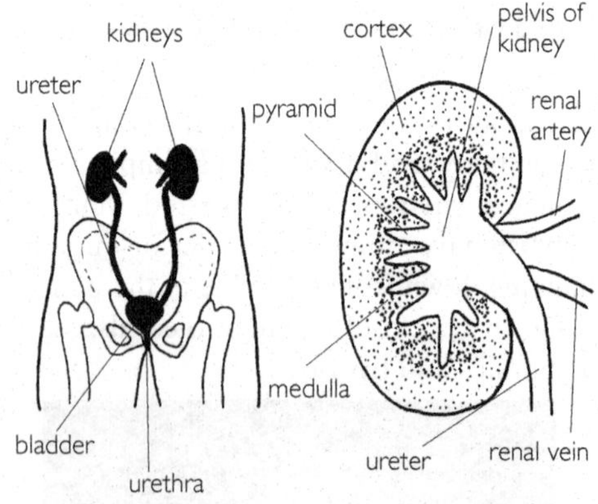

The main parts of the human excretory system (left) and a kidney in cross-section (right).

kinetic energy

Kinetic energy is the energy that an object has because of its motion.

The faster an object moves, the more kinetic energy it has. An object that is not moving has no kinetic energy.

kingdom

See **classification of living things**

knee

The knee is the main **joint** in the leg or hind leg of a **vertebrate** animal.

The human knee is unusual because it involves four bones. They are the femur, fibula, tibia and patella.

See also **skeleton**

krill

Krill are small **crustaceans** that live in huge numbers in the cold parts of the oceans.

Krill look like shrimps and feed on **plankton**. They are eaten by many whales, penguins, seals, sea birds and squids.

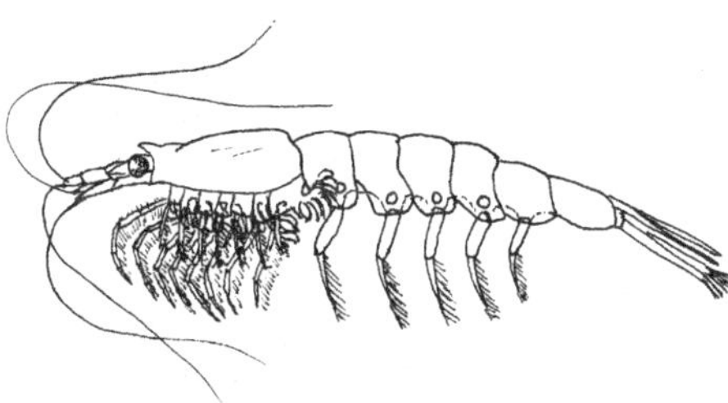

A krill from the Southern Ocean.

Ll

landform

A landform is any feature of the Earth's surface, such as a cliff, mountain, plain, plateau, valley or volcano.

La Niña

La Niña is the normal wind patterns and ocean currents in the Pacific Ocean. It is the opposite of the **El Niño**.

large intestine

The large intestine is a part of the **alimentary canal** and consists of the caecum, colon and rectum.

The large intestine absorbs water from the wastes that pass through it, and forms the solid wastes called faeces.

larva

A larva is the second stage in the life cycle of an **insect** or other **invertebrate** animal, after the egg stage. The plural of larva is larvae.

A larva can usually feed by itself, but it does not look like the adult insect and cannot reproduce.

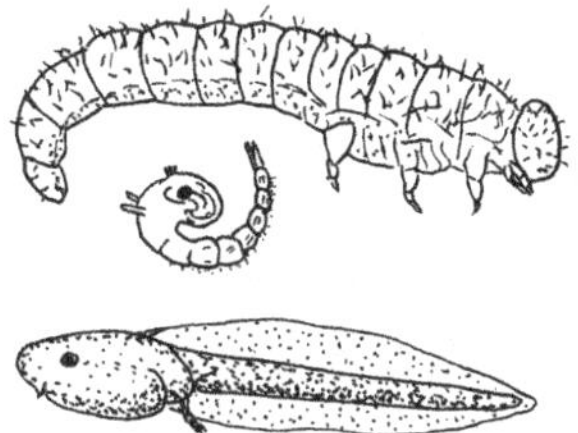

Some examples of larvae (top to bottom): sawfly larva, mosquito larva and tadpole.

See also **metamorphosis**, **nymph**

larynx

The larynx is the upper part of the trachea in the throat of some vertebrate animals.

In humans and other mammals, the larynx contains the vocal cords that enable us to speak.

laser

Laser stands for **l**ight **a**mplification by the **s**timulated **e**mission of **r**adiation. It is a device for producing a very powerful, narrow beam of light.

A laser beam consists of only one **frequency** of light, and emits the light in only one direction. Lasers are used for welding metal, printing, communications, medical surgery, to read and write CDs and for special effects in shows.

latitude

See **lines of latitude and longitude**

launch vehicle

A launch vehicle is a rocket that is designed to lift a spacecraft into orbit around the Earth.

lava

Lava is molten rock that flows out of a volcano. When it cools, it becomes a **volcanic rock**.

See also **magma**

leaf

A leaf is the part of a plant where most of the **photosynthesis** happens.

Leaves are usually flat and green, but there are many different sorts.

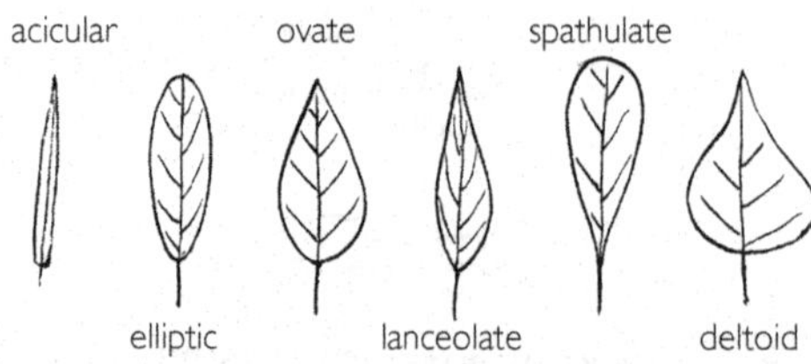

Some common leaf shapes.

lens

A lens is any **transparent** object that can bend light.

A lens that makes a beam of light narrower is called a converging lens. A lens that makes a beam of light wider is called a diverging lens.

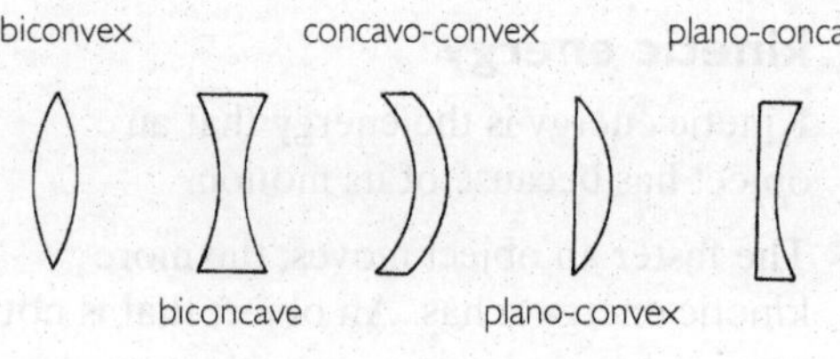

Some common lens shapes.

level

A level is any device that is used to check that two points are at the same height. Some levels can also be used to measure the difference in height between two points.

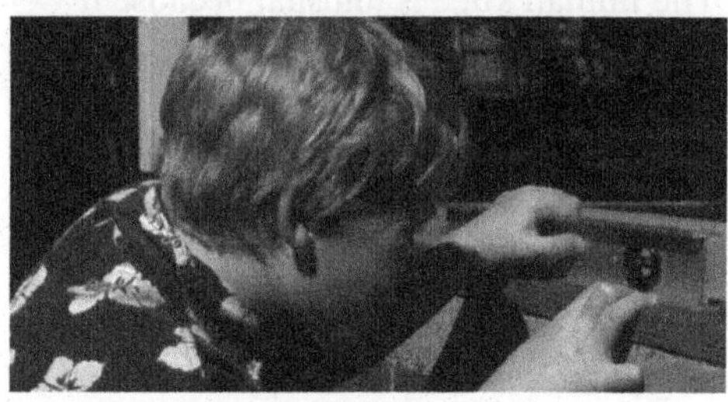

Using a builder's level.

lever

A lever is a sort of simple **machine**. It is a stiff object that can move an object by turning about one point, which is called the fulcrum.

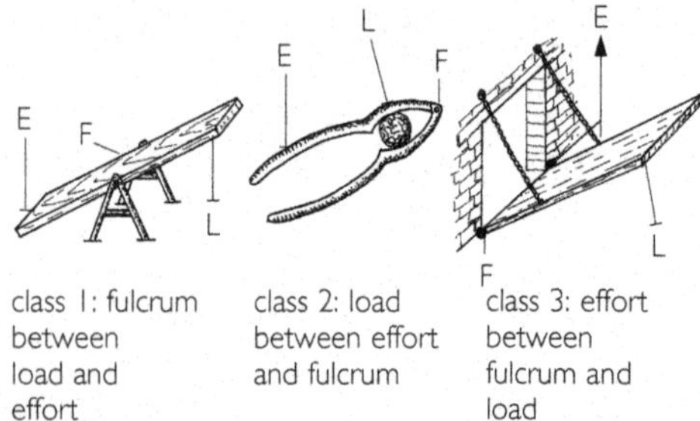

Three classes of levers (E = effort, F = fulcrum and L = load).

lichen

A lichen is an organism made up of a **fungus** and an **alga** living together. The alga makes food for the fungus, and the fungus protects the alga from drying out.

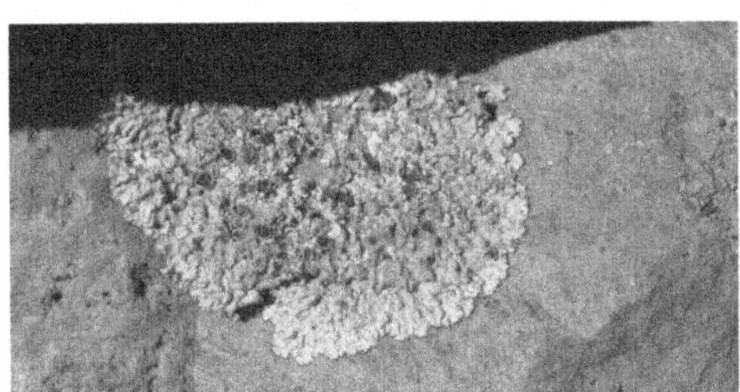
A lichen growing on a rock.

life cycle

A life cycle is all the stages that an organism goes through during its life. It is called a cycle because, even though the organism eventually dies, it can reproduce so that the cycle continues.

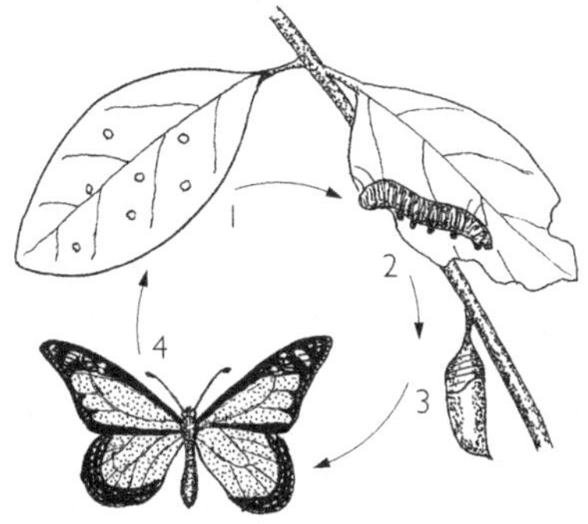

Life cycle of a butterfly (clockwise, from top left): 1. eggs laid on underside of leaf; 2. larva (caterpillar); 3. pupa (where metamorphosis happens); 4. adult butterfly.

lift

In physics, lift is the upward **force** produced by the flow of air over a wing or other smooth object.

Lift is caused because there is more air pressure under the wing than on top of it when it moves through the air.

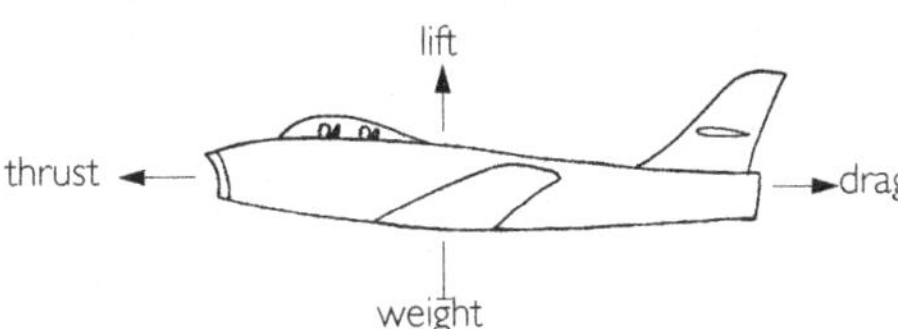

Lift and other forces that act on an aeroplane.

See also **drag**, **thrust**

light

Light is a sort of energy that can be seen by the eye. It is made up of a range of different colours that are mixed together to produce white light.

See also **photon**, **spectrum**

lightning

Lightning is a huge **electric current** that flows through the air, causing a bright flash of light.

Lightning is caused when **static electricity** that has built up in clouds is suddenly released.

See also **thunder**

light year

A light year is the distance that light travels in one year. It is almost 10 million million kilometres (9.465 million million to be exact).

limestone

Limestone is a sort of **sedimentary rock** that is formed mainly from **minerals** containing **calcium**.

Most limestone is made up of the shells of dead **molluscs** and other organisms.

lines of latitude and longitude

The lines of latitude and longitude are lines drawn on a map that help to describe the exact position of a place on the Earth's surface.

The lines of latitude show the distance from the **equator** in degrees. The lines of longitude show the distance from the **prime meridian**, also in degrees. For example, the city of Sydney is at about latitude 34 degrees south and longitude 151 degrees east.

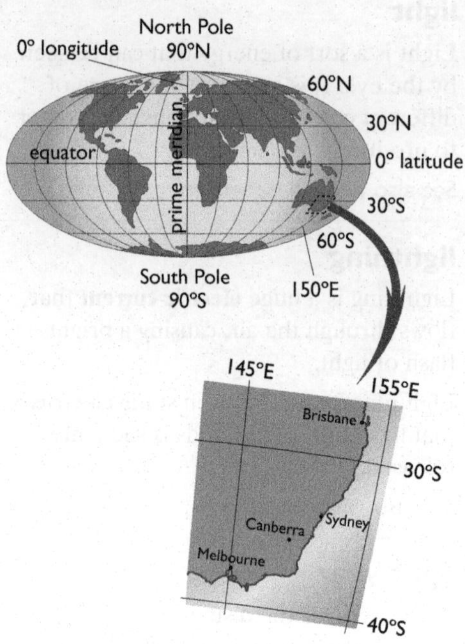

liquid

A liquid is a **state of matter** between a solid and a gas. It has a fixed volume but not a fixed shape. If it is put into a container, it changes shape to fit the container but does not change its volume.

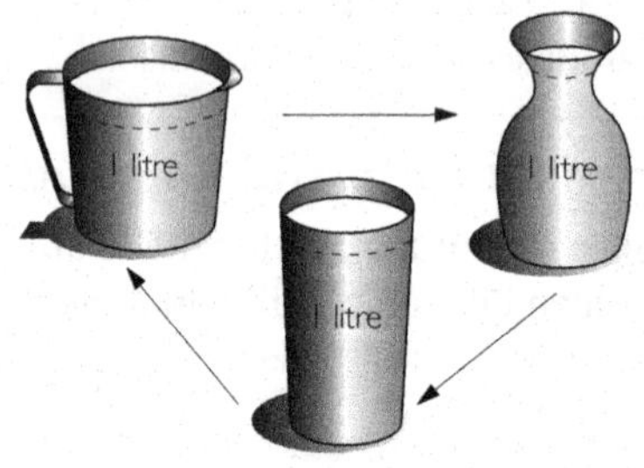

A liquid can change shape but does not change volume.

liquid crystal

A liquid crystal is a substance that acts like a liquid but has its **molecules** organised in a regular way, almost like a solid. Light passing through the crystal changes colour if an electric current is applied to it.

liquid crystal display

A liquid crystal display is a sort of **monitor** used with some computers. It is made of thousands of tiny cells containing a liquid crystal, sandwiched between two thin layers of glass with a special coating. It is usually called an LCD screen.

litmus paper

Litmus paper is paper that is specially treated so that it changes colour when it is dipped into an **acid** or a **base**. It is either red or blue.

litter

Meaning 1 Litter is the remains of plants, such as leaves and twigs, lying on top of the soil.

Meaning 2 Litter is rubbish that has been scattered about in the environment, instead of being put into rubbish bins or recycled.

liver

The liver is a large and important **organ** in the abdomen of a **vertebrate** animal. It controls the amount of nutrients in the blood, and helps to store extra food and iron. It also breaks down and removes wastes.

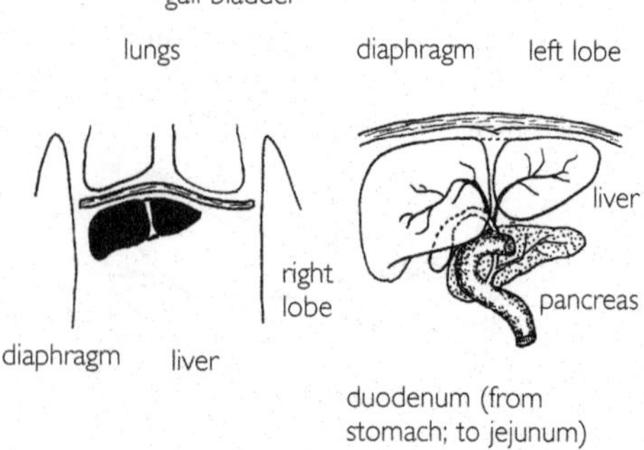

The human liver and related organs.

liverwort

Liverworts are small plants that reproduce by **spores**. They are either thin and leafy, like a moss, or thick and flat.

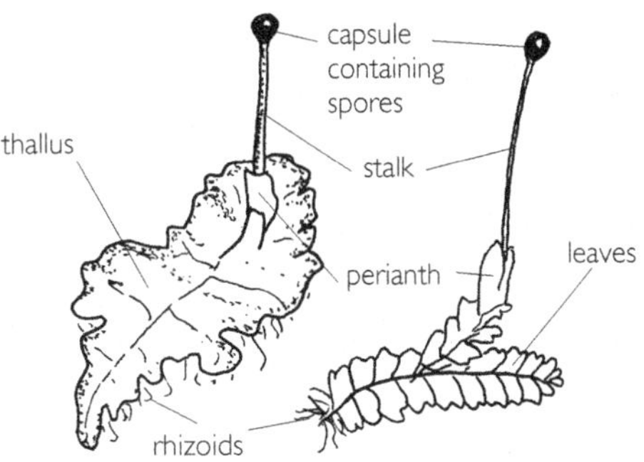

Thallose liverwort (left) and leafy liverwort (right).

lizard

Lizards are **vertebrate** animals that are covered in scales and have four legs, movable eyelids, a long tail and eardrums. Lizards are **cold-blooded** animals, so they are active only when the weather is warm. They are closely related to **snakes**.

load

The load is the force (usually weight) that must be overcome by a machine, such as a **lever**, or held up by a structure, such as an arch.

loam

Loam is soil made up of sand, silt and clay. It is usually very good soil for growing crops.

locomotion

Locomotion is the way that something moves around. It can be applied to machines as well as animals.

lodestone

A lodestone is a piece of black iron **ore** called magnetite that is a natural **magnet**. Lodestones were used as **compasses** in ancient times.

longitude

See **lines of latitude and longitude**

luminous

Something that is luminous produces its own light.

lunar eclipse

A lunar eclipse is when the Earth's shadow stops sunlight from reaching the Moon. The shadow is seen to pass across the Moon. A total lunar eclipse happens when the Moon goes through the centre of the shadow. A partial lunar eclipse happens when the Moon goes through the edge of the shadow.

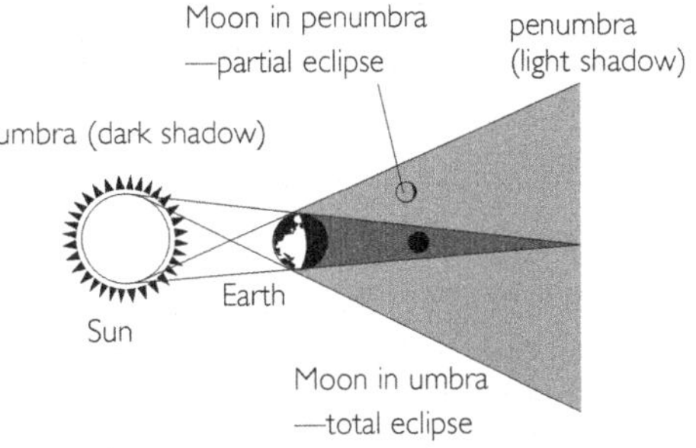

lung

The lungs are two inflatable organs in the thorax of air-breathing vertebrates. They add oxygen to the blood and remove carbon dioxide. The lungs are inflated and deflated by the movement of the diaphragm.

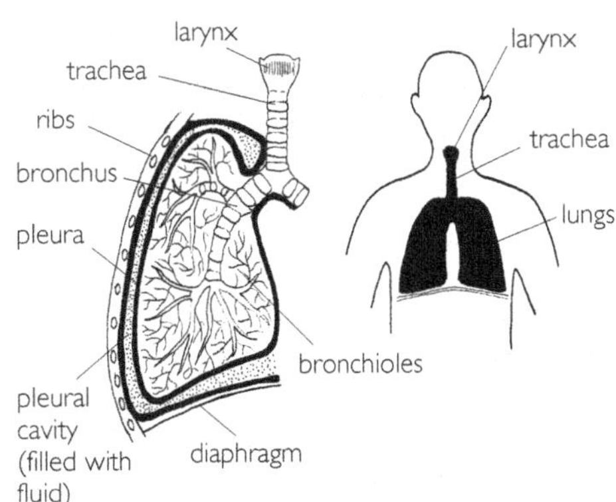

Mm

machine

A machine is anything that makes it easier to do **work**. Machines can be simple or complex. Simple machines include axles, gears, inclined planes, levers, pulleys, screws, wedges and wheels. Complex machines include computers, washing machines, vacuum cleaners, motor cars and aeroplanes.

Magellanic clouds

The Magellanic clouds are two galaxies that are close to our own **galaxy** (the Milky Way). On a dark night they look like fuzzy patches near the South Celestial Pole.

magma

Magma is very hot liquid rock beneath the Earth's crust. It can be pushed up through cracks in the crust.

See also **lava**, **mantle**

magnet

A magnet is a piece of metal that can attract metal objects by **magnetism**. Most magnets are made of iron, but nickel or cobalt can also be used.

See also **electromagnet**, **pole**

magnetism

Magnetism is a sort of **force** that causes some metals to be attracted to each other. It is caused by electrons moving in the atoms of the metal or through the metal.

magnetite

See **lodestone**

magnitude

Meaning 1 Magnitude is a number that represents the brightness of a star. Brighter stars have smaller magnitudes.

The apparent magnitude of a star is its brightness as seen from the Earth. The absolute magnitude of a star is its brightness as seen from a distance of 33 light years.

Meaning 2 Magnitude is a number that represents the strength of an **earthquake**.

See also **Richter scale**

mammal

A mammal is a **vertebrate** animal with **lungs**, mammary glands and a well-developed **nervous system** connected to a large brain. Mammals include humans, chimpanzees, dogs, cats, horses, kangaroos, whales, dolphins, mice and rats.

Examples of some mammals (clockwise, from left to right): Doll's Sheep, Marmoset, Grey-headed Flying-fox, Giraffe, Blue Whale, Bottlenose Dolphin, Spiny Mouse, Eastern Grey Kangaroo and Snowshoe Rabbit.

mammoth

Mammoths were elephant-like mammals that became extinct during the last ice age. There were several species, including the Northern Woolly Mammoth, which had a long coat of fur, and the Imperial Mammoth, which was the largest of all the mammoths.

Northern Woolly Mammoth.

mantle

The mantle is the part of the Earth between the core and the **crust**. The upper part of the mantle is mostly solid rock, but the rest is mostly molten rock.

See also **magma**

marble

Marble is a very hard sort of **metamorphic rock** that is formed when limestone rock melts and then cools again. It is used in building and for stone ornaments.

Mars

Mars is the fourth planet from the Sun. It is a rocky planet made up mainly of iron and silicates. Mars is 227.4 million kilometres from the Sun and takes 687 days to make one orbit. It rotates once every 24 hours 37 minutes. Its diameter is 6794 kilometres, and the gravity at the surface is about $\frac{2}{5}$ of the gravity on Earth. Mars has two moons called Phobos and Deimos. The surface is mainly dry and dusty, with craters, extinct volcanoes and ancient water channels, and there is an ice cap at each pole. The atmosphere is mainly carbon dioxide.

See also **moon, solar system**

marsupial

Marsupials are **mammals** that carry their young in a pouch until they are old enough to look after themselves. Only female marsupials have this pouch, which is called a marsupium and contains the mammary glands that produce milk.

mass

Mass is the amount of **matter** in an object. It is a measure of the **inertia** of the object. The more mass the object has, the more inertia it has.

materials

The materials are all the things you need to do an **experiment**.

matter

Matter is anything made up of **atoms** or parts of atoms. It can be a solid, a liquid or a gas.

See also **states of matter**

medium

Meaning 1 In physics, a medium is a solid, liquid or gas through which light, sound or other sort of energy passes.

Meaning 2 In biology, a medium is a substance on which a **micro-organism** is grown in a laboratory. A common medium is agar, which is made from seaweed.

megafauna

Megafauna means large animals, especially the species that are now extinct. Australia once had many species of megafauna, including giant kangaroos and wombats.

Diprotodon optatum, an example of megafauna from Australia.

melting point

The melting point is the temperature at which a solid turns into a liquid. For example, the melting point of ice is 0 degrees Celsius.

mercury

Mercury is a silvery metal that is found in small amounts in **ores** on the Earth. Pure mercury is the only metal that is a liquid at ordinary temperatures on Earth. It is very poisonous and must be used with great care. The symbol for mercury is Hg.

See also **barometer, thermometer**

Mercury

Mercury is the closest planet to the Sun. It is a small, rocky planet made up of nickel-iron and silicates. Mercury is about 58 million kilometres from the Sun and takes 88 days to make one orbit. It rotates once every 59 days. Its diameter is 4880 kilometres, and the gravity at the surface is a little less than half the gravity on Earth. Mercury has no moons and no atmosphere. Its surface is covered with craters.

See also **solar system**

mesosphere

See **atmosphere**

Mesozoic era

See **geological time**

metabolism

Metabolism means all the different **chemical reactions** that happen inside the body to make new tissues, to produce energy for cells and to get rid of wastes.

metal

A metal is a shiny substance that is a good conductor of heat and electricity. Metals usually have a regular arrangement of atoms. All metals except **mercury** are solid at ordinary temperatures on Earth.

metamorphic rock

Metamorphic rock is a rock that has been formed from other rocks by great heat or pressure. For example:

gneiss	is formed from	granite
slate	is formed from	shale
quartzite	is formed from	sandstone
marble	is formed from	limestone
greenstone	is formed from	basalt

See also **igneous rock, sedimentary rock**

metamorphosis

Metamorphosis is the change of an animal from one form into another, such a caterpillar into a butterfly or a tadpole into a frog. It is used especially for the change of an insect **larva** into an adult.

See also **pupa**

meteor

A meteor is a piece of rock that enters the Earth's atmosphere.

Meteors are usually travelling so fast that they burn up in the atmosphere, producing a bright streak of light as they shoot across the sky. Most of them are less than the size of a pea, but some are huge chunks of rock.

A meteor (arrowed) streaks across the sky.

meteorite

A meteorite is a **meteor** that hits the surface of the Earth. Some meteorites are so big that they make craters in the surface.

meteorology

Meteorology is the study of weather, climate and other changes that happen in the Earth's **atmosphere**.

methane

Methane is a gas that has no smell or colour. Each **molecule** has one carbon atom and four hydrogen atoms, so its formula is CH_4.

Methane is the main gas in **natural gas**, and is also made when organic matter decomposes.

microchip

A microchip is a small electronic **chip**.

micro-organism

A micro-organism is an organism or virus that is too small to see without a lens or microscope. Micro-organisms include bacteria and most one-celled organisms.

microprocessor

Meaning 1 A microprocessor is a **microchip** that controls the way that a **personal computer** works. Microprocessors are usually described by how fast they can work. For example, a 1 GHz (gigahertz) microprocessor can do one thousand million operations every second.

Meaning 2 A microprocessor is a small computer designed to do one particular job, such as counting the number of cars that go past on a road.

microscope

A microscope is an instrument that allows you to see small objects clearly.

A light microscope uses glass lenses to let you see objects using normal light. An electron microscope lets you see objects using a beam of electrons instead of light and electromagnets instead of lenses.

microwaves

Microwaves are a sort of **electromagnetic radiation** with a short **wavelength** between 1 millimetre and 500 millimetres. They are used to send television, radio and telephone signals, and for heating food in a microwave oven.

migration

Migration is the movement of many animals from one place to another. Most animals migrate before winter to avoid cold weather, to find more food or to go to places where they can breed.

Milky Way

See **galaxy**

mineral

Meaning 1 A mineral is a natural substance with a particular chemical composition and structure. Some examples are quartz, ruby and rutile.

Meaning 2 A mineral is an **ore** that is dug out of the ground.

minor planet

A minor planet is another name for an **asteroid**.

mirage

A mirage is a sort of optical illusion that makes an object that is far away seem to float in the air. It is caused when light from the object is reflected off a layer of cold air near the ground.

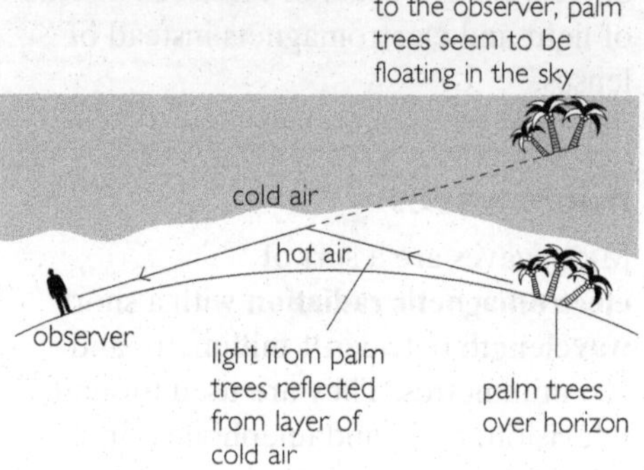

mirror

A mirror is a very shiny surface that reflects almost all the light that shines on it. Mirrors can be any shape, but they are usually flat, concave or convex. Mirrors used in telescopes and other scientific instruments are usually made of special glass that has a thin coating of silver or aluminium on the front. Ordinary mirrors can be made of other materials and have the coating on the back.

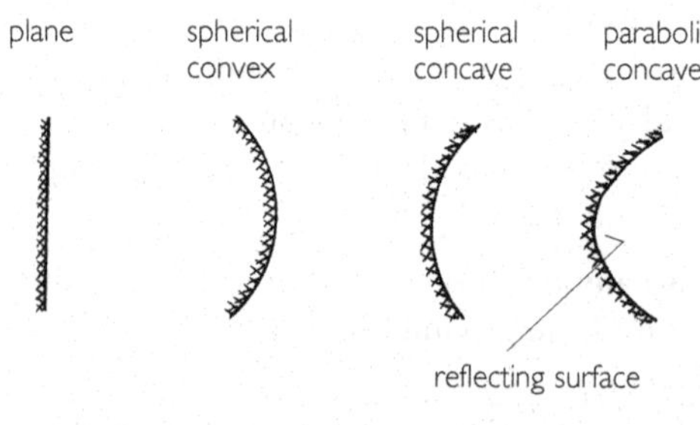

Some common mirror shapes.

mixture

A mixture is made when a substance is added to another substance, but they do not change. For example, salt added to pepper makes a mixture that is still salt and pepper.

See also **solution**

modem

A modem is an electronic device that connects a computer to a **server** through a telephone line or optical cable. A modem can be a part of the computer, or it can be a **peripheral device**.

Mohs scale

The Mohs scale is a scale of numbers used to describe how hard something is.

To work out the hardness of something, you simply scrape it against different minerals listed in the scale. If it scrapes away the mineral, it is harder than that mineral, so it will have a higher hardness number. If it is scraped by the mineral, it is softer than the mineral, so it will have a lower hardness number.

Mohs scale of hardness

Hardness	*Mineral*
1	Talc
2	Gypsum
3	Calcite
4	Fluorite
5	Apatite
6	Orthoclase
7	Quartz
8	Topaz
9	Corundum
10	Diamond

molecule

A molecule is the smallest particle of a **compound** that can exist. Molecules are made up of two or more **atoms**. The atoms can be from the same **element** or from different elements.

mollusc

Molluscs are **invertebrate** animals that usually have a hard shell protecting their soft tissues and organs.

Some molluscs have spiral shells, some have shells made of two parts joined together, and others hardly have any shell at all. Most molluscs live in the oceans, but some live in lakes and streams or on land.

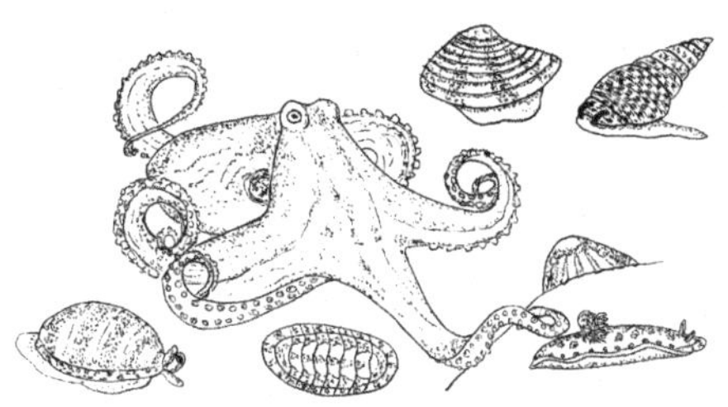

Some examples of molluscs (clockwise, from top left): octopus, bivalve, prosobranch (snail), limpet, opistobranch (sea slug), chiton and cowrie.

monitor

Meaning 1 A monitor is a screen attached to computer to display images or text. It is also called a VDU, or visual display unit.

Meaning 2 Monitors are large lizards found in Africa, Australia and Indonesia.

The lace monitor of Australia.

monotreme

Monotremes are **mammals** that lay eggs. The only monotremes that are still living are platypuses and echidnas.

A Short-beaked Echidna.

monsoon

The monsoon is a strong wind that blows across the Indian Ocean and southern Asia. It brings heavy rains from April to October in the northern hemisphere, and from December to February in the southern hemisphere.

moon

A moon is a natural **satellite** that orbits around a planet. All the planets in the **solar system** have moons except Mercury and Venus.

Planetary moons

Planet	*Moons (in order from closest to farthest from the planet)*
Earth	Moon (Luna)
Mars	Phobos, Deimos
Jupiter	J3, J1, Amalthea, J2, Io, Europa, Ganymede, Callisto, Leda, Himalia, Elara, Lysithea, Ananke, Carme, Pasiphae, Sinope
Saturn	S15, S14, S13, S11, S10, Mimas, Enceladus, Tethys, S16, S17, Dione, Dione B, Rhea, Titan, Hyperion, Iapetus, Phoebe
Uranus	Muriel, Ariel, Umbriel, Titania, Oberon
Neptune	Triton, Nereid
Pluto	Charon

M

Moon

The Moon is the natural **satellite** that orbits the Earth. It is 3476 kilometres in diameter, and 382 000 kilometres from the Earth on average. It rotates once every 27 days and 7 hours, which is the same time it takes to orbit the Earth once.

Morse code

Morse code is a code invented by Samuel Morse to send messages by telegraph or radio. Each letter and number has its own combination of dots and dashes. The message is sent using a simple tapping machine. A short tap is a dot and a long tap is a dash.

The Morse code

A	.–	N	–.
B	–...	O	–––
C	–.–.	P	.––.
D	–..	Q	––.–
E	.	R	.–.
F	..–.	S	...
G	––.	T	–
H	....	U	..–
I	..	V	...–
J	.–––	W	.––
K	–.–	X	–.––
L	.–..	Y	–.––
M	––	Z	––..
1	.––––	6	–....
2	..–––	7	––...
3	...––	8	–––..
4	....–	9	––––.
5	.....	0	–––––

moss

A moss is a small plant that produces **spores** inside a capsule. The capsule and its stalk are usually brown. Mosses have leaves and make their food by **photosynthesis**. They are found in all sorts of places: from rainforests to deserts.

See also **liverwort**

motor

A motor is a machine that changes electrical energy or chemical energy into mechanical energy. Some examples are the motor of a car and the motor in an electric drill.

mould

Meaning 1 In biology, a mould is a sort of tiny fungus that grows on vegetables and other organic matter.

Meaning 2 A mould is a hollow shape that is used to form a solid shape. A liquid is poured into the mould and when it turns into a solid, the mould is taken off, leaving the solid shape.

A plastic mould for making jelly.

multimedia

In computing, multimedia means using different sorts of information or effects.

For example, a multimedia computer game might use video pictures, graphics, sound and text.

multimeter

A multimeter is an instrument that can be used to measure different features of an electric circuit, such as voltage, current and resistance.

muscle

In animals, muscle is a strong tissue that can stretch and shorten without being damaged.

Muscles enable animals to move, and also help with blood flow, breathing, eyesight and other functions in the body.

See also **joint**

mushroom

A mushroom is a **fungus** that has a stalk and a cap with gills.

Nn

nadir

The nadir is the point on the **celestial sphere** that is directly below you. It is in the opposite direction to the **zenith**.

natural gas

Natural gas is a gaseous mixture of **hydrocarbons** that is found underground in sedimentary rocks.

Natural gas is usually found where there is underground oil, and can be obtained by drilling down into the rock. It was formed by the decay of dead plants millions of years ago. It usually contains flammable gases, such as methane, ethane, propane and butane, so it can be used as a fuel.

Neanderthals

Neanderthals were humans who lived in Europe and western Asia between 100 000 and 35 000 years ago. They were shorter than modern humans and had a different head shape. Neanderthals no longer exist.

nebula

A nebula is a fuzzy-looking object in the night sky. The plural of nebula is nebulae.

Some nebulae are galaxies, so they are called galactic nebulae. Others are huge clouds of gas and dust called gaseous nebulae.

A gaseous nebula.

N

Neocene period

See **geological time**

Neptune

Neptune is the eighth planet from the Sun. It was discovered in 1846 by a French astronomer, J. C. Galle, after mathematicians worked out its position from wobbles that were observed in the orbit of Uranus. It is made up mainly of ice, hydrogen and helium, probably with a rocky core. It is 4497 million kilometres from the Sun and takes 165 years to make one orbit. Neptune is about 50 000 kilometres in diameter and rotates once every $18\frac{1}{2}$ hours. The surface is probably an ocean of liquid hydrogen. The gravity at the surface is a little more than the gravity on the Earth. Neptune has two moons called Triton and Nereid, and it seems to have a ring system.

See also **moon**, **solar system**

nerve

Nerves are the tissues that pass messages between the brain or spinal cord and the different parts of the body.

Nerves send information about our body and our surroundings to the brain. They also send instructions from the brain to the rest of the body.

See also **nervous system**

nervous system

In animals, the nervous system is made up of the brain, spinal cord (or nerve cord) and nerves.

The nervous system receives, interprets and responds to information about the body and its surroundings. It also coordinates the various parts of the body.

network

In computing, a network is a group of computers that are connected together. Computers in a network can share files and software, as well as a connection to the Internet.

See also **server**

neutral

A neutral substance is not an acid and not a base. It has a **pH** of exactly 7.

neutron

A neutron is a particle in an atomic **nucleus** that does not have an **electric charge**.

See also **atom**

neutron star

A neutron star is a very dense **star** that consists mainly of neutrons. It is probably the last stage in the life of a large star.

See also **black hole**

nitrogen

Nitrogen is an **element** that is important for all life on Earth. It makes up about 78% of the air we breathe. Nitrogen is also found in **proteins**, amino acids and other substances needed by living things. The symbol for nitrogen is N. Nitrogen has many uses in technology: as a gas and as a liquid.

nitrogen cycle

The nitrogen cycle is the continuous movement of **nitrogen** through the environment. It is one of the main chemical cycles on the Earth. The main part of the nitrogen cycle is shown in the diagram on following page.

See also **carbon cycle**, **oxygen cycle**, **water cycle**

nocturnal

Nocturnal means being active at night or happening at night. It is the opposite of **diurnal**.

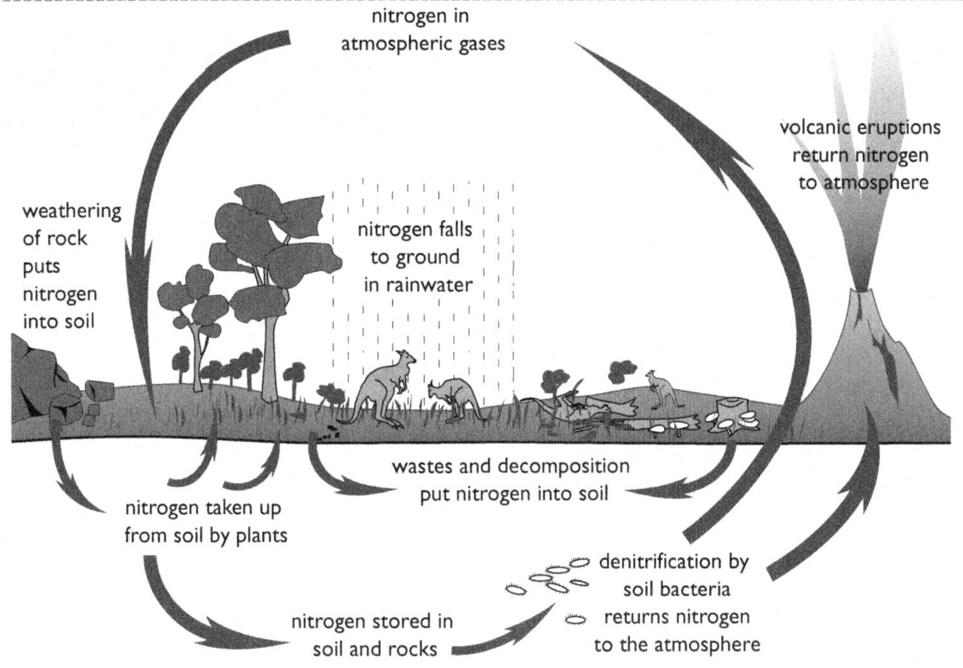

The nitrogen cycle. Nitrogen in the soil is used by plants, and animals take in the nitrogen when they eat the plants. When the plants and animals die, their bodies are broken down by bacteria, and the nitrogen is returned to the soil.

noise

Noise is any unwanted sound. It is measured on the decibel scale.

The quietest sound that can be heard measures 0 decibels. A whisper measures about 20 decibels, passing traffic is about 70 to 80 decibels, and a rock band measures about 110 to 130 decibels.

North Celestial Pole

The North Celestial Pole is the point on the **celestial sphere** that is directly above the Earth's geographic North Pole. In the northern hemisphere, the stars seem to turn around this point.

See also **South Celestial Pole**

nova

A nova is a **star** that suddenly becomes much brighter than normal.

A nova can happen when an old star blows away its outer layers. It can also happen in a **binary star**, when one star pulls matter from the other one.

nuclear energy

Nuclear energy is energy that comes from nuclear fission or fusion.

Nuclear fission is when the **nucleus** of an atom splits into two parts. Nuclear fusion is when the nucleus of one atom joins with the nucleus of another atom to form a new, larger nucleus. Both of these events produce a large amount of energy.

nuclear reactor

A nuclear reactor is a device designed to produce **nuclear energy** from nuclear fission.

Nuclei of radioactive **uranium** or plutonium atoms are split by firing **neutrons** into them. This produces energy, as well as more neutrons, which can split other atoms.

nucleus

Meaning 1 In biology, the nucleus is the part of the cell that controls the functions of the cell and contains the **DNA**.

Meaning 2 In physics, the nucleus is the central core of an **atom**, consisting mainly of protons and neutrons.

Meaning 3 In astronomy, a nucleus is the dense part of the head of a **comet**.

---

nutrient

A nutrient is a substance that an organism uses to stay alive and grow. Nutrients for humans include carbohydrates, proteins, fats, minerals and vitamins.

---

nutrition

Nutrition means the way that organisms get the food they need to stay alive and grow, and how they turn the food into living tissue.

---

nymph

A nymph is a young dragonfly, grasshopper or other **insect** that does not have a pupal stage. A nymph has mouth parts and eyes like the adult insect, but it does not have adult wings or reproductive organs. Nymphs develop directly into adults.

See also **larva**

Oo

objective lens

An objective lens is the main lens in an optical instrument, such as a **telescope**, binoculars or a **microscope**.

See also **eyepiece**

---

observatory

An observatory is a place where there is a telescope for looking at the sky. The telescope could be an optical **telescope** or a **radiotelescope**.

---

octopus

An octopus is a **mollusc** that has eight arms with suckers on them, a large sack-like body, a biting mouth with strong jaws and well-developed eyes. Octopuses are closely related to squids and cuttlefish.

---

odour

An odour is something you can smell. Odours are caused by chemicals in the air. We can smell them if they are breathed into the nose.

---

offspring

Offspring is another name for a young animal.

---

oil

An oil is a **viscous** liquid that does not mix very well with water.

Many different oils are found in plants and animals, and mineral oils are found in underground rocks. We can also make **synthetic** oils. We use oils mainly for cooking and to make moving parts in machines work smoothly.

omnivore

An omnivore is an animal that eats plants and animals.

See also **carnivore**, **herbivore**

on-line

On-line means connected to a computer or a computer **network**, especially the Internet. It also refers to things that can be done on a computer network, such as on-line banking.

opaque

Opaque means not letting light pass through. For example, a solid piece of metal is opaque.

See also **translucent, transparent**

operating system

An operating system is the **program** that tells a computer how to work. There are many different operating systems. For example, in **personal computers** there are Windows XP and Mac OS X.

optical fibre

Optical fibres are very fine threads of glass that can be used to send light waves over long distances.

Optical fibres are used in telecommunications because information, such as TV pictures, can be sent as light waves. Optical fibres are also used in medical instruments for looking into the lungs, stomach, intestines, bladder and other parts of the body.

orbit

An orbit is the path followed by an object that is going around another object. Examples of orbits are the path of a moon or spacecraft around a planet, a planet around a star and an electron around the nucleus of an atom. The orbits of moons and planets are usually ellipses.

Ordovician period

See **geological time**

ore

An ore is a **mineral** that can be processed to obtain a **metal**.

Some examples of ores are argentite (for silver), bauxite (for aluminium), cassiterite (for tin), cuprite (for copper), galena (for lead), haematite (for iron), pitchblende (for uranium) and rutile (for titanium).

organ

In living things, an organ is any part that has a special function. In mammals, organs include the brain, eyes, lungs, stomach, kidneys and liver. In flowering plants, organs include leaves, roots and flowers.

organic

Meaning 1 In chemistry, organic means containing carbon atoms. Organic substances include all the substances made by living organisms.

See also **inorganic**

Meaning 2 Organic means made only from natural things, without the use of **synthetic** chemicals.

ovum

Meaning 1 In animals, an ovum is an egg cell in female animals. This cell can be fertilised by male **sperm** to create a new animal.

Meaning 2 In plants, an ovum is the cell in the female part of a flower that can be fertilised by a **pollen** grain to create a **seed**.

oxygen

Oxygen is an **element** that is in rocks, water and air. It is the most common element on the Earth.

In the air, oxygen is a gas. Each molecule of oxygen gas is made up of two oxygen atoms, so its formula is O_2. The symbol for oxygen is O.

See also **ozone**

ozone

Ozone is a colourless gas made up of oxygen atoms.

Ozone exists naturally in the atmosphere, mainly in the **ozone layer**. It is made naturally when **ultraviolet radiation** from the Sun hits oxygen molecules. It is also made by an electric spark in the air, such as in a lightning stroke or in some electrical equipment. Each molecule of ozone has three atoms of oxygen, so its formula is O_3.

ozone hole

An ozone hole is an area of the ozone layer where there is less ozone than normal.

Because there is less ozone, **ultraviolet radiation** can pass through an ozone hole more easily. Ozone holes occur naturally, but they are made worse by some chemicals we use that destroy ozone.

ozone layer

The ozone layer is a layer of the **atmosphere** where there is a lot of ozone. It is between 15 and 20 kilometres above the Earth's surface.

The ozone absorbs most of the ultraviolet light in sunlight, and stops it reaching the Earth's surface.

See also **ozone hole**

Pp

palaeontology

Palaeontology is the study of fossils and extinct organisms.

Palaeontology helps us to understand **evolution** and **geology**, and tells us what the Earth's climate might have been like in ancient times.

Palaeozoic era

See **geological time**

pancreas

The pancreas is a **gland** in the body that makes the enzymes used in digestion. It also makes chemicals called insulin and glucagon that control the sugar levels in the blood.

parallax

Parallax is the change in the position of an object against the background when you look at the object from two different places.

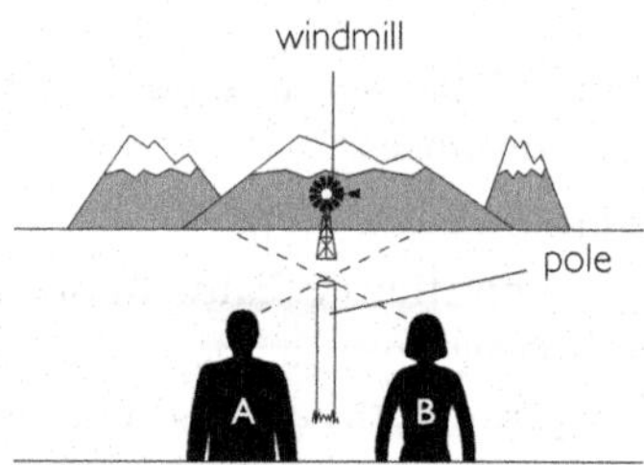

The effect of parallax. Observer A sees the windmill on the left side of the pole. Observer B sees the windmill on the right side of the pole.

parasite

Parasites are organisms that live in or on other organisms (called hosts) and feed off them.

Some examples of parasites are mistletoes on trees, fleas on dogs and tapeworms in humans. Parasites do not always harm the host, but many of them do.

pasteurise

Pasteurise means to heat milk or other liquid foods to destroy harmful bacteria.

Milk is usually pasteurised by first heating it to 65 degrees Celsius for 30 minutes and then quickly cooling it to less than 10 degrees Celsius.

pendulum

A pendulum is an object that swings freely back and forth about one point.

The simplest pendulum is a rod or string with a weight attached to one end. The time taken for one swing depends only on the length of the string. Because the time of the swing can be set exactly by changing the length of the rod, pendulums are often used in clocks.

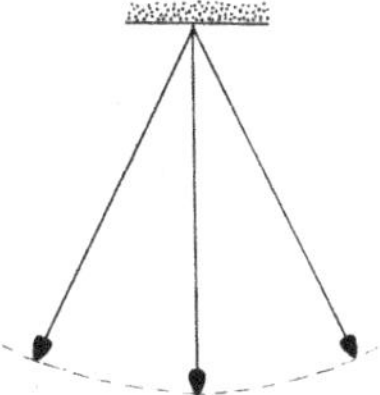

A simple pendulum consists of a mass attached to a string. The mass swings the same distance each side of vertical, and every swing takes exactly the same time.

PC

See **personal computer**

penumbra

A penumbra is the part of a shadow that is not completely dark.

See also **eclipse**

percussion

Percussion is the striking of one object against another, such as a hammer against a nail.

The sound or vibration produced by percussion is called a percussion wave. Musical instruments that are played by hitting one part against another, such as a drum, a triangle or a xylophone, are called percussion instruments.

perennial

Meaning 1 A perennial plant is a plant that lives for more than two years.

See also **annual**

Meaning 2 Perennial refers to a stream that flows all year round. Streams that flow for only part of the year are called intermittent.

peripheral device

A peripheral device is any part of a computer that is connected to the **CPU** of a computer but is not needed for the CPU to operate. Some examples are a modem and a printer.

Permian period

See **geological time**

personal computer

A personal computer is a computer that is designed to be used by one person at a time. A personal computer is often called a PC.

See also **server**

pesticide

A pesticide is a chemical that is used to kill pest organisms.

Petri dish

A Petri dish is a small glass dish with straight sides and a lid.

Petri dishes are used in laboratories to grow bacteria and other micro-organisms, or to hold things to look at under a microscope.

pH

pH is a measure of how acidic or alkaline something is, such as water or soil. **Acids** have a pH of less than 7 and **bases** have a pH of more than 7. A pH of exactly 7 is called neutral.

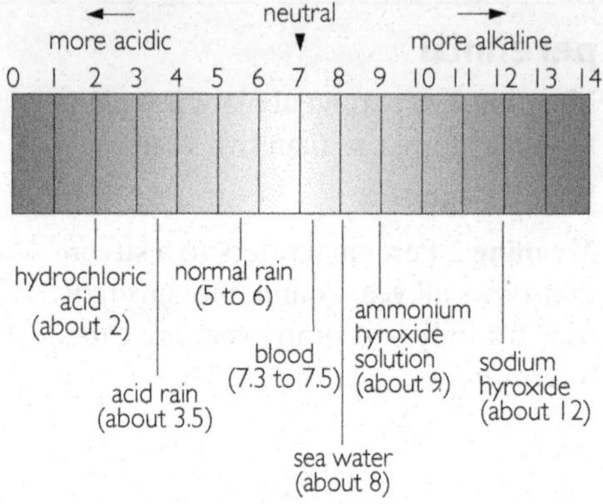

phases of the Moon

In astronomy, a phase is the shape of the bright part of the **Moon** that we see from the Earth. This shape changes as the Moon orbits the Earth.

The planets also have different phases as they orbit the Sun, but they are not so obvious.

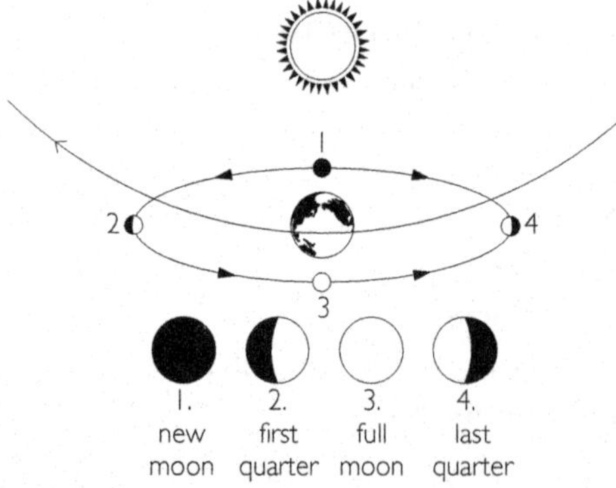

phloem

Phloem is the tissue that carries food from the leaves to the other parts of a plant. It is made up of bundles of hollow tubes that are joined at their ends.

See also **wood, xylem**

photosynthesis

Photosynthesis is the way that plants and algae use sunlight, carbon dioxide and water to make new tissues.

The energy in sunlight is trapped by a **pigment** in cells called chlorophyll. This energy is then used in a chemical reaction between carbon dioxide and water to make simple **organic** chemicals. These chemicals can be used by the plant to make new tissues.

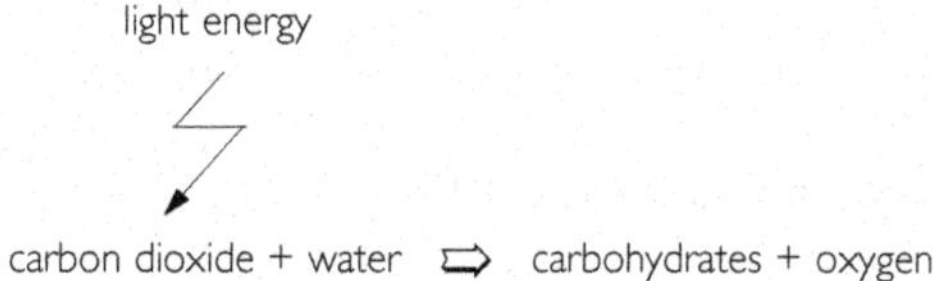

phylum

See **classification of living things**

physics

Physics is the study of energy and matter, especially the forces involved when they interact or change.

Physics can be divided into the study of heat, light, sound, electricity and magnetism, mechanics, quantum physics and relativity.

pig iron

Pig iron is the sort of iron that is obtained by heating iron **ore**.

It is called pig iron because it is made into blocks called pigs. It is used to make cast iron and alloys, such as **steel**.

pigment

A pigment is anything that gives colour to a substance.

Pigments are found in living things, and can also be made artificially for use in paints and dyes.

pipette

A pipette is a fine glass tube that is used to transfer small amounts of liquid into a container.

A pipette often has a scale marked on it so that the amount can be measured exactly.

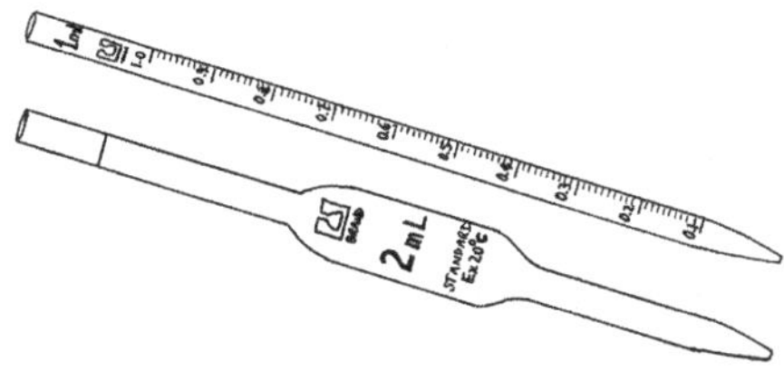

Two types of pipette.

pitch

Meaning 1 In **acoustics**, pitch is the main **frequency** of a sound heard by a listener. A high-pitched sound has a high frequency and a low-pitched sound has a low frequency.

Meaning 2 In engineering, pitch is the angle of the blade of a propeller.

Meaning 3 Pitch is the distance between the grooves of a **screw**. One turn of the screw makes a movement equal to the pitch.

Meaning 4 Pitch is an up or down movement of the ends of an aeroplane or boat.

Meaning 5 In chemistry, pitch is a very **viscous** liquid that is left over from making tar or bitumen. It is used mainly for making roads and for waterproofing buildings.

planet

A planet is a large, solid, globe-shaped object that orbits a star.

Planets do not produce their own light. They can be seen only because they reflect light from the star.

See also **asteroid**, **comet**, **moon**, **solar system**

planetarium

A planetarium is a machine that can show what the night sky looks like. It does this by projecting beams of light onto the inside of a dome.

The building that houses a planetarium is sometimes called a planetarium too.

plankton

Plankton are tiny organisms that drift about in oceans or lakes. They include animals, **algae** and **protozoa**.

Plankton are the main food for many animals, including whales.

See also **krill**

plant

A plant is an organism that can make new tissues by **photosynthesis**, and is made of many different sorts of cells.

See also **classification of living things**

plasma

Meaning 1 In biology, plasma is the watery part of **blood**. It is mainly water but also contains salts, proteins, sugars, fats, hormones, vitamins and wastes from cells.

☞

Meaning 2 In physics, plasma is a sort of matter that exists in stars. It can also exist in the space between stars, and can be made in a laboratory. It contains free **electrons** and an equal number of positively charged **ions**. Plasma is sometimes called the fourth **state of matter**.

plastic

Plastics are solid materials that can be shaped or moulded by heat or pressure. Most plastics are made from **synthetic** substances.

plate tectonics

Plate tectonics is the idea that the Earth's crust is made up of different pieces called plates that can move around very slowly. Plate tectonics explains how mountains are built, why **earthquakes** and **volcanoes** happen, and how continents have moved about on the Earth.

See also **continental drift**

Pluto

Pluto is the ninth planet from the Sun, and it is the most distant planet that we know of. It was discovered in 1930 by an American astronomer, Clyde Tombaugh. It is probably made up mostly of ice and rock. Pluto is 5900 million kilometres from the Sun on average, but sometimes it is closer to the Sun than Neptune because it has an unusual orbit. It takes 248 years to make one orbit around the Sun. Pluto is about 3100 kilometres in diameter, and the gravity at the surface is only about $\frac{1}{20}$ of the gravity on the Earth. Pluto has one moon called Charon, which is so big that some astronomers call Pluto and Charon a "double planet".

See also **moon, solar system**

plutonic rock

Plutonic rocks are **igneous rocks** formed when **magma** cools far below the surface of the Earth. Plutonic rocks are also called intrusive rocks.

The most common sort of plutonic rock is **granite**.

See also **volcanic rock**

pole

Meaning 1 In geography, a pole is the very north or south point on the Earth's surface. The geographic pole is where the Earth's axis passes through the surface. The magnetic pole is the place on the Earth that one end of a compass needle points to. It is in a slightly different place to the geographic pole.

Meaning 2 In physics, a pole is the end of a **magnet**. Every magnet has two poles: one called the north-seeking (or north) pole and the other called the south-seeking (or south) pole.

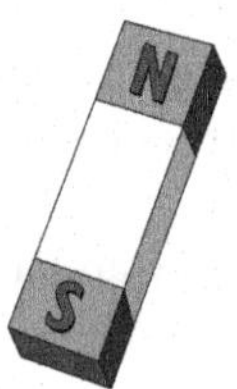

The poles of the Earth (left) and the poles of a bar magnet (right).

pollen

Pollen is the powdery grains released by the male part of a flower.

A pollen grain contains a cell that can join with an **ovum** to make a seed.

pollutant

A pollutant is anything that causes **pollution**.

pollution

Pollution is something in the environment that has been produced by humans, and is harmful to living things.

Pollution can be in the air, water or soil, and includes noise and heat.

See also **greenhouse effect**, **smog**

population

Meaning 1 In ecology, a population is a group of organisms that belong to the same species, living in a particular area or **community**.

Meaning 2 Population is the number of organisms in an area.

porous

Porous means that a liquid or a gas can pass through.

potential energy

Potential energy is the energy that an object has stored inside it because of its position or shape.

For example, a spring gains potential energy if you squash it. Potential energy can be turned into kinetic energy. For example, you can make the spring jump by letting it go.

powder

A powder is a small, dry solid, such as dust or talcum powder.

Precambrian era

See **geological time**

precipitation

Precipitation means any sort of water that falls from the atmosphere, such as rain, sleet, hail or snow.

predator

A predator is an animal that kills and eats other animals (called their prey).

Some examples are eagles, dolphins, snakes and spiders.

See also **parasite**, **scavenger**

prey

Prey is an animal that is killed and eaten by a predator.

primate

Primates are a group of mammals that includes humans, apes, monkeys and lemurs. They have large brains, excellent eyesight and well-developed hands and feet.

prime meridian

The prime meridian is the line of longitude that passes through the Royal Observatory at Greenwich, in England. It has a longitude of 0 degrees.

See also **lines of latitude and longitude**

prism

A prism is a block of glass or other **transparent** material, usually with a triangular base.

Prisms are used to change the direction of a beam of light, and also to separate it into a **spectrum**.

processor

A processor is the integrated circuit that controls all the operations of a **computer**.

See also **microprocessor**

producer

Producers are organisms that produce their own food. They include plants, algae, plankton and fungi.

See also **consumer, food chain**

program

A program is a set of instructions that tell a computer how do something. Programs are written in special codes that computers can follow.

See also **software**

projectile

A projectile is anything that is thrown or shot into the air.

Because of gravity, projectiles follow a curved path until they hit the ground.

propeller

A propeller is a spinning shaft with twisted blades attached.

Propellers are used to drive some aeroplanes and boats along. The engine of the aeroplane or boat spins the propeller. Because the blades are twisted, the air or water is pushed back behind the propeller. This produces a forward **thrust** that pushes the aeroplane or boat along.

propulsion

Propulsion means the way in which an object is made to move.

See also **jet propulsion, rocket**

protein

A protein is a large **molecule** made up of chains of smaller molecules called amino acids. Proteins contain atoms of carbon, hydrogen, oxygen and nitrogen and sometimes also sulfur.

Proteins are an important part of all living things, especially animals. Some examples are haemoglobin in blood, actin in muscles, albumen in eggs and silk made by silkworms.

protist

Protists are simple organisms that live in wet places. They include **algae**, diatoms, **protozoa** and slime moulds.

Most protists consist of one cell, but some can be very large, especially algae. Protists are grouped together in the kingdom Protoctista.

See also **classification of living things**

proton

A proton is a positively charged particle in the **nucleus** of an atom.

Every atom has protons in its nucleus, although a hydrogen atom has only one. A proton has the same charge as an electron, but it has a much greater mass.

protozoa

Protozoa are tiny one-celled organisms. They were once thought to be simple animals, but they are really **protists**.

Most protozoa reproduce by simply splitting into two. Some protozoa cause diseases in humans, including malaria, sleeping sickness and dysentery.

See also **classification of living things**

pulley

A pulley is a machine consisting of ropes and wheels for lifting heavy objects.

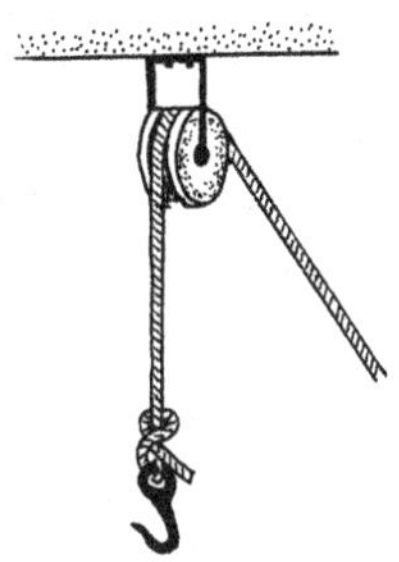

A simple pulley.

pulsar

A pulsar is a star-like body that emits regular pulses of radio waves. Astronomers think that pulsars are rotating **neutron stars**.

pulse

A pulse is when the amount or value of something changes suddenly and then goes back to what is was before.

Pulses can happen in liquids and gases, and also in light, sound and electricity. The pulse you can feel in your arteries is caused by your **heartbeat**.

pupa

A pupa is the third stage in the life cycle of some **insects**. It comes after the egg and **larva** stages. A pupa does not move or eat. It is slowly turns into the adult insect. The plural of pupa is pupae.

See also **metamorphosis**

pupil

The pupil is the hole in the middle of the **iris** that lets light into the eye.

Qq

quantum

A quantum is the smallest amount by which some things, such as energy, can change.

This idea of a quantum means that energy can only have particular values. This is the basic idea behind quantum theory and quantum physics.

quark

A quark is believed to be one of the smallest possible parts of matter. Electrons and protons, for example, might be made up of quarks.

Scientists think that there are different sorts of quarks, but there is still a lot of argument about what they are really like.

quartz

Quartz is a mineral made up of a special form of a compound called silicon dioxide.

Quartz is one of the commonest minerals on Earth. There are many different sorts of quartz, including agate, amethyst, chalcedony and jasper.

See also **Mohs scale**

quasar

A quasar is a very distant, star-like object that seems to be moving away from the Earth at a very high speed. The name is an abbreviation for quasi-stellar. Quasars are believed to be the most distant objects in the universe. Some scientists believe that a quasar is the centre of a galaxy, where there is a **black hole**.

Quaternary period

See **geological time**

Rr

rack and pinion

A rack and pinion is a sort of **gear** in which a toothed wheel (the pinion) moves a toothed rod (the rack).

Most microscopes use a rack-and-pinion system for focusing.

radar

Radar is short for **ra**dio **d**etection **a**nd **r**anging. It is a way of finding a distant object, such as an aeroplane, by sending out radio waves and listening to the echo that bounces back off the object.

radiation

Meaning 1 Radiation means **electromagnetic radiation.**

Meaning 2 Radiation is the movement of energy (especially heat) through air or another **medium** from a hot object to a cooler one.

See also **infra-red radiation, ultraviolet radiation**

Meaning 3 In biology, radiation is the spread of organisms into a new area.

radioactivity

Radioactivity is energy or particles emitted by a substance when the nuclei of its **atoms** are disintegrating (breaking down).

Radioactivity occurs naturally in some rocks, but it can also be made to happen by shooting neutrons into substances, such as uranium or plutonium.

See also **nuclear energy**

radiotelescope

A radiotelescope is a radio receiver that can pick up very faint radio signals from space. Most radiotelescopes are dish-shaped so that they can focus the signals into a detector above the centre of the dish.

The famous radiotelescope at Parkes in New South Wales.

rainbow

A rainbow is a curved **spectrum** caused by sunlight shining through millions of raindrops.

When the light enters a drop, it is reflected off the inside of the drop. When it goes back out of the drop, it is refracted, so it spreads out into its different colours.

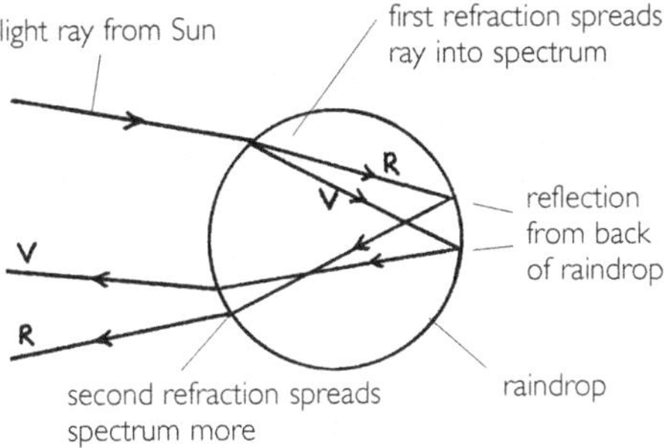

V = violet light
R = red light

See also **refraction, reflection**

rainforest

A rainforest is a forest that grows in places where it rains a lot. The plants in rainforests are **evergreen** and usually have big leaves, so not much light gets down to the ground.

A cool temperate rainforest in south-eastern Australia.

rare species

A rare species is a species that is not found in many places, or a species that has only a small number of living members. A rare species is not necessarily **endangered**.

ray

Meaning 1 A ray is a narrow beam of light, such as the light produced by a **laser**.

Meaning 2 A ray is a fish with wing-like fins and a long, thin tail. Rays are closely related to sharks.

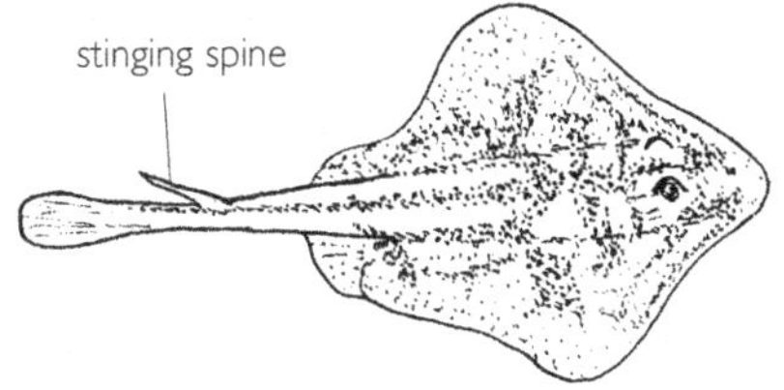

A typical ray.

reaction

Meaning 1 In chemistry, a reaction is any change that happens when two substances are mixed together.

See also **chemical reaction**

Meaning 2 In physics, a reaction is a force that is produced when an object hits or pushes on something. The reaction force is always in exactly the opposite direction to the force of the object.

real time

Meaning 1 In computing, real time means that the output happens at the same speed as the input. For example, when you type words on a computer keyboard, they appear on the screen at exactly the same speed that you type.

Meaning 2 Something that happens in real time is happening as you watch it. For example, if you go to a football game, you are watching the game in real time. But if you watch it on a videotape, you are watching it in “delayed time”.

recycling

Recycling means using waste materials like old paper, bottles, timber or scrap metal to make new products.

red blood cell

Red blood cells are the cells in the **blood** that carry oxygen around the body. They contain a red substance called haemoglobin. Red blood cells are also called erythrocytes.

See also **white blood cell**

red dwarf

A red dwarf is a normal **star** that shines with a red light because its surface is cool compared with other stars.

red giant

A red giant is an old **star** that has used up most of its hydrogen. It is made up mostly of helium, and its energy is produced by nuclear fusion.

reflection

Reflection is when something bounces back off the boundary between one **medium** and another.

For example, light waves are reflected when they hit a mirror, sound waves are reflected when they hit a wall, and water waves are reflected when they hit a cliff.

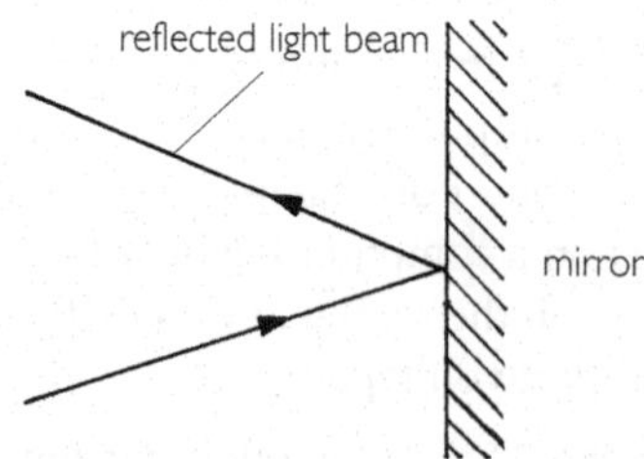

When light hits a flat mirror at an angle, it is reflected at exactly the same angle.

See also **refraction**

reflex

A reflex is the movement of a muscle caused by a **stimulus**.

An example is the change in the size of the iris of the eye that is caused by a change in the amount of light entering the eye.

refraction

Refraction is when a wave changes direction as it passes from one **medium** into another at an angle. Water waves are refracted when they go into deeper or shallower water. Light waves are refracted when they go into a different substance. Sound waves are refracted when they go into colder or warmer air.

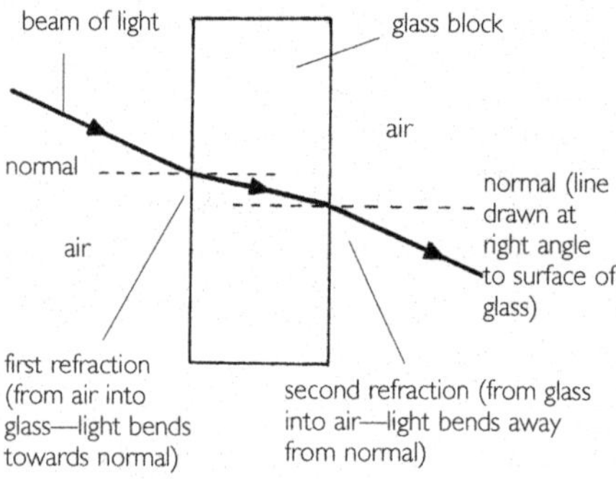

Refraction of light passing through a glass block.

See also **reflection**

refrigeration

Refrigeration means keeping something cold, even when the weather is hot.

regolith

Regolith is the layer of broken and weathered rock and soil that lies on top of the solid rock of the Earth's crust.

remote access

Remote access means being able to connect to a device, such as a computer, from another place.

Connecting to the Internet from your home computer is an example of remote access.

renewable energy

Renewable energy is energy that will not run out, no matter how much we use. The two main sorts of renewable energy are **solar energy** and **wind energy**.

Hydroelectricity and geothermal power are sometimes also called renewable energy, but they can eventually run out.

repel

To repel means to push away.

See **repulsion**

reproduction

Reproduction is the making of a new organism by one or two parents.

When only one parent is involved in reproduction, it is called asexual reproduction. When two parents are involved, it is called sexual reproduction.

See also **fertilisation**

reproductive system

The reproductive system is the part of an animal that is involved in reproduction. In humans, it consists mainly of the testes, vas deferens and penis (in males), or the ovaries, oviducts, uterus and vagina (in females).

reptile

A reptile is a **cold-blooded** animal that has scales covering the skin and breathes air. Reptiles breathe air using lungs, and the females produce eggs. Reptiles include lizards, snakes, turtles, tortoises and dinosaurs.

Some examples of reptiles (top to bottom): chameleon, sea snake and crocodile.

repulsion

Repulsion is when electricity or magnetism causes two objects to be pushed away from each other.

In electricity, repulsion happens between two negative charges or two positive charges. In magnetism, repulsion happens between two north **poles** or two south poles.

resistance

Resistance is a measure of how hard it is for electrons to flow through a device in an electric circuit. It is measured in units called ohms.

See also **current**, **voltage**

resolving power

Resolving power is a measure of how well an optical instrument can distinguish between two objects that are close together. It is often used to describe the quality of microscopes, telescopes and binoculars. Resolving power is sometimes wrongly called resolution.

respiration

Meaning 1 Respiration means taking in oxygen from air or water. In humans, this is the same meaning as breathing.

Meaning 2 Respiration is the **chemical reactions** that happen inside a cell to produce energy.

respiratory system

The respiratory system is the part of an animal that is used to obtain oxygen from the air or water.

In humans, the respiratory system consists of the nasal and mouth cavities, trachea, lungs and diaphragm.

retina

The retina is the part of the **eye** that detects light and changes it into nerve impulses. The retina is made up of two layers. The first layer stops light from being reflected back into the eye. The second layer consists of **rods and cones**.

revolution

A revolution is a single orbit of one object around another, such as the Moon around the Earth.

revolve

To revolve means to go around something in a circular or elliptical path. For example, the Moon revolves around the Earth.

See also **rotate**

rib

A rib is one of the thin, curved bones that protect the heart and lungs of a vertebrate animal. Ribs are connected to the vertebrae of the **backbone**.

Richter scale

The Richter scale is a scale of numbers used to describe the strength of an **earthquake**. It is based on how strong the earthquake would feel 100 kilometres from its **focus**. The smallest earthquake that can be felt measures about 2, and the strongest earthquake recorded so far measured 8.9. Earthquakes above 6 on the scale can cause great damage.

ring system

A ring system is a series of rings of rocks, ice and dust around a planet. Jupiter, Saturn, Uranus and Neptune all have ring systems.

Saturn and its ring system.

robot

A robot is a machine designed to do jobs without the need for a human operator.

Robots are used in modern manufacturing for jobs such as welding metal or assembling electrical instruments. They are usually controlled by computer systems or microprocessors. Almost all robots look nothing like humans.

rock cycle

The rock cycle is the continuous process involving the formation of rocks and their breakdown into rock fragments and soil.

See also **deposition, erosion**

rocket

A rocket is something that moves by pushing gas out of a narrow opening at enormous speed. The gas is produced by burning fuel and oxygen that are carried inside the rocket.

Solid-fuel rockets have the fuel and oxygen combined in a solid mass that is burnt to produce gas. Liquid-fuel rockets carry liquid fuel and oxygen in separate tanks, and burn them together in an engine.

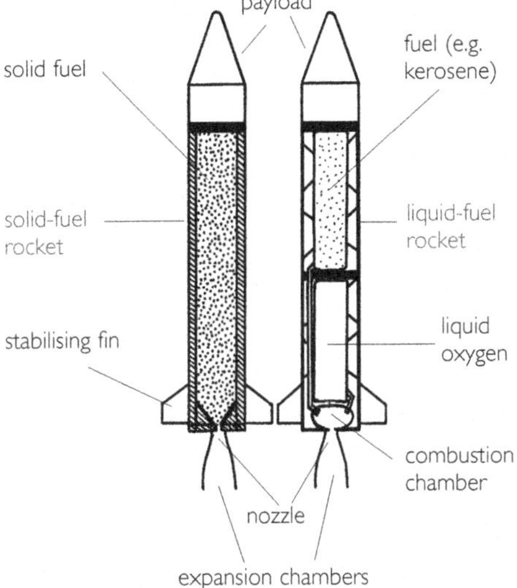

rods and cones

Rods and cones are special structures in the **retina** of the eye that change light energy into nerve impulses. The rods help us to see in dim light, and cones help us to see colours.

root

A root is the part of a plant that anchors it in the soil and draws in water and nutrients from the soil. In some plants, such as potatoes, the roots also store food for the plant.

rotate

Rotate means to spin around. The line that an object spins around is called the axis of rotation.

See also **revolve**

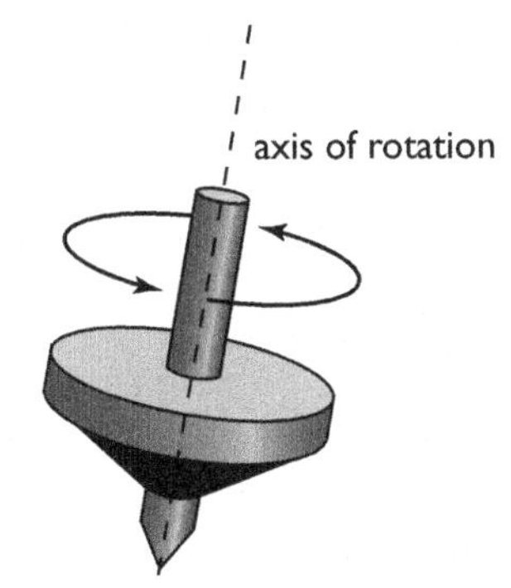

rubber

Rubber is a bouncy, elastic substance made from the sap of rubber trees or synthetic materials. Rubber made from sap is called natural rubber or latex. Other sorts of rubber are called synthetic rubber.

rust

Rust is a brownish, powdery substance called iron oxide that is made when iron is exposed to air. The oxygen in the air combines with the iron to form the rust.

See also **corrosion**

Ss

saline

Saline means salty.

Saline is used to describe water that is too salty to drink. It is also used to describe soil that is too salty for most plants.

See also **desalination**

saliva

Saliva is a mixture of water, **enzymes** and other substances made in the mouth of many animals.

Saliva helps food to slip easily down the oesophagus, and also starts the process of **digestion**.

salt

Meaning 1 Common salt is a **compound** of sodium and chlorine. Each molecule of salt is made up of one atom of sodium and one atom of chlorine, so its formula is $NaCl$.

Meaning 2 A salt is a compound made by mixing an **acid** with a **base**. When they are solid, salts usually have a crystalline form, like common salt.

sand

Sand is loose particles of rock that are up to 2 millimetres wide. Sand is made up mostly of **quartz**.

sap

Sap is a sticky liquid that carries nutrients from the roots to the leaves of a plant.

satellite

A satellite is any object that orbits around a planet. For example, the Moon is a satellite of the Earth. Spacecraft can also be satellites. They are called artificial satellites.

Saturn

Saturn is the sixth planet from the Sun, and the second largest planet in the solar system. It is about 1430 million kilometres from the Sun and takes 29.5 years to make one orbit. Like Jupiter, Saturn is made up mainly of hydrogen and frozen gases, and probably has a rocky core. Saturn is 120 000 kilometres in diameter and rotates once every 10 hours and 40 minutes. The gravity at the surface is just a bit more than the gravity on Earth. The surface is probably liquid hydrogen, and the atmosphere is made up mostly of hydrogen. Saturn has 17 moons and a very large **ring system** made up of rocks, ice and dust.

See also **moon**, **solar system**

scale

Meaning 1 Scale is another word for **balance**.

Meaning 2 A scale is a set of marks or lines that can be used to measure something, such as distance on a map or volume in a pipette.

Method 3 Scales are thin, hard plates that cover the bodies of fish and reptiles.

Meaning 4 Scales are tiny plates that cover the wings of some insects, such as butterflies and moths.

scavenger

A scavenger is an animal that feeds on the remains of animals that have died naturally, or that have been killed by a **predator**.

science

Science is the knowledge about the universe that we get from observations, measurements and **experiments**. Science helps us to understand how the universe works. It also helps us to predict what will happen when we try something new.

See also **hypothesis, scientific method, theory**

scientific method

The scientific method is the way that scientists gain knowledge. It involves making observations, working out **theories**, and doing **experiments** that can be repeated by someone else.

scientific name

A scientific name is a name that scientists use to identify a **species**. Every species has its own scientific name, and the name has two parts. The first part is the name of the **genus** that the species belongs to. The second part is the name of the particular species. Scientific names are written in italics. For example, the Giant Panda's scientific name is *Ailuropoda melanoleuca.*

scorpion

A scorpion is an **arachnid** with a long tail that carries a sting on the end. The front legs have claws for grasping prey. Scorpions are found in many parts of the world, and many of them are dangerous to humans.

A small scorpion, showing the tail sting.

screw

Meaning 1 A screw is a sort of simple **machine**. It consists of a rod with a spiral thread, running through a lifting platform. When a handle on the end is turned, the screw lifts the platform. Many car **jacks** are screws.

Meaning 2 A screw is a sort of fastener for wood and metal, with a spiral thread that helps drive the screw into position.

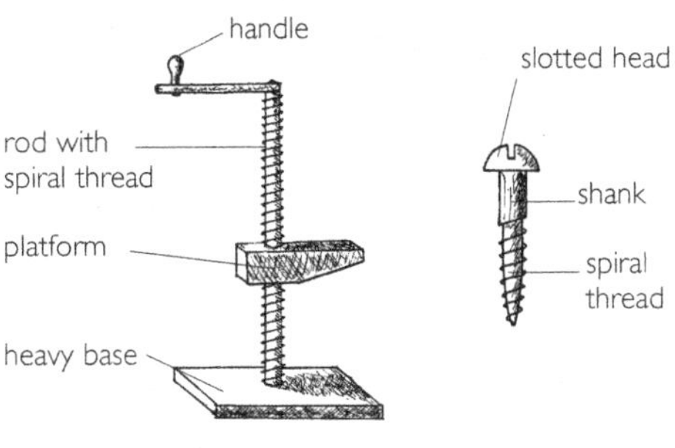

season

A season is a part of the year when the weather has particular features.

Most parts of the Earth have four seasons: winter, spring, summer and autumn (or fall). In most countries, each season is considered to begin at a **solstice** or an **equinox**. Some areas of the Earth, especially the **tropics** and the **poles**, do not have four seasons.

The four seasons

Season	*Northern hemisphere*	*Southern hemisphere*
Spring	21 Mar. to 20 June	22 Sept. to 20 Dec.
Summer	21 June to 21 Sept.	21 Dec. to 20 Mar.
Autumn (fall)	22 Sept. to 20 Dec.	21 Mar. to 20 June
Winter	21 Dec. to 20 Mar.	21 June to 21 Sept.

S

sea star

Sea stars are star-shaped crustaceans that live in the sea. They are also called starfish.

Sea stars crawl around on rocks or on the sea bottom, using tiny tube-like feet on their "arms". Many sea stars eat other sea animals, such as molluscs. Other sea stars eat algae or plankton.

A small sea star.

sea urchin

Sea urchins are round, spiny animals that live in the shallow parts of the oceans.

Sea urchins are closely related to sea stars. Most sea urchins feed on algae that get caught on the spines or on tiny tube-like feet that cover the body.

A sea urchin feeding.

seaweed

A seaweed is a large **alga** that grows in the sea. Seaweeds are found close to the coast. Most of them grow on rocks or mud, but some float in the water. The biggest seaweeds are kelps, which can grow up to 60 metres long.

Neptune's Necklace is a common seaweed on rocky shores.

sediment

Sediment is pieces of sand, soil, clay or other matter that have sunk to the bottom of a stream, lake or ocean.

sedimentary rock

Sedimentary rock is rock that has been formed from sediment. This is the sort of rock in which most fossils are found.

seed

A seed is a plant part that contains the **embryo** and the food that it needs to grow.

The embryo in a seed does not grow until **germination**. Some seeds have features that help them to be carried to new places by the wind, animals or water.

seismic

Seismic means relating to movements in the Earth's crust, such as earthquakes.

semiconductor

A semiconductor is a material that is halfway between a **conductor** and an **insulator**. The two most common semiconductors are germanium and silicon.

Semiconductors can be made to conduct electricity sometimes, but not at other times. This makes them very useful in electronics because they can be used to change how electrons flow in an electric circuit.

sense

A sense is a way that an animal detects things that are outside its body.

Humans have five main senses. They are taste, touch, hearing, sight and smell. Another sense is the sense of balance. In humans, the sense organs are the skin, eyes, ears, nose and tongue.

server

A server is a computer that is used to help a group of computers work together in a **network**.

sewage

Sewage is a mixture of solid and liquid wastes from households and other places. It includes wastes from toilets, sinks and laundries.

Sewage is pumped to a treatment plant where most harmful substances are removed. The wastes are allowed to decompose, and then the cleaned sewage is usually pumped into a stream, into the ocean or into a special pit in the ground.

sewerage

Sewerage is a system of pipes that carries **sewage** and wastewater to a sewage treatment plant.

shooting star

Shooting star is another name for a **meteor**.

SI units

SI units are the units of measurement used in science. SI stands for Système International d'Unités (International System of Units).

See **table of commonly used SI units** (page 122)

sight

Sight is the **sense** that we use to detect light.

silicon

Silicon is an **element** that forms crystals when it is a pure solid.

Silicon is the third most common element in the Earth's crust, after oxygen and aluminium. It forms crystals, and it is a **semiconductor**. Silicon is used in electronics and for making silicone. The symbol for silicon is Si.

silicone

Silicone is a soft, waxy substance made up mainly of long chains of silicon atoms and oxygen atoms that are joined to other atoms.

Silicone is very useful because it is not affected much by cold, heat, chemicals or water. It is used in oils, waxes, synthetic rubber and waterproof fillers.

silk

Silk is long, fine threads made by the **larvae** of some moths. It is made up of **proteins** in long chains. The best silk is made by the Chinese Silkmoth. It is made into long threads that can be woven into a soft, shiny fabric that we also call silk.

silt

Silt is a layer of very small particles of soil and other matter that sits on the bottom of a stream, lake or ocean. Silt occurs only where the water is still or moving very slowly because the small particles take a long time to sink to the bottom.

Silurian period

See **geological time**

S

siphon

A siphon is a tube that moves liquid from one place to another using only gravity.

One end of the tube is dipped into the liquid and the other end is held below the liquid level. Once the liquid is sucked up into the tube, gravity keeps it flowing down to the outlet.

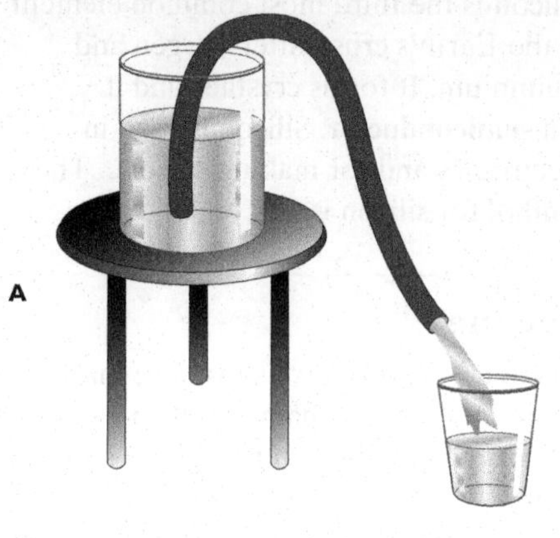

simple siphon.

skeleton

A skeleton is a group of bones or bony parts that are connected together to support the body.

In humans and other **vertebrate** animals, the skeleton is inside the body. This is called an endoskeleton. There are more than 200 bones in the human skeleton.

Some **invertebrate** animals also have a skeleton, but it is on the outside of the body. This is called an exoskeleton.

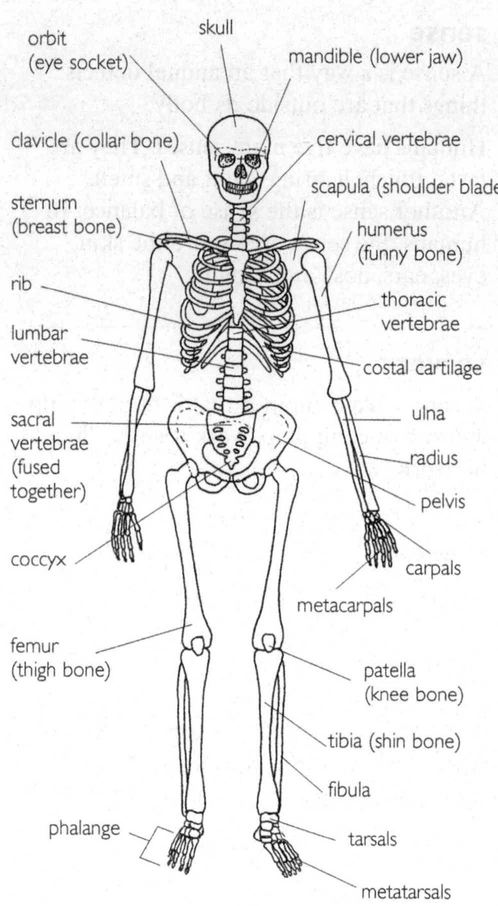

The human skeleton.

skin

Skin is the soft, tough covering on the body of a **vertebrate** animal. Human skin is made up of two layers called the epidermis and the dermis. The epidermis is the outside layer, and the dermis is underneath. Skin is attached to the body by fatty tissue that allows the skin to move over the muscles and other tissues underneath.

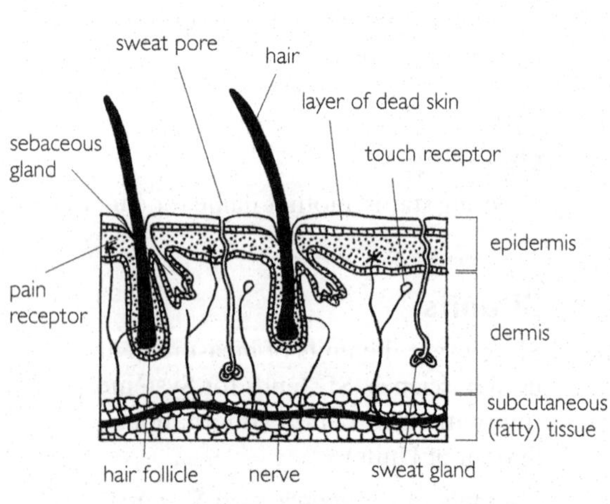

Human skin.

skull

The skull is the solid set of bones that surrounds the brain of a vertebrate animal.

See also **skeleton**

slate

Slate is a fine, hard **metamorphic rock** formed when **clay** is heated inside the Earth's crust.

Slate splits easily into thin sheets, so it is used a lot in buildings for floors and roofing tiles.

small intestine

The small intestine is a part of the **alimentary canal**. It is where most of the food mixture from the stomach is digested and absorbed into the blood. It consists of the duodenum, jejunum and ileum.

smell

Smell is the sense that is detected by the nose. It detects chemicals in the air when they are breathed in and land on the wet lining inside the nose called the mucous membrane.

smog

Smog is a cloud-like pollution that is common in large cities.

Smog originally meant a mixture of smoke and fog, but now it means any sort of cloudy pollution in the air.

snail

A snail is a **mollusc** with a spiral shell and eyes on long stalks. Some snails live on land, but most live in water, especially along the coast. Snails eat soft plant parts or algae.

A European Garden Snail with its eye stalks extended.

snake

Snakes are **reptiles** that have no legs or eardrums. They have a special jaw that can allow them to swallow animals whole. Snakes also have a forked tongue, and most snakes also have two sharp fangs in their mouth.

Most snakes eat small vertebrate animals, such as frogs and mice, which they can swallow whole. Pythons do not have fangs. They kill their prey by squeezing them, and they can eat larger animals.

A Carpet Python from Queensland.

snow

Snow is light, feathery ice crystals that form in clouds when water vapour condenses at very low temperatures.

Snowflakes are made up of these crystals stuck together.

See also **condensation**

soap

Soap is a **detergent** made by boiling up natural animal or plant fats with sodium hydroxide (which makes a hard soap) or potassium hydroxide (which makes a liquid soap).

Soaps include bar soaps for washing your skin. Most washing powders, shampoos and washing-up liquids are not soaps because they are made entirely from **synthetic** substances.

sodium

Sodium is an **element** that occurs in common salt.

Pure sodium is a silvery substance that reacts violently with water, so it is usually kept in oil. Sodium is needed in large amounts by animals, mainly for controlling the movement of water into and out of cells. The symbol for sodium is Na.

software

Software is any **program** that can be loaded onto a computer, such as the system and the applications.

soil

Soil is a mixture of rock particles and dead plant and animal matter that sits on top of the Earth's crust. It is where most plants grow and get their nutrients.

See also **regolith**

solar cell

A solar cell is a device that converts **solar energy** directly into electricity. Solar cells are also called photovoltaic cells.

solar eclipse

A solar eclipse is when all or part of the Sun is blocked out by the Moon. This happens when the Earth goes into the shadow cast by the Moon.

A total solar eclipse is seen from any place that is in the umbra, or centre, of the shadow. A partial solar eclipse is seen from anywhere that is in the penumbra.

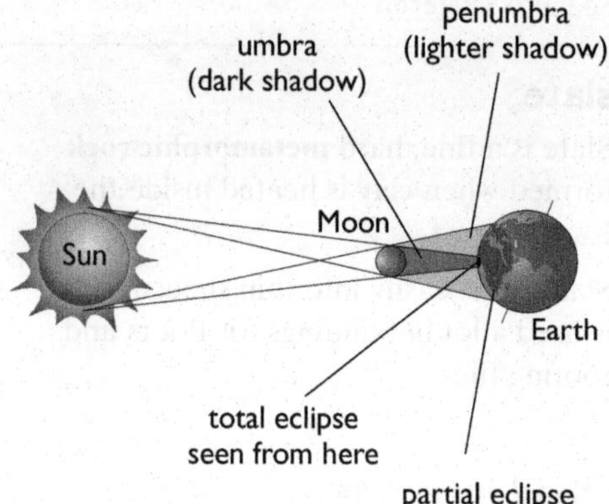

solar energy

Solar energy is the energy in sunlight.

Solar energy is used directly by plants for photosynthesis. It can also be used directly to produce electrical energy or to heat water.

solar system

The solar system is the Sun, its planets, their moons, comets, asteroids and all the other matter that orbits around the Sun.

solid

A solid is a **state of matter** in which the atoms or molecules are held tightly together so that they cannot move around. This means that a solid cannot flow like a liquid or a gas. It has a fixed volume and a fixed shape.

solstice

A solstice is a time of the year when the Sun is farthest north or south in the sky.

The solstices happen on about 21 June and 21 December every year. At these times, the length of daylight is either longest or shortest, depending on whether you are in the northern hemisphere or the southern hemisphere.

See also **equinox**, **season**

soluble

Soluble means able to be **dissolved** in a liquid.

See also **dissolve**

solution

A solution is a **mixture** made when something is **dissolved** in liquid. The liquid part is called the solvent, and the dissolved substance is called the solute.

sonic boom

A sonic boom is a loud booming sound that is heard when an aircraft flies past you at more than the speed of sound. It is caused by a huge shock wave that is pushed ahead of the aircraft.

sound

Sound is waves that can be detected by the human ear.

In the air, a sound wave is an area of high pressure that moves through the air. It travels at about 344 metres per second. Its speed in another **medium**, such as water, is different.

sound barrier

The sound barrier is the speed of sound in air, which is about 1238 kilometres per hour. This speed is sometimes called Mach 1. It is called a barrier because it used to be hard for an aircraft to pass through it. Today, some aircraft can fly several times faster than the speed of sound.

See also **sonic boom**

South Celestial Pole

The South Celestial Pole is the point on the **celestial sphere** that is directly above the Earth's geographic South Pole. In the southern hemisphere, the stars seem to turn around this point.

See also **North Celestial Pole**

space

Meaning 1 Space is another name for **volume**.

Meaning 2 Space is how we define the position of an object. To do this, we say that there are three directions or dimensions of space. These three directions are often called the *x*, *y* and *z* dimensions.

Meaning 3 Space is everything that is beyond the Earth's atmosphere. It includes the galaxies, stars, planets and moons, as well as the gas, dust and empty areas in between.

species

A species is a group of organisms that have exactly the same features. Members of a species can breed and produce **fertile** organisms that look like the parents.

spectrum

The spectrum is all the different types of radiation that exist. It includes light, infra-red radiation, ultraviolet radiation, X-rays and radio waves. The part of the spectrum that we can see is called the **visible spectrum**.

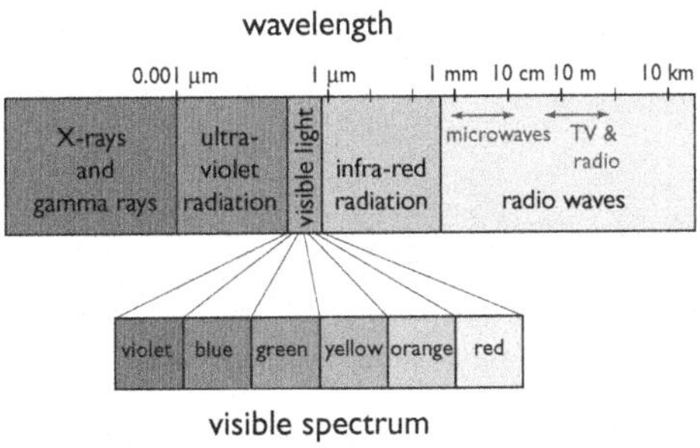

speed

Speed is how fast something is moving. It is the distance moved divided by the time it takes to move that far.

We usually measure speed in metres per second or kilometres per hour. Speed does not include the direction of movement, so it is different from **velocity**.

speed of light

The speed of light is how fast light travels in a **vacuum**, which is 299 792 metres per second.

The speed of light is probably the fastest speed that can be reached in the universe. It is a little less when light travels through other materials, such as air or glass.

sperm

A sperm is the sex cell produced by a male animal, plant, fungus or alga. In animals, its proper name is spermatozoon. In other organisms, it is called a spermatozoid.

See also **fertilisation**

spider

A spider is an **arachnid** with eight legs and a pair of fangs near the mouth for injecting venom into its prey.

Most spiders spin webs to catch their prey, but some are hunters that live in holes in the ground or in trees.

A huntsman spider hunting its prey.

spine

Meaning 1 The spine is another name for the **backbone**.

Meaning 2 A spine is a long, sharp part on the outside of a plant or animal.

sponge

A sponge is an animal without any organs or other tissues.

Most sponges live in the sea. The larvae swim about in the water, but adults live attached to a rock or anchored in sand or mud. They feed on the **plankton** that drift past in the water.

spore

A spore is a large cell that can grow into a new organism without **fertilisation**.

Spores are produced by ferns, mosses, liverworts, hornworts, fungi, lichens and algae.

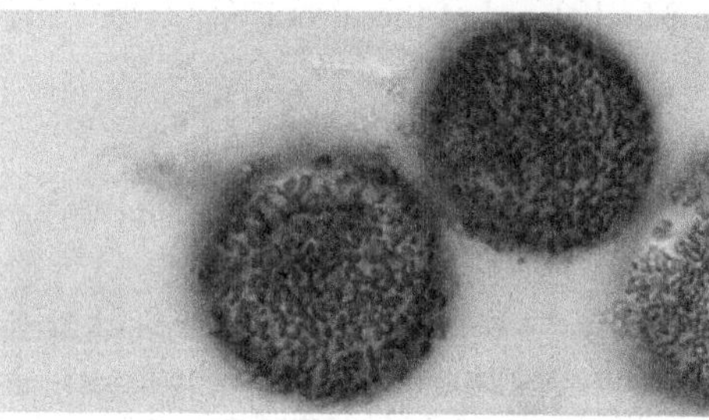

Spores of a liverwort viewed under a microscope, shown about 400 times actual size.

spring

Meaning 1 A spring is a piece of coiled wire that can be used to store and then release energy.

Energy is stored by squashing the spring, It is released when the spring is released.

Meaning 2 Spring is the **season** that follows winter. It is the time when plants begin to grow again, and animals become more active.

spur gear

A spur gear is a **gear** made up of wheels with teeth around the edge.

stalactites and stalagmites

Stalactites and stalagmites are rocks formed by water dripping from the roof of a limestone cave.

The water contains a substance called calcium carbonate. When the water evaporates on the floor or the roof of the cave, solid calcium carbonate is left behind. This gradually builds up into long, pointy rocks. The one hanging from the roof is called a stalactite. The one rising up from the floor is called a stalagmite.

Stalactites and stalagmites in an underground cave.

star

A star is an object in space that shines with its own light.

The nearest star is our own Sun. Stars are made up mostly of hydrogen, and the light is produced by **nuclear energy**. Stars are formed when a cloud of hydrogen gradually comes together into a ball of matter. When the ball is large enough, nuclear reactions start and the star begins to shine.

Some star facts

The 10 brightest stars as seen from Earth (except the Sun)

Star	*Magnitude*	*Constellation*
Sirius	−1.45	Sirius Major
Canopus	−0.72	Carina
Arcturus	−0.06	Boötes
Vega	0.00	Lyra
Rigel Kent	0.01	Centaurus
Capella	0.06	Auriga
Rigel	0.15	Orion
Procyon	0.35	Canis Minor
Betelgeuse	0.40	Orion

The 10 stars closest to Earth

Star	*Distance in light years*
Proxima Centauri	4.1
Rigel Kent (Alpha Centauri)	4.3
Sirius A	8.7
Procyon A	11.3
Tau Ceti	12.0
Altair (Alpha Aquilae)	16.5
Eta Cassiopeiae A	18.0
Beta Hydri	21.0
Fomalhaut (Alpha Piscis Austrini)	22.6
Pi Orionis	26.0

See also **blue giant**, **brown dwarf**, **constellation**, **galaxy**, **nebula**, **neutron star**, **nova**, **red dwarf**, **red giant**, **supernova**, **white dwarf**

starch

Starch is a substance made by plants to store energy. It is a **carbohydrate** and occurs in roots, seeds and fruits.

Starch is an important source of energy for plant-eating animals.

star cluster

A star cluster is a group of stars that are close together in the sky.

Most star clusters are made up of stars that look close but are really far apart in space. Some clusters are large **star systems**.

star system

A star system is a group of stars that formed from the same cloud of gas. They are close together in space.

Star systems may be only two or three stars, or they may be clusters of hundreds or millions of stars. A **galaxy** is also a sort of star system.

See also **binary star, star cluster**

states of matter

The states of matter are the three forms that matter can have. They are the **solid** state, the **liquid** state and the **gaseous** state. Scientists sometimes call **plasma** the fourth state of matter.

static electricity

Static electricity is a build-up of **electrical charge** in a substance, without any flow of electrons.

Lightning is caused by static electricity, but the lightning itself is actually an **electric current**. You can make static electricity easily by rubbing a glass or acetate rod on a cloth.

Static electricity produced by a van de Graaff generator has charged this girl's hair, causing it to be repelled from their body.

steam

Steam is water in the form of tiny droplets or gas, which is produced when water boils.

See also **water vapour**

steel

Steel is an **alloy** made up mainly of iron and carbon.

There are different sorts of steel, depending on what other elements they contain. For example, stainless steel contains a large amount of chromium.

stimulus

A stimulus is something in the surroundings that makes an organism change.

For example, a cold wind is a stimulus that gives you goose bumps, and a bright light is a stimulus that makes the pupil in your eye smaller.

stomach

The stomach is the organ in the body where the main **digestion** of food begins.

In humans an empty stomach is quite small, but it can expand to hold a large amount of food. The inside of the stomach produces **enzymes** that start to break down the food. The partly digested food is emptied into the duodenum for further digestion.

See also **alimentary canal**

stratocumulus

See **cloud**

stratosphere

See **atmosphere**

stratus

See **cloud**

subject

In an **experiment**, the subject is the thing that you are investigating.

For example, if you are measuring how the mass of a block of ice changes as it melts, the subject is the block of ice.

sublimation

Sublimation is when a solid changes straight into a gas, without becoming a liquid first.

See also **evaporation**

suction

Suction is a push caused by a difference in **air pressure**.

For example, when you put a straw into a drink and suck on the end, what you are really doing is reducing the air pressure in your mouth. This means that the air pressure pushing on the drink can push the drink up the straw and into your mouth.

sugar

A sugar is a sweet-tasting substance that can dissolve in water.

There are many sorts of sugars. For example, common table sugar is called sucrose. Sugars are **carbohydrates**, so a lot of energy can be obtained from them.

sulfur

Sulfur is an **element** that occurs naturally in many **ores**. It also occurs as a pure, yellow substance.

Sulfur is needed by most living things. It is used to build **proteins** and is needed for some cell functions. It is taken up into plants from the soil. Sulfur is also important in industry, especially for making hard, tough rubber. Sulfur is the correct spelling in science, but it can also be spelt *sulphur*. The symbol for sulfur is S.

Sun

The Sun is the star at the centre of the **solar system**.

sundial

A sundial is an outdoor clock that shows the time by the position of a shadow.

A sundial consists of two main parts: the gnomon and the dial. As the Sun crosses the sky, the position of the gnomon's shadow changes. The time is read on the dial.

A bronze sundial in the Royal Botanic Gardens, Melbourne.

sunspot

A sunspot is a dark spot on the surface of the Sun where the temperature is lower than normal.

Sunspots are probably caused by huge magnetic fields that stop hot gases coming up to the surface of the Sun.

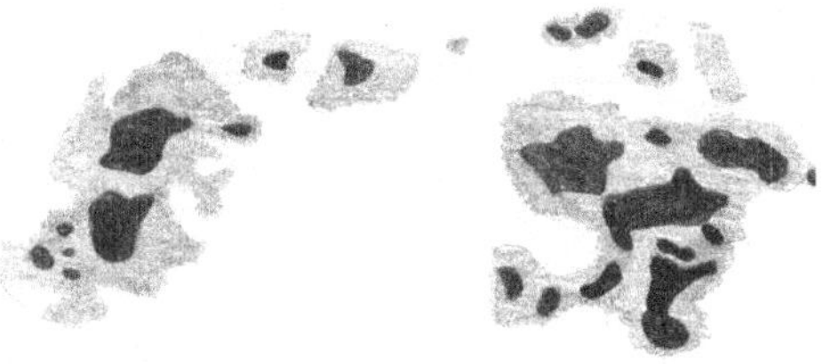

A drawing of sunspots.

supernova

A supernova is an old **star** that explodes suddenly, after it has used up all its fuel and collapsed. It is much brighter than a **nova**.

S

switch

A switch is a part of an **electric circuit** that can be used to turn the flow of electrons on or off. Some switches can also change the electron flow from one part of a circuit to another.

symbiosis

Symbiosis is when two different organisms live together and help each other.

For example, a lichen is a symbiosis between a fungus and an alga. The alga makes food for the fungus, and the fungus protects the alga from too much heat and light.

See also **parasite**

synthetic

Synthetic means made in a laboratory or factory from chemicals, instead of from things found in nature.

For example, nylon is a synthetic material made by mixing chemicals called diamines and dicarboxylic acid.

system

A system is a group of things that work together.

For example, an ecosystem is a group of organisms and non-living things working together, and the digestive system is a group of organs that work together to digest food.

Tt

A tadpole is the **larva** of a frog or toad.

Tadpoles hatch from jelly-like eggs laid in the water or on the stream bank. As a tadpole grows into an adult, its tail grows smaller, legs appear and lungs develop to replace the gills.

taste

Taste is the **sense** that uses the tongue. The tongue is covered with tiny sensors called taste buds. Hairs on the end of each taste bud react to the substances on the tongue, and send a signal to the brain.

technology

Meaning 1 Technology is the use of science and engineering knowledge for practical purposes. It is also called applied science.

Meaning 2 Technology is equipment and machines developed by scientists and engineers.

See also **bioengineering**, **genetic engineering**, **information technology**

tectonic plate

A tectonic plate is one of the large pieces of the Earth's crust that can move around.

See also **continental drift**

Teflon

Teflon is a sort of **plastic** called polytetrafluoroethene, or PTFE. It is very smooth, does not react with most chemicals and stays solid at very high temperatures.

Teflon is used to coat cooking utensils to make them "non-stick". It is also used in joints that cannot use oil or grease to keep them working smoothly, such as artificial hip joints and moving parts in spacecraft.

telescope

A telescope is a system of lenses that can be used to make distant objects look bigger.

There are many different sorts of telescopes. The two common ones are the refractor and the reflector.

In a refracting telescope, two sets of lenses (the objective and the eyepiece) are used to form the image of the object.

In a reflecting telescope, a mirror is used instead of an objective lens.

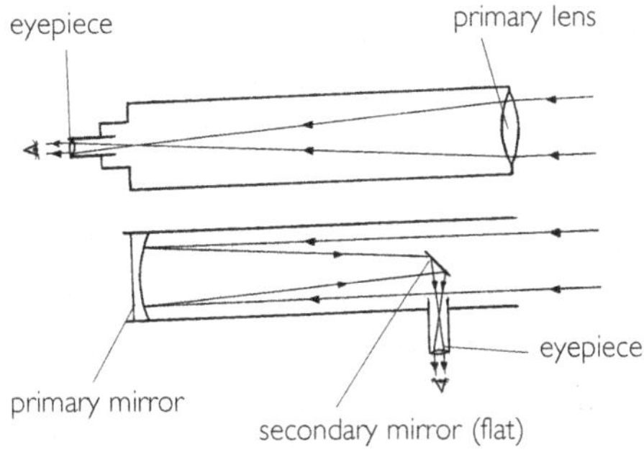

Two common types of telescopes: (top) a Keplerian refractor; (bottom) a Newtonian reflector.

See also **radiotelescope**

temperature

Temperature is how hot or cold something is. The temperature of an object tells us how much heat energy can flow between the object and its surroundings.

There are different ways of measuring temperature, but the most common is the **Celsius scale**.

See also **thermometer**

tendon

A tendon is a piece of tissue that connects a muscle to a bone. It helps to stop the end of the muscle from tearing when it contracts. Tendons cannot stretch much because they are made from a stiff substance called collagen.

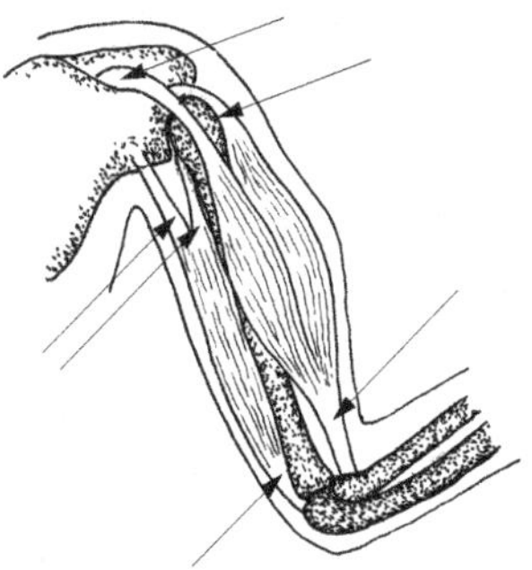

Tendons in the upper arm (arrowed).

tension

Tension is a **force** that causes something to stretch. It is the opposite of **compression**.

tentacle

A tentacle is a long, flexible body part that is found in many invertebrates, such as octopuses, snails and jellyfish.

Tentacles are usually used for picking up food or holding onto objects. Some sorts of tentacles have other parts, such as eyes or stinging cells, attached to them.

A sea anemone with its tentacles extended to feed.

terrestrial

Terrestrial means living or happening on the ground.

See also **aquatic**, **arboreal**

Tertiary period

The Tertiary period is an old name for the Neogene and Palaeogene periods.

See also **geological time**

theory

A theory is an idea about why something happens, or how something works.

Theories are used to explain all the things we observe in the universe, such as gravity, chemical reactions and species. In science, a theory is tested by doing an **experiment**.

thermal

Thermal means relating to heat or temperature.

For example, thermal clothing is designed to keep your body warm in cold weather.

thermometer

A thermometer is an instrument that is used to measure **temperature**.

The most common sort of thermometer is a liquid thermometer. This is a thin tube with liquid, such as mercury or alcohol, inside. When the temperature changes, the level of the liquid in the tube changes. The temperature is read from a scale on the tube.

Electronic thermometers use the fact that the **resistance** of substances changes when the temperature changes. They have a digital readout.

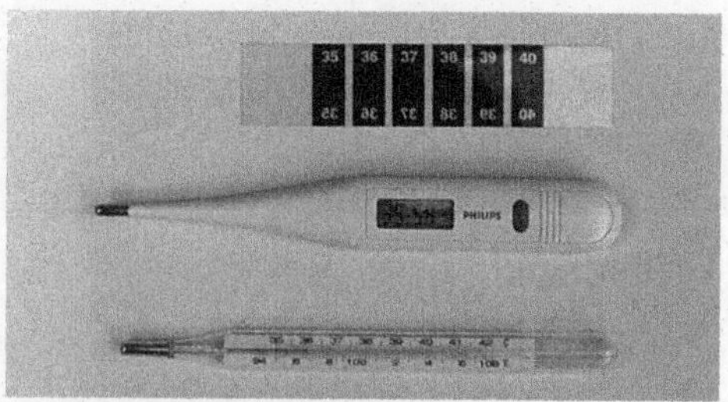

Three types of thermometer (from top to bottom): liquid crystal, digital and mercury.

thermosphere

See **atmosphere**

thorax

The thorax is the front part of the body of an animal between the head and the **abdomen**.

In vertebrates, it is the upper part of the body containing the heart and lungs inside the ribcage.

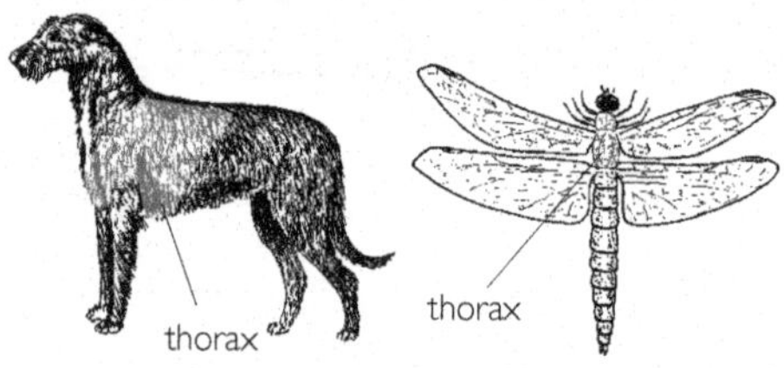

thrust

Thrust is a **force** produced when an object ejects liquid or gas in one direction. Thrust always acts in the direction of motion.

Thrust is one of four main forces that affect an aircraft flying through the air. The others are drag, weight and lift. Thrust is produced by the engine and pushes the aircraft forwards. Drag is produced by friction between the the aircraft and the air, and pulls the object backwards. Lift is produced by the flow of air over the wings and tail, and causes the aircraft to rise.

See also **aerodynamics**, **lift**

thunder

Thunder is the sound produced by **lightning**.

Thunder happens when the air is suddenly heated by the lightning, causing it to expand quickly. This produces a large pressure wave in the air, which we hear as thunder.

tibia

The tibia is the larger of the two bones in the lower leg of a vertebrate animal.

The top of the tibia is connected to the femur at the knee and the bottom is connected to the ankle.

See also **skeleton**

tidal energy

Tidal energy is electrical energy that is produced from the rise and fall of water caused by **tides**.

Water running into a storage dam during the flood tide is used to turn a **turbine** that produces electricity. When water flows out during the ebb tide, the generator turns the other way and also produces electricity.

tidal wave

Meaning 1 Tidal wave is another name for a **tsunami**.

Meaning 2 A tidal wave is a small wave in a stream caused by the tides.

tide

Tides are the constant, regular rising and falling of water. They are caused by the pull of gravity of the Moon, and by the rotation of the Earth around the centre of the Earth–Moon system. The Sun's gravity also has an effect on the tides.

A high tide occurs when the Moon is overhead because the water is pulled out towards the Moon. Another high tide occurs when the Moon is on the other side of the Earth. A low tide occurs when the Moon is just over the horizon because the water is pulled away in a different direction.

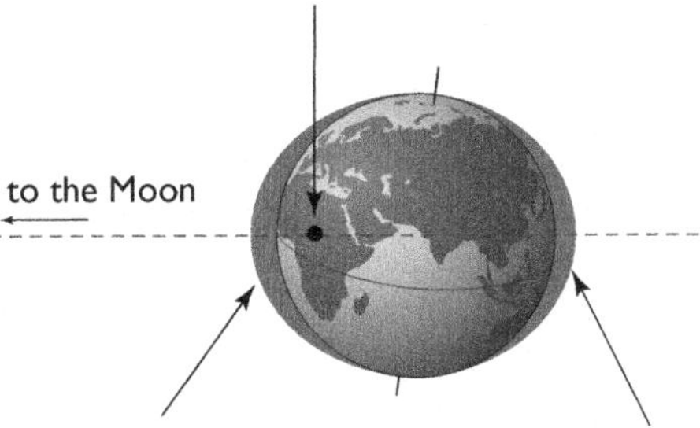

time

Time is the fourth dimension, after the three dimensions of **space**. To a scientist, it is one way of measuring the difference between two events.

Time is a very difficult thing to understand, but it is easy to measure. We can measure it using something that happens at the same speed all the time, such as the rotation of the Earth. Scientists use the vibration of atoms to measure time.

See also **day**, **year**

toad

Toads are **amphibians** that look like frogs. Unlike frogs, toads have fully webbed toes, ridges above the eyes and a slightly different skeleton.

The Cane Toad.

toadstool

Toadstool is an old word for some sorts of fungi, especially poisonous **mushrooms**. It is not used in science any more.

tongue

In vertebrate animals, the tongue is a large muscular organ attached to the inside of the mouth.

In humans, the tongue is used for tasting, moving food while chewing, swallowing and speech.

tooth

Meaning 1 In vertebrates, a tooth is a bony structure attached to the jaws, and used for biting and chewing food.

Human teeth are divided into molars, premolars, canines and incisors.

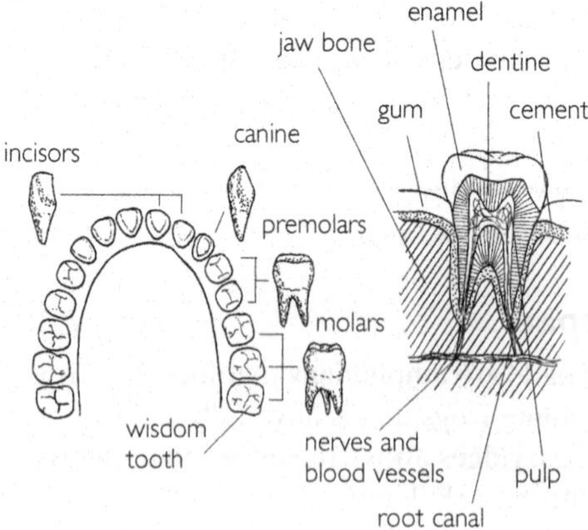

Meaning 2 A tooth is another name for a cog on a wheel in a **gear**.

topsoil

Topsoil is the topmost layer of soil. It consists mainly of organic matter, including decaying leaves.

See also **horizon**, **humus**, **litter**

tornado

A tornado is a funnel-shaped mass of air up to 100 metres wide that is spinning very quickly.

The air in a tornado can be moving at than 400 kilometres per hour. Most tornadoes occur in the USA and Australia. A tornado over the sea is sometimes called a waterspout.

An approaching tornado.

tortoise

A tortoise is a **turtle** that lives on land or in fresh water.

touch

Touch is the **sense** produced when you feel something with your skin.

toxic

Toxic refers to a substance that can cause harm to an organism.

For example, chlorine gas is toxic to most plants and animals and funnel-web spider venom is toxic to humans.

translucent

Translucent means allowing most of the light to pass through but scattering some of it.

Translucent materials usually look white or frosty, and you can usually only see fuzzy shapes through them. An example of a translucent material is frosted glass.

See also **opaque**, **transparent**

transparent

Transparent means letting most of the light through without absorbing it or scattering it. You can see quite clearly through transparent materials.

See also **opaque**, **translucent**

transpiration

In plants, transpiration is the loss of water into the atmosphere.

Transpiration happens mainly through tiny holes (stomata) in the leaves. The water is replaced by drawing up water through the roots or directly into the leaves.

Triassic period

See **geological time**

trilobite

Trilobites were small, flat **arthropods** that lived in the oceans during the Permian period.

Trilobites were covered by overlapping plates. They lived on the ocean floor and ate the remains of other animals.

tropical cyclone

A tropical cyclone is a **cyclone** in the **tropics** that develops into a huge storm over the ocean.

A tropical cyclone can cover tens of thousands of square kilometres, and can have winds of more than 200 kilometres per hour. If it passes over land, it can be very destructive.

Tropical cyclones (arrowed) over Australia and the Pacific Ocean.

tropics

The tropics are the areas on the Earth where the Sun can be directly overhead. These regions are between the latitudes $23\frac{1}{2}$ degrees north (called the tropic of Cancer) and $23\frac{1}{2}$ degrees south (called the tropic of Capricorn.

The tropics are hot and often wet. They are the areas where **monsoons** and **tropical cyclones** occur, and most of the Earth's rainforests grow there.

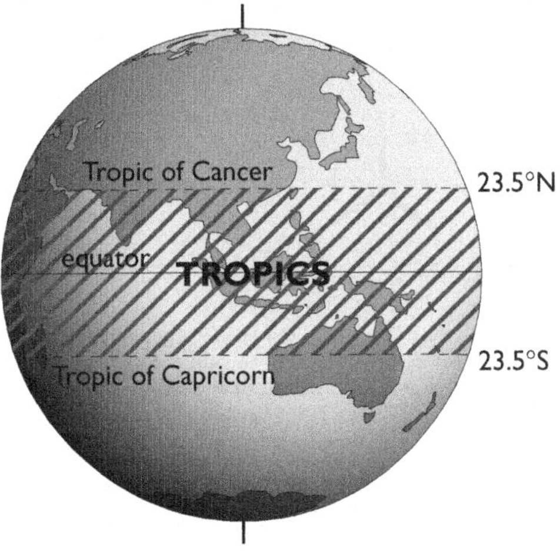

troposphere

See **atmosphere**

T

tsunami

A tsunami is a huge ocean wave caused by an **earthquake**.

Tsunamis can travel for thousands of kilometres across the ocean. When they meet land, they can cause great destruction. Tsunamis are sometimes also called tidal waves.

tundra

The tundra is a huge cold, treeless area covering the far northern parts of Europe and Canada. Most plants there are small because the summer growing time is very short.

turbine

A turbine is a machine in which a liquid or a gas is used to turn a wheel or propeller.

Large turbines are used to produce electricity. Steam, air or water turns a huge wheel or propeller, which is connected to a large electricity generator.

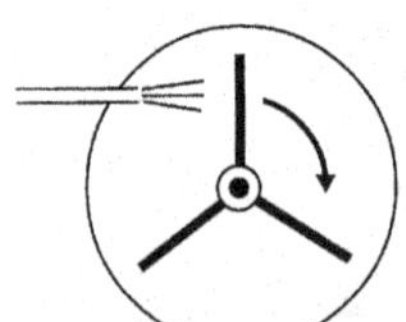

Two types of turbine. On the left, gas or steam is blown through the wall to spin the propeller. On the right, it blows out of the rotor, causing the rotor to spin.

turtle

Turtles are **reptiles** that have a bony shell protecting the body.

Turtles that live in the sea have flippers and are often called sea-turtles. Turtles that live in fresh water or on land have toes and are often called tortoises.

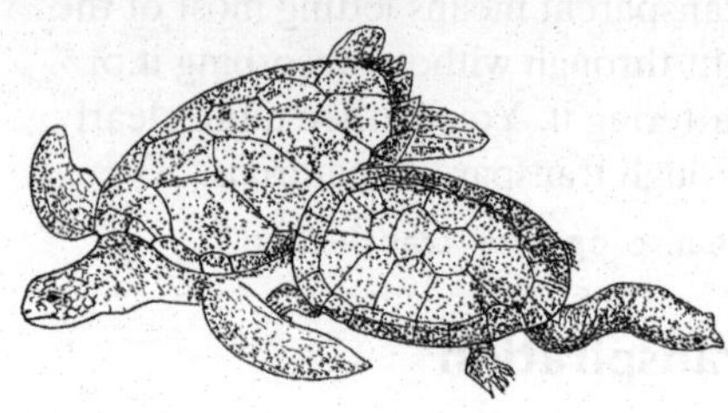

Green Sea-turtle (left) and Eastern Long-necked Tortoise (right).

typhoon

Typhoon is the name used in the China region for a **tropical cyclone**.

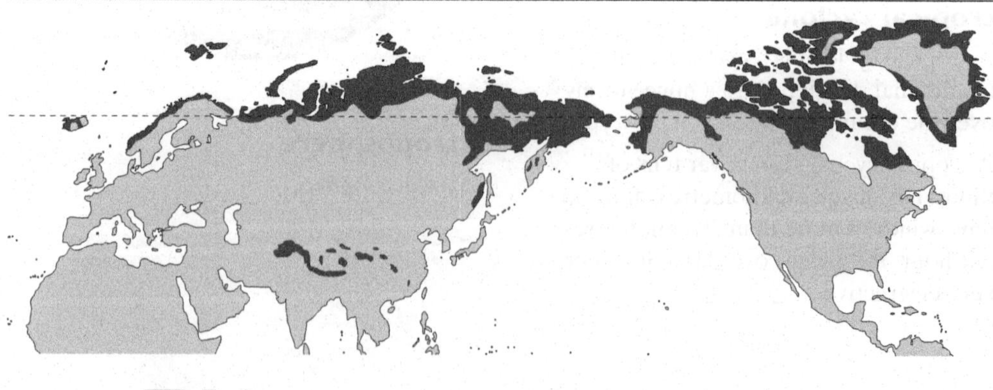

The tundra region.

Uu

ulna

The ulna is the larger of the two bones in the forearm of a **vertebrate**. It connects the elbow with the outside part of the wrist.

See also **skeleton**

ultrasound

Ultrasound is **sound** that is too high-pitched to be heard by humans. It has a **frequency** of more than 20 000 hertz.

Many other animals, such as bats and dolphins, can hear ultrasound and use it to detect prey and other objects. Ultrasound is used in medicines to look at internal organs without damaging them, and also to treat kidney stones and other problems. In manufacturing, it is used to check metal parts for defects and to clean surfaces and small parts.

ultraviolet radiation

Ultraviolet radiation is radiation that is past the violet end of the **spectrum**. It is also called UV.

Ultraviolet radiation can damage the tissues of plants and animals. Ultraviolet light can be produced artificially by a mercury vapour lamp. It has many uses in chemistry and technology.

See also **ozone layer**

umbilical cord

Meaning 1 The umbilical cord is the cord of tissue that connects an animal foetus to the mother through the placenta. It contains two **arteries** and a **vein** that keep up a supply of blood to the foetus. Your navel is the scar left when your umbilical cord was cut after you were born.

Meaning 2 An umbilical cord is a tube that connects an astronaut's spacesuit to the spacecraft, and provides the spacesuit with air and electrical power.

Meaning 3 An umbilical cord is a cable or pipe that is connected to a rocket before it is launched.

umbra

The umbra is the very dark part of a shadow.

See also **lunar eclipse, solar eclipse**

universe

The universe is all the matter, space and energy that exists. Some scientists think that there is more than one universe.

See also **big bang**

uranium

Uranium is a white **metal** that is very radioactive. It is found in the **ore** pitchblende, which is mined to obtain the metal.

Uranium is used as a fuel in nuclear reactors and also in nuclear weapons. The symbol for uranium is U.

See also **radioactivity**

Uranus

Uranus is the seventh planet from the Sun. It was discovered in 1781 by an English astronomer, William Herschel. It is made up mainly of ice, hydrogen and helium, probably with a rocky core. It is about 2880 million kilometres from the Sun and takes 84 years to make one orbit. Uranus is 51 800 kilometres in diameter and rotates once every $15\frac{1}{2}$ hours. The gravity at the surface is just less than the gravity on Earth. The surface is probably an ocean of liquid hydrogen, and the atmosphere is made up mainly of hydrogen and methane. Uranus has five **moons**, as well as a ring system.

See **solar system**

UV

See **ultraviolet radiation**

Vv

vaccine

A vaccine is a substance prepared from dead, harmless micro-organisms that helps the body defend itself against attack by the same live micro-organisms.

The vaccine makes the body produce substances that will attack the micro-organism if it invades the body. Vaccines are commonly used to help protect us against viruses and bacteria. Some have to be injected, but others can be taken by mouth.

vacuum

It is possible to have a space without anything in it at all. This sort of space is called a vacuum.

vacuum flask

A vacuum flask is a container that uses a vacuum to stop heat moving in or out.

The main part of the flask is a glass or metal bottle with two walls. When the bottle is made, the air is pumped out from between the walls to make a **vacuum**. This vacuum is an **insulator**, so the temperature inside the bottle stays the same for a long time. A plastic stopper is used to stop heat moving in or out of the mouth of the bottle, and the whole bottle is protected inside a plastic or metal case.

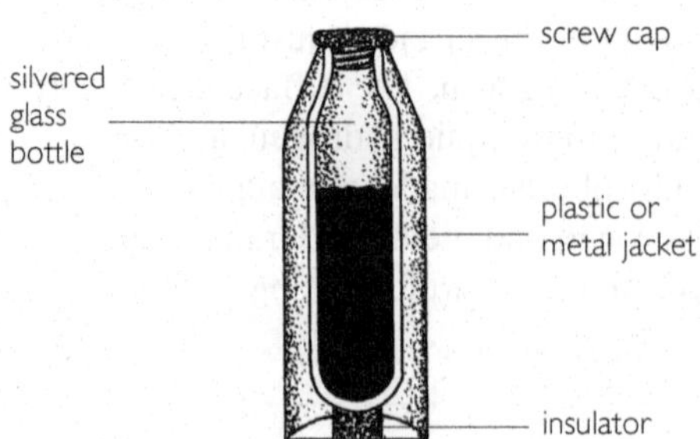

A cross-section of a vacuum flask.

valve

A valve is a device that can control the flow of something, such as a liquid, a gas or electricity.

A glass valve in a laboratory pipette.

vapour

Vapour is another name for **gas**. It is not the same as fog, mist or steam.

vegetation

Vegetation means all the plants growing in an area, including trees, shrubs, grasses, herbs, ferns, mosses and liverworts.

vein

Meaning 1 In animals, veins are the small tubes that carry **blood** back to the **heart** after the oxygen has been removed. Veins are also the fine tubes in the wings of insects that support the wings and carry nerves and blood vessels.

Meaning 2 In plants, veins are bundles of tubes in a leaf that carry fluids to and from the leaf cells.

Meaning 3 In geology, a vein is a crack or gap in a rock that has been filled up with **minerals**. Gold, silver and many other important minerals are found in veins.

velocity

Velocity is the **speed** and direction that something is moving in.

For example, if a car is going north along a road at a speed of 40 kilometres per hour, its velocity is 40 kilometres per hour north. If it then turns around and drives back the other way at the same speed, its velocity would be 40 kilometres per hour south.

venom

A venom is a poison produced by an animal to kill or stun its prey. Many different sorts of animals have venoms. They include snakes, spiders, scorpions, bees, wasps, ants, ticks, fish, octopuses, jellyfish and cone shells.

See also **antivenom**

Venus

Venus is the second closest planet to the Sun. It is a rocky planet made up mainly of nickel-iron and silicates. Venus is about 108 million kilometres from the Sun and it takes 225 days to make one orbit. Venus rotates once every 243 days. Its diameter is 12 100 kilometres, and the gravity at the surface is just less than the gravity on Earth. Venus has no moons. Its atmosphere is mainly carbon dioxide, with clouds of sulfuric acid. The surface is very hot and is probably covered with active volcanoes and ancient craters.

See also **solar system**

vertebra

A vertebra is any of the bones or cartilage that make up the **backbone** of an animal. The plural of vertebra is vertebrae.

vertebrate

A vertebrate is an animal with a **skeleton** of **bone** or **cartilage**, and also a well-developed **brain** inside a **skull**.

Vertebrates include mammals, fish, reptiles, amphibians and birds.

See also **invertebrate**

VHF

VHF is an abbreviation for very high frequency. It is a radio **frequency** between 3 and 300 megahertz.

vibration

A vibration is a motion back and forth around one position. Vibrations include sounds, radio waves, earthquake tremors and the motion of a pendulum.

video

In computing, a video is a moving picture that is stored as an electronic file.

See also **graphic**

virtual

In computing, virtual means something that is done on a computer instead of in real life.

Examples are a virtual game of golf and a virtual tour of a famous building.

virus

Meaning 1 In biology, a virus is a tiny non-living particle that can multiply inside a living cell.

Viruses can multiply only when they are inside the right sort of cell. Many viruses cause diseases in plants and animals. Some examples of viral diseases in plants are potato blight and tobacco mosaic. In humans, some examples are AIDS, hepatitis, influenza, polio, rabies and smallpox.

See also **bacteria**, **vaccine**

Meaning 2 In computing, a virus is a program that can spread from one computer to another, and interferes with the operation of a computer. A virus can cause damage by changing the operating system, causing software to stop working properly or even destroying files and programs.

viscosity

Viscosity describes how easily a substance flows. A low viscosity means that the substance flows easily. A high viscosity means that it hardly flows at all.

viscous

Something that is viscous acts like a **liquid** but does not flow easily. Some examples are honey, golden syrup and pitch.

visible spectrum

The visible spectrum is the range of colours of light that the human eye can see.

The visible spectrum is divided into six main colours: red, orange, yellow, green, blue and violet. The colour indigo is no longer included.

See also **spectrum**

vitamin

A vitamin is an **organic** compound that a living thing needs in very small amounts to stay healthy.

Humans need many different vitamins, but the most important ones are vitamins A, B complex, C, D, E and K.

The main vitamins and their sources

Vitamin	*Common sources*
A	Fish, yellow fruits and vegetables, dairy products, eggs
B1 (thiamine)	Fish, meat, soybeans, milk, cereals, green vegetables
B2 (riboflavin)	Meat, soybeans, dairy products, green vegetables, eggs, yeast
Niacin	Meat, fish, nuts, potatoes, cereals, tomatoes, leafy vegetables
C	Citrus fruits, tomatoes, leafy vegetables
D	Fish, liver, also formed by sunlight on skin
E	Leafy vegetables, dairy products, vegetable oils

volcanic rock

Volcanic rock is **igneous rock** that forms when **magma** cools above the ground. It is also called extrusive rock.

The most common volcanic rock is **basalt**, which is formed when **lava** cools. Some other volcanic rocks are formed from magma, pieces of rock or cinders that are blown out of a volcano. These are called pyroclastic rocks.

See also **plutonic rock**

volcano

A volcano is an opening in the Earth's **crust** from which **lava** flows. Steam, smoke, ashes and pieces of rock can also come out of a volcano.

Volcanoes are often cone-shaped hills or mountains with a **caldera** in the top. A volcano that can still eject lava or other material is "active". One that cannot is "extinct".

An active volcano in the Hawaiian Islands.

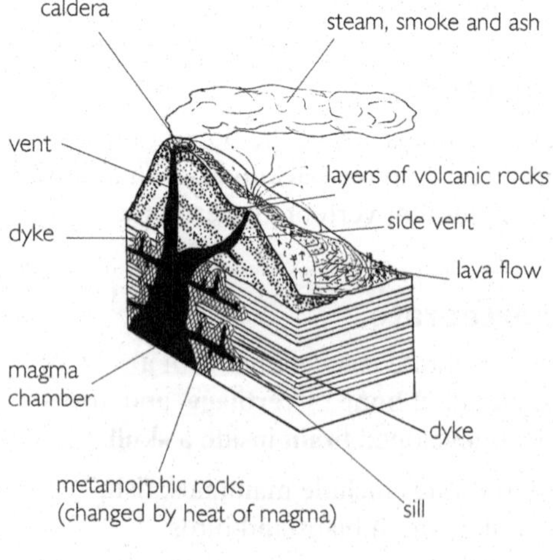

A cross-section of a volcano.

volt

Volt is the unit used to measure **voltage**. The symbol for volts is V.

voltage

Voltage is a measure of the electric force that can be produced by a cell or battery. It is also a measure of the amount of **work** needed to move **electrons** from one part of an **electric circuit** to another. Voltage is measured using a voltmeter.

volume

Meaning 1 Volume means the amount of space taken up by something.

Volume can be described in different units, such as litres, cubic centimetres or cubic metres.

Meaning 2 Volume means how loud a sound is.

warm-blooded

Warm-blooded refers to an animal that can keep the temperature inside its body steady when the temperature outside changes. All mammals and birds are warm-blooded.

See also **cold-blooded**

wasp

Wasps are insects that are closely related to bees and ants. They have four wings, a narrow "waist" between the thorax and abdomen and a sting at the tip of the abdomen.

A mud-dauber wasp.

waste

Waste is anything that is left over after something has been made or used. It can be a solid, liquid or gas.

See also **biodegradable**, **recycling**

watchglass

A watchglass is a small glass dish used for evaporating small amounts of liquids when you are doing experiments.

water

Water is a liquid that is found just about everywhere on Earth.

Water is a **compound** of oxygen (O) and hydrogen (H), and its chemical formula is H_2O. That means that every molecule of water contains two atoms of hydrogen and one atom of oxygen.

See also **ice**, **steam**, **water vapour**

W

water cycle

The water cycle is the never-ending movement of water through the environment.

Water in the atmosphere forms clouds and eventually falls to the ground as rain, snow, sleet or hail. This water may be used by plants, animals and other living things. Eventually the water runs into the oceans, where it evaporates back into the atmosphere again. The water cycle is also called the hydrological cycle.

water energy

The energy that flowing water has because it is moving is called water energy.

Water energy can be used to run machines (such as in a water mill), or turned into electrical energy for homes and factories (such as in a hydroelectric power station).

See also **hydroelectricity**

water table

The water table is the top of water that is lying under the ground.

In some places the water table is close to the surface, and in other places it is a long way down. If you dig a hole down to the water table, you have made a well.

water vapour

Water vapour is gaseous water in the air.

Water vapour is produced by evaporation, not by boiling, so it is colder than **steam**.

See also **condensation**

watt

The watt is the unit we use to measure how much power a machine produces. A scientist would say that one watt of power is equal to one joule of **energy** operating for one second. The symbol for watts is W.

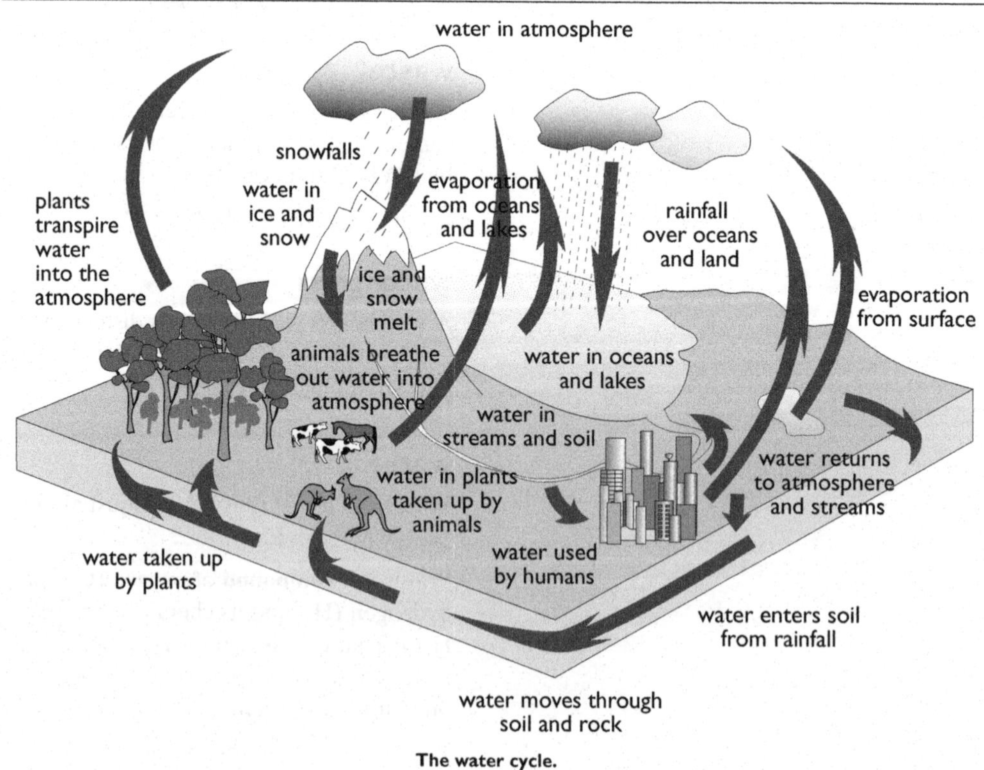

The water cycle.

wave

A wave is a moving series of crests and troughs that make a smooth curve, like a sideways *S*.

Many things move in waves or have a wave-like form, including sound, light and radio transmissions. Waves in oceans are caused mainly by winds blowing over the surface, not by tides.

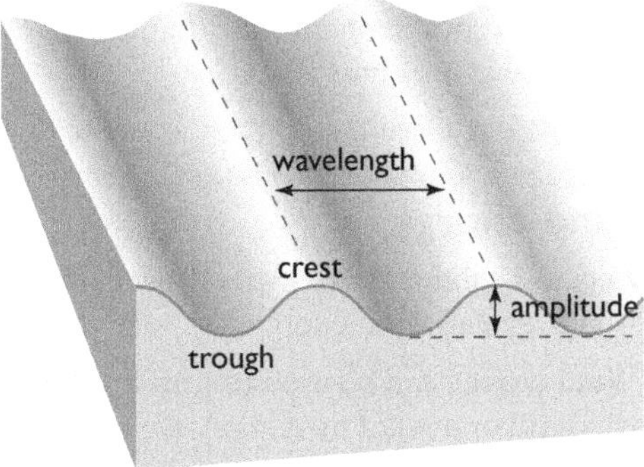

wavelength

Wavelength is the distance from one crest of a wave to the next one.

weather

The weather is what the atmosphere is like at one time and place.

We often describe the weather in terms of the temperature, humidity, cloudiness, rainfall, wind speed and air pressure.

See also **climate**

weathering

Meaning 1 Weathering is when rocks are broken down into smaller and smaller pieces. Some things that cause weathering are extreme heat or cold, chemicals in water and tree roots pushing into cracks.

Meaning 2 Weathering is a change in the surface of a material, such as wood. This can be caused by things such as sunlight, wind, water or chemicals.

See also **erosion**

website

A website is the information you can see on the **World Wide Web** at a web address.

wedge

A wedge is a solid object with two flat sides that meet at a sharp angle.

Wedges are used to split materials, such as wood and rock, to stop things moving and to fill gaps of different sizes.

A wedge being used to stop a door from closing.

See also **inclined plane**

weed

When a plant is growing where it is not wanted it is called a weed.

Weeds on farms are called agricultural weeds. Weeds in natural areas are called environmental weeds.

weight

Everything on Earth has weight. It is the downward pull or force on an object that is caused by the Earth's **gravity**. Your weight would be different on another planet because the gravity there would be different.

Weight is a **force**, and in science it is measured in units called newtons (symbol N). If someone says their weight is 50 kilograms, they are really talking about their **mass**. On Earth, the weight of an object is about 10 times its mass, so a 50 kilogram person actually weighs about 500 newtons.

weightlessness

Imagine you are in a stationary spaceship far from any star or planet. There is no gravity pulling on you, so you just float about. This is called weightlessness.
It is when the pull of **gravity** is almost zero.

It is possible to feel weightless even when there is still a pull from gravity. It happens to a skydiver before the parachute opens. It also happens to an astronaut in a spacecraft orbiting the Earth. This sort of weightlessness is called apparent weightlessness.

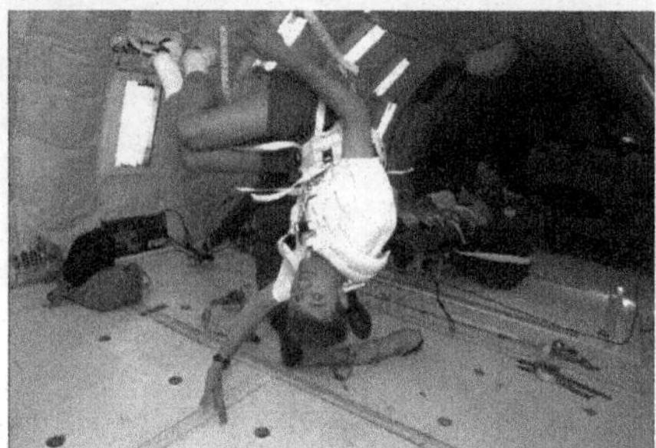

Astronauts training on Earth can experience apparent weightlessness in an aircraft that flies in a downward parabolic curve.

wetland

A wetland is any large area covered by still or slow-moving water for a large part of the year. Wetlands include lakes, swamps, estuaries and salt marshes.

whale

Whales are ocean-living **mammals** that have a blowhole on the top for breathing, flippers instead of arms, no hind limbs and a flat, horizontal tail fin.

The Blue Whale can grow up to 34 metres long and is the largest living animal.

wheel

A wheel is any flat, circular object that can turn on an **axle** that passes through its centre.

white blood cell

White blood cells are cells in human blood that help to protect the body against infections and poisons. White blood cells are also called leucocytes.

See also **red blood cell**

white dwarf

A white dwarf is very small and dense stars that shines with a white light. It is really the core of an old star that has blown away its outer layers.

See also **nova**, **red giant**

wind energy

Because wind is moving air, it has energy, which we call wind energy.

Wind energy can be used to turn a windmill or a wind turbine. A windmill turns wind energy into mechanical energy (for example, to operate a pump), and a wind turbine produces electrical energy for homes and factories.

wind turbine

A wind turbine is a huge propeller that is designed to turn wind energy into electricity. A large group of wind turbines is called a wind farm.

These wind turbines supply much of the electricity for a town in Queensland.

See also **turbine**

wood

Wood is the strong, dense part of a tree or shrub and is made of **xylem**.

Hardwood is the wood of leafy trees, such as eucalypts and oaks. Softwood is the wood of conifers, such as pine trees.

work

When an object is moved by a force, we say that work is done on the object. The amount of work done is equal to the amount of extra **energy** that the object gets.

Work can also be done on an object by changing things, such as its size, electrical charge, temperature or chemical nature.

World Wide Web

The World Wide Web is a world-wide information system that you can find on the **Internet**. World Wide Web sites have web addresses beginning with "www".

worm

Worms are soft-bodied **invertebrates** with a body made up of many segments. They include **earthworms**, leeches and marine worms.

worm gear

A worm gear is a **gear** made up of a grooved rod and a toothed wheel. It is used mainly to change the direction of the turning motion.

Guitar strings are tuned using worm gears.

See also **gear**, **rack and pinion**

Xx

xerophyte

A xerophyte is a plant that can live in places where there is almost no water. One example is a **cactus**.

X-ray

X-rays are a sort of energy that can pass right through things without destroying them. An X-ray photograph is made by passing X-rays through an object, and then onto a photographic film. In a medical X-ray, bones show up as white and the soft organs and muscles as grey or black.

See also **electromagnetic radiation**

xylem

Xylem is the special tissue in plants that carries water all the way from the roots to the leaves.

See also **phloem**, **wood**

Yy

year

One year is the time taken for a planet to travel once around the Sun.

The Earth's year is $365\frac{1}{4}$ days. Because it would be confusing to have an extra one-quarter of a day every year, we have three years of 365 days in a row and then a "leap year" of 366 days.

See also **solar system**

yeast

A yeast is a very small fungus that can convert **sugars** into carbon dioxide and alcohol.

Some yeasts are very important in baking, brewing and winemaking.

yolk

Yolk is the yellowish material in the centre of an egg. It contains all the food needed by a growing **embryo**.

See also **albumen**

Zz

zenith

The zenith is the point on the **celestial sphere** that is directly above you.

See also **nadir**

zero error

A zero error is when an instrument, such as a spring scale, does not read zero when it should. This causes every measurement made using that instrument to be wrong.

A zero error can usually be fixed by turning an adjusting screw.

zinc

Zinc is a shiny metal that is used to coat other metals to stop them from corroding.

Zinc is also used to make metal **alloys**, such as brass, and in paints, electrical parts and TV screens. A tiny amount of zinc is needed by plants and animals to keep their cells healthy. The symbol for zinc is Zn.

See also **corrosion**

zodiac

The zodiac is the part of the sky that the Sun, Moon and planets seem to move through. It is a narrow band that passes through 12 large **constellations**, which are called the zodiacal constellations.

zoology

Zoology (pronounced *zo-ology*) is the scientific study of animals. It includes how they have evolved, how they are classified, their structure, how they function, their role in ecosystems and where they live.

zooplankton

Zooplankton are animals that are part of **plankton**. They are mostly tiny crustaceans and the larvae of other animals, such as fish.

See also **krill**

zygote

A zygote is the new cell that is formed when an **ovum** is fertilised. It is the cell that can eventually grow into a new organism.

See also **fertilisation**

Acknowledgments

The authors would like to thank Jane Pitkethly for reviewing the Maths section.

The publisher wishes to thank David Meagher for providing the majority of the photographs, as well as the following copyright holders for granting permission to reproduce their material:

Andi Cairns: p. 154 (left); Corbis (Picture Australia): p. 203; M. C. Escher Foundation: p. 34; Bruce Fuhrer: pp. 143 (bottom left), 158, 168 (right), 181 (bottom left), 216 (top, bottom left), 223, 230 (top left); Michael Gill, Leonids 2001 Meteor Gallery website: p. 190; Dennis Kunkel Microscopy, Inc.: p. 165 (right); Zoë Meagher: p. 109 (middle right); NASA Human Space Flight Gallery: p. 240 (right); NOAA Photo Library: pp. 132 (left), 230 (right), 231; NSSDC Photo Gallery: pp. 149 (bottom left), 169 (right), 195, 212; Oz Images/Mark Dixon (Portland, Oregon): p. 159; Picture Source: p. 224; Fiona Smith: pp. 173 (left, bottom right), 177, 193 (right).

Every effort has been made to trace the original source of copyright material contained in this book. The publisher would be pleased to hear from copyright holders to rectify any errors or omissions.

Printed in Australia
09 Jan 2019
695097